LAB MANUAL TO AC

Automotive Service: Inspection, Maintenance, and Repair

Second Edition

Tim Gilles
Chuck Rockwood

THOMSON
DELMAR LEARNING

Australia Canada Mexico Singapore Spain United Kingdom United States

Lab Manual to Accompany Automotive Service: Inspection, Maintenance, and Repair

Second Edition

Tim Gilles and Chuck Rockwood

Vice President, Technology and Trades SBU:
Alar Elken

Editorial Director:
Sandy Clark

Acquisitions Editor:
David Boelio

Development Editor:
Christopher Shortt

Channel Manager:
Fair Huntoon

Marketing Coordinator:
Sarena Douglass

Production Director:
Mary Ellen Black

Production Editor:
Barbara L. Diaz

Art/Design Coordinator:
Rachel Baker

Technology Project Manager:
David Porush

Technology Project Specialist:
Kevin Smith

Editorial Assistant:
Jill Carnahan

Printed in the United States of America
1 2 3 4 5 XX 05 04 03

For more information contact
Delmar Learning
Executive Woods
5 Maxwell Drive, PO Box 8007,
Clifton Park, NY 12065-8007
Or find us on the World Wide Web at
www.delmarlearning.com

Library of Congress Cataloging-in-Publication Data:
Card Number: [2002034992]

ISBN: 1-4018-1235-X

NOTICE TO THE READER

Publisher does not warrant or guarantee any of the products described herein or perform any independent analysis in connection with any of the product information contained herein. Publisher does not assume, and expressly disclaims, any obligation to obtain and include information other than that provided to it by the manufacturer.

The reader is expressly warned to consider and adopt all safety precautions that might be indicated by the activities herein and to avoid all potential hazards. By following the instructions contained herein, the reader willingly assumes all risks in connection with such instructions.

The publisher makes no representation or warranties of any kind, including but not limited to, the warranties of fitness for particular purpose or merchantability, nor are any such representations implied with respect to the material set forth herein, and the publisher takes no responsibility with respect to such material. The publisher shall not be liable for any special, consequential, or exemplary damages resulting, in whole or in part, from the readers' use of, or reliance upon, this material.

Contents

Preface

The *Lab Manual to Accompany Automotive Service, Inspection, Maintenance and Repair* is designed to help the student build automotive skills. It contains two parts. Part I is made up of activity sheets to reinforce the theory learned in the core text. Activity sheet exercises include parts identification, matching exercises, and fill-in sheets designed to help reinforce the student's understanding of the operation of the automobile and its systems.

In Part II a wide variety of hands-on lab worksheets emphasizes practical real-life skills needed to service today's automobiles. Worksheets are presented in an increasing order of difficulty. The student should complete one task before progressing to the next one. Each project or lab assignment is built upon the next in a logical sequence in much the same manner as lab science instructional programs are constructed. Worksheets include these features:

Objective: A description of what the student will have learned after completion of the lab assignment and the ASE task or work-related activity that applies to the new skill acquired.

Directions: Specific instructions for completing the lab assignment.

Tools and Equipment Required: A list of the most important hand tools and equipment needed to complete the lab assignment.

Parts and Supplies: A list of the most important automotive components and miscellaneous materials needed to complete the lab assignment.

Procedure: A step-by-step description of the work to be performed in the lab assignment.

Notes: Inserted where helpful hints will facilitate the completion of the lab assignment.

Cautions: Warnings about potential hazardous situations that could cause personal injury or damage to the vehicle.

Shop Tips: Suggestions and shortcuts with careful instructions for implementing them.

Environmental Notes: Proper disposal methods for hazardous materials and product expiration information.

Illustrations: Fully illustrated to enhance lab discussion.

Figure Credits

Activity Sheets: AS-7 Courtesy of Tim Gilles, AS-9 Part I Courtesy of Tim Gilles, AS-9 Part 2 Courtesy of Ford Motor Company, AS-9 Part 3 Courtesy of Hunter Engineering Company, AS-9 Part 3 Courtesy of Baldor Electric Company, AS-9 Part 3 Courtesy of Tim Gilles, AS-9 Part 3 Courtesy of DaimlerChrysler

Corporation, AS-9 Part 3 Courtesy of Snap-On Tools Company, AS-9 Part 4 Courtesy of Tim Gilles, AS-9 Part 4 Courtesy of Delta International Machinery Corp. AS-62 Courtesy of Tim Gilles, AS-76 Courtesy of General Motors Corporation, AS-113 Courtesy of NAPA Institute of Automotive Technology, AS-113 Courtesy of Ford Motor Company, AS-113 Courtesy of General Motors Corporation, AS-115 Courtesy of American Honda Motor Co., Inc.

Lab Prep Worksheets: WS I-3 Courtesy of Tim Gilles, WS I-4 Courtesy of Tim Gilles, WS I-7 Courtesy of Tim Gilles, WS 1-1 Courtesy of Chek-Chart Publications, WS 1-2 Courtesy of General Motors Corporation, Service Technology Group, WS 1-3 Courtesy of Automotive Lift Institute, WS 2-5 Courtesy of The Gates Rubber Company, WS 2-7 Courtesy of GE Lighting, WS 2-12 Courtesy of Ford Motor Company, WS 3-3 Courtesy of Tim Gilles, WS 3-4 Courtesy of Tim Gilles, WS 3-7 Courtesy of American Honda Motor Co., Inc., WS 3-8 Courtesy of American Honda Motor Co., Inc., WS 4-3 Courtesy of Plews/Edelmann Division, A Gates Group Company, WS 4-3 Courtesy of Tim Gilles, WS 4-4 Courtesy of Rubber Manufacturers Association, WS 4-5 Courtesy of Tim Gilles, WS 4-5 Courtesy of Rubber Manufacturers Association, WS 4-6 Courtesy of Tim Gilles, WS 4-7 Courtesy of Tim Gilles, WS 6-1 Courtesy of Tim Gilles, WS 6-3 Courtesy of Tim Gilles, WS 6-5 Courtesy of Tim Gilles, WS 6-9 Courtesy of The Gates Rubber Company, WS 6-12 Courtesy of The Gates Rubber Company, WS 6-15 Courtesy of Tim Gilles, WS 6-16 Courtesy of Ford Motor Company, WS 6-19 Courtesy of Ford Motor Company, WS 6-19 Courtesy of The Gates Rubber Company, WS 6-23 Courtesy of DaimlerChrysler Corporation, WS 7-2 Courtesy of Ford Motor Company, WS 7-8 Courtesy of DaimlerChrysler Corporation, WS 7-14 Courtesy of DaimlerChrysler Corporation, WS 7-23 Courtesy of Tim Gilles, WS 8-1 Courtesy of General Motors Corporation, Service Technology Group, WS 8-8 Courtesy of Sun Electric Corporation, WS 8-10 Courtesy of Tim Gilles, WS 8-11 Courtesy of Tim Gilles, WS 8-12 Courtesy of General Motors Corporation, WS 8-13 Courtesy of Ford Motor Company, WS 8-14 Courtesy of DaimlerChrysler Corporation, WS 8-18 Courtesy of DaimlerChrysler Corporation, WS 8-18 Courtesy of Tim Gilles, WS 8-21 Courtesy of Tim Gilles, WS 8-23 Courtesy of DaimlerChrysler Corporation, WS 8-26 Courtesy of Tim Gilles, WS 8-27 Courtesy of Heli-Coil, WS 8-28 Courtesy of Tim Gilles, WS 9-2 Courtesy of Tim Gilles, WS 9-3 Courtesy of DaimlerChrysler Corporation, WS 9-4 Courtesy of Tim Gilles, WS 9-4 Courtesy of Federal-Mogul Corporation, WS 9-5 Courtesy of Federal-Mogul Corporation, WS 9-9 Courtesy of DaimlerChrysler Corporation, WS 9-9 Courtesy of Tim Gilles, WS 9-10 Courtesy of Timken Company, WS 9-10 Courtesy of Tim Gilles, WS 9-11 Courtesy of Tim Gilles, WS 9-13 Courtesy of American Honda Motor Co., Inc., WS 9-14 Courtesy of Ford Motor Company, WS 9-17 Courtesy of Snap-On Tools Company

Part I

Activity Sheets

Section One

The Automobile Industry

Instructor OK ______________ Score ____________

Activity Sheet #1

IDENTIFY FRONT- AND REAR-WHEEL DRIVE COMPONENTS

Name______________________________ Class ____________________

Directions: Identify the front- and rear-wheel drive components in the drawing. Place the identifying letter next to the name of the component.

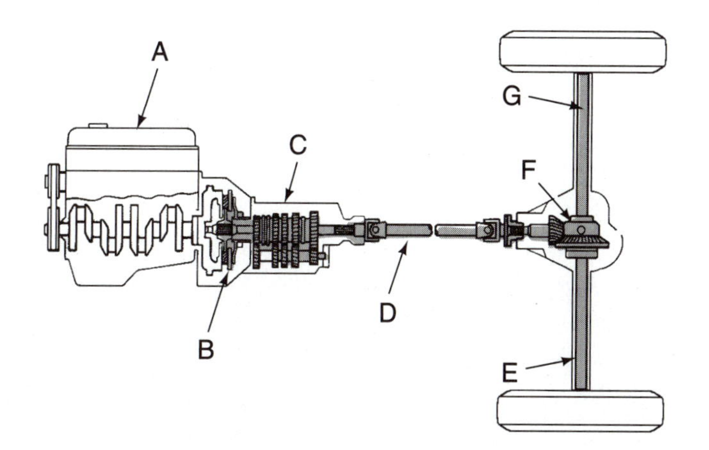

Axle Housing	E	Propeller Shaft	F	Rear Axle Shaft	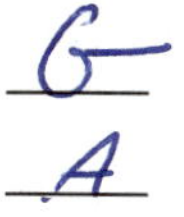
Clutch	B	Transmission	C	Engine	A
Differential	D				

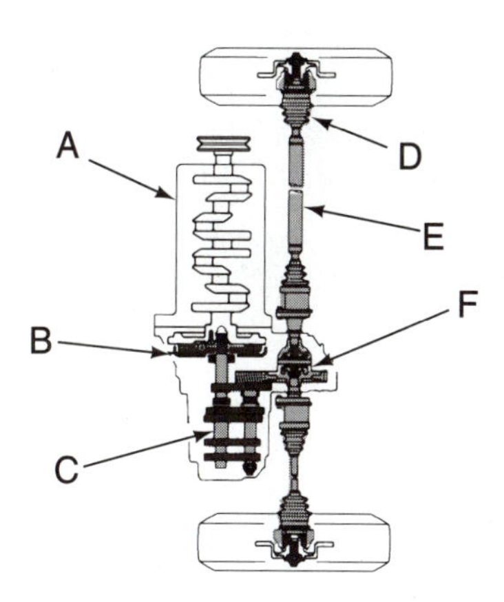

Differential	____	Drive Axle	F	Transaxle	____
CV Joint	D	Engine	A	Clutch	____

STOP

Instructor OK ______________ Score ____________

Activity Sheet #2

THE AUTOMOTIVE SERVICE INDUSTRY

Name______________________________ Class ____________________

Directions:

1. List four careers in the automotive repair industry.
   a. ______________________________
   b. ______________________________
   c. ______________________________
   d. ______________________________

2. List four types of automotive repair shops.
   a. ______________________________
   b. ______________________________
   c. ______________________________
   d. ______________________________

3. List the eight ASE areas of specialization required to become a Master Auto Technician.
   a. ____________________ e. ____________________
   b. ____________________ f. ____________________
   c. ____________________ g. ____________________
   d. ____________________ h. ____________________

4. List two additional ASE certifications.
   a. ______________________________
   b. ______________________________

5. Are there any special licenses required to perform certain types of work in your area?
   a. ______________________________
   b. ______________________________

STOP

Part I

Activity Sheets

Section Two

Shop Procedures, Safety, Tools, Equipment

Instructor OK ______________ Score ____________

Activity Sheet #3
SERVICE INFORMATION

Name________________________ Class ______________

Directions: Match the words on the left to the descriptions on the right. Write the letter for the correct word on the line provided. For the terms that you are not certain of, use the glossary in your textbook.

A. Owner's manual	____	Information on an underhood label
B. Microfiche	____	Automotive Engine Rebuilders Association
C. CD-ROM	____	Technical service bulletin
D. R&R	____	Number to identify a particular vehicle
E. Flat rate	____	Repair shop that repairs older vehicles
F. Flat-rate manual	____	Vacuum diagram label
G. Estimates repair cost	____	Magazine for the automotive professional
H. TSB	____	Found in a lubrication service manual
I. VIN	____	Sells and repairs new vehicles
J. 10th digit	____	Manual that is the most comprehensive one for a particular vehicle
K. Emission requirements	____	Sells parts for most vehicles
L. Manufacturers	____	VIN number year identifier
M. Lubrication fittings	____	Automatic Transmission Rebuilders Association
N. Vacuum hose routing	____	Parts and time guide
O. Trade journal	____	A service writer
P. AERA	____	Plastic film containing service information
Q. ATRA	____	Booklet that comes with a new car
R. Dealership	____	Service information accessed by a personal computer
S. Independent	____	Estimated time to perform a repair
T. Parts store	____	Remove and replace
U. Dealership parts department	____	Sells parts for only one make of vehicle

STOP

Instructor OK ______________ Score ____________

Activity Sheet #4

IDENTIFY MEASURING INSTRUMENTS

Name______________________________ Class ____________________

Directions: Identify the measuring instruments in the drawing. Place the identifying letter next to the name of the tool.

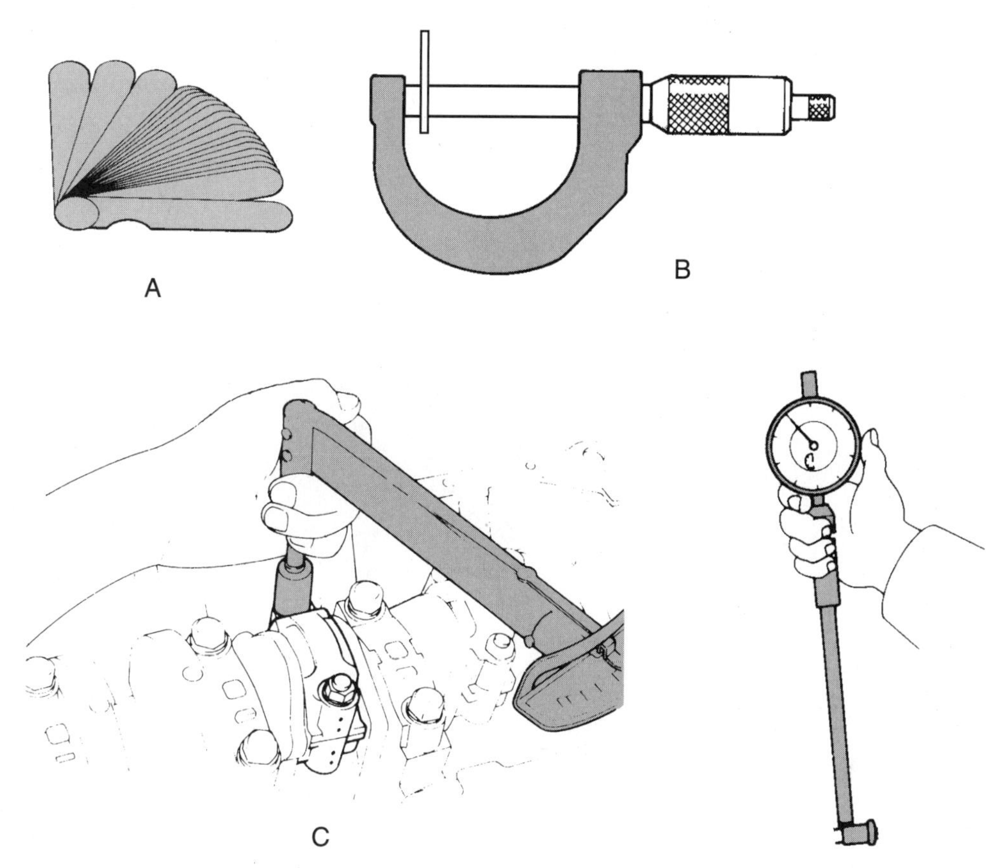

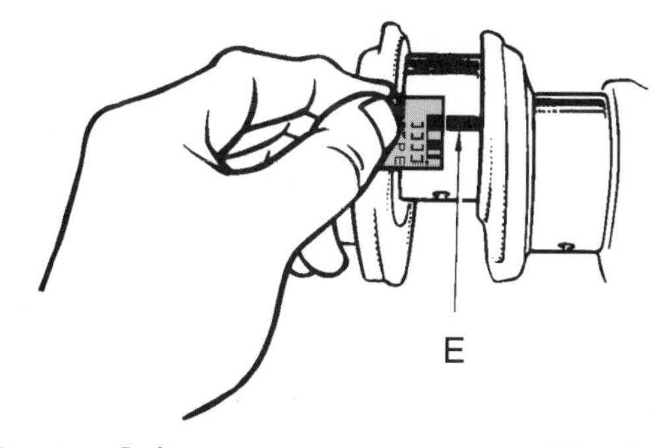

F

Vernier Caliper	____	Plastigage	____
Dial Indicator	____	Micrometer	____
Feeler Gauge	____	Torque Wrench	____

Instructor OK ______________ Score __________

Activity Sheet #5
MEASURING

Name______________________________ Class ____________________

Directions: Match the words on the left to the descriptions on the right. Write the letter for the correct word on the line provided. For the terms that you are not certain of, use the glossary in your textbook.

A. British Imperial system	____	Inside diameter
B. Metric system	____	Used to calibrate a micrometer
C. Foot-pounds	____	Cubic inches equals 2.7 liters
D. Newton-meters	____	How 1/1000″ is expressed with decimals
E. Plastigage	____	5 inches converts to how many millimeters?
F. O.D.	____	Measuring system that uses fractions and decimals, based on inches, feet, and yards
G. I.D.	____	Used to measure small holes
H. LCD	____	A strip of plastic that deforms when crushed; used to measure oil clearance
I. 0.001″	____	How torque readings are expressed in the metric system
J. Gauge block	____	Extra gauge that long-range dial indicators have
K. Vernier	____	Outside diameter
L. Ball gauge	____	One revolution of the micrometer thimble measures ______________ .
M. Dial indicator	____	Scale on the micrometer used to measure to 0.0001″
N. Feeler gauge	____	How torque readings are expressed in the English system
O. Revolution counter	____	One revolution of the micrometer thimble equals what fraction of an inch?
P. 127	____	Liquid crystal display
Q. 60	____	One liter equals ______ cubic centimeters.
R. 162	____	Gauge used to measure valve clearance
S. 1000	____	Approximate cubic inches are in 1 liter

TURN

T. Vernier caliper	____	Instrument that would be used to measure endplay
U. 1/40 of an inch	____	Measuring system based on the meter
V. 0.025″	____	One tool that can be used to measure I.D., O.D., and depth

Instructor OK ______________ Score ______________

Activity Sheet #6

FASTENER GRADE AND TORQUE

Name______________________________ Class ______________

Directions:

1. The size of the following bolt is 3/8″ × 16 × 3. List the dimensions in the correct blanks. Locate a bolt of this size and list the size of the wrench that fits the bolt head correctly.

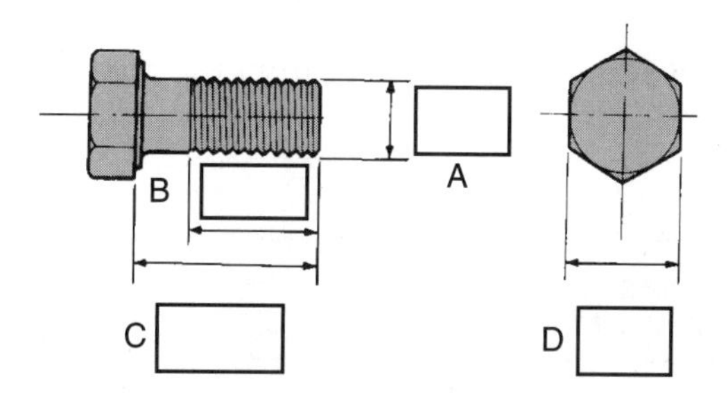

2. Locate fasteners of the following sizes and determine which size wrench fits the fastener head.

   A. 1/2″ ____ C. 6 mm ____

   B. 1/4″ ____ D. 10 mm ____

3. List the grade under each of these SAE inch standard bolts.

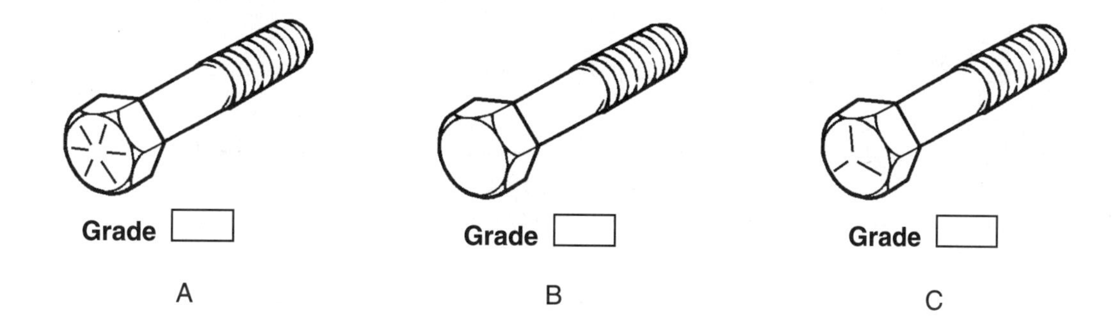

4. Place an X under the metric bolt shown here that is closest in strength to an SAE grade 8 bolt.

TURN

5. A bolt is usually properly torqued to____% of its elastic limit.

6. A nut can be turned easily onto a bolt for the first few threads and then begins to turn hard. What is the most likely cause?

____

7. Refer to the bolt torque chart below. What is the proper torque for a 3/8″ × 16 grade 5 fastener?

____

Material	SAE 2 Mild Steel		SAE 5		SAE 8	Socket Head Cap Screws
Minimum Tensile P.S.I. Strength	74,000	60,000	120,000	105,000	150,000	160,000
Proof P.S.I. Load	55,000	33,000	85,000	74,000	120,000	136,000
Steel Grade Symbols						
Bolt Diameter Inches	Torque: pound foot					
1/4	7	—	10	—	14	16
5/16	14	—	21	—	30	33
3/8	24	—	37	—	52	59
7/16	39	—	60	—	84	95
1/2	59	—	90	—	128	145

8. List the grade under each of these SAE inch standard nuts.

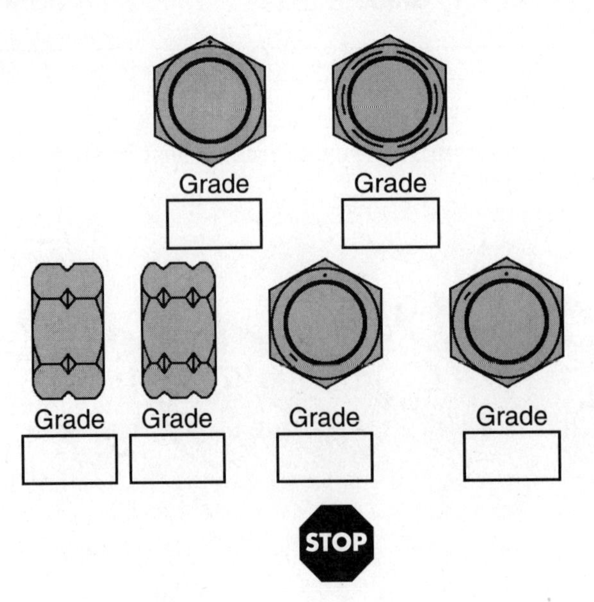

Instructor OK ______________ Score ____________

Activity Sheet #7

TAPS AND DRILLS

Name______________________________ Class ________________

Directions:

1. List the name next to the correct tap.

a. ___________

b. ___________

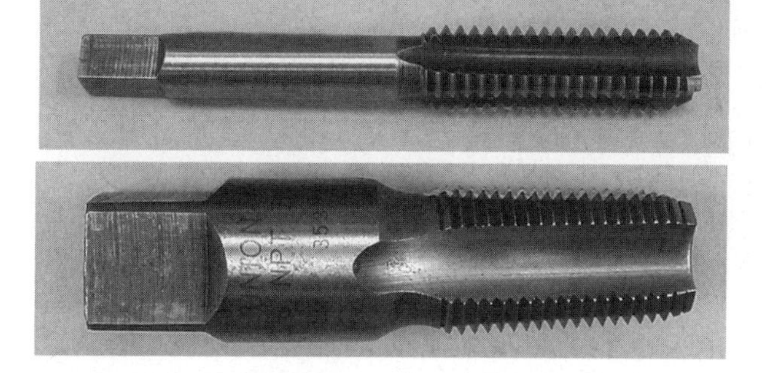

c. ___________

d. ___________

2. Refer to the tap drill chart.

❏ What is the correct size tap drill for a 1/4 × 20 screw thread? ___________

❏ What is the closest fractional drill in a 1/64″ increment drill index that could safely be used as a tap drill for a 1/4 × 20 thread? A decimal equivalent chart is located in the Appendix of the core text. ___________

❏ When referring to a screw thread, what does the 20 in 1/4 × 20 mean? ______________

____

Thread diameter	Threads per inch			Decimal equivalent	Tap drill Approx. 75% full thread	Decimal equivalent of tap drill
	NC	NF	NS			
12	...	28	...	.2160	14	.1820
12	...	...	32	.2160	13	.1850
1/4	20	...	...	.2500	7	.2010
1/4	...	28	...	.2500	3	.2130
5/16	18	...	...	.3125	F	.2570
5/16	...	24	...	.3125	I	.2720
3/8	16	...	...	.3750	5/16	.3125
3/8	...	24	...	.3750	Q	.3320
7/16	14	...	...	.4375	U	.3680
7/16	...	20	...	.4375	25/64	.3906
1/2	13	...	...	.5000	27/64	.4219
1/2	...	20	...	.5000	29/64	.4531

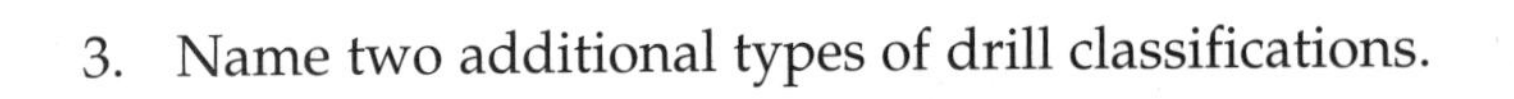

3. Name two additional types of drill classifications.

❑ fractional

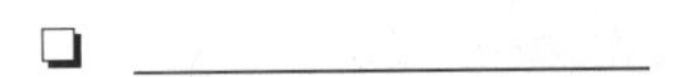

❑ ____________________

❑ ____________________

4. List the trade name for this type of replaceable thread insert.

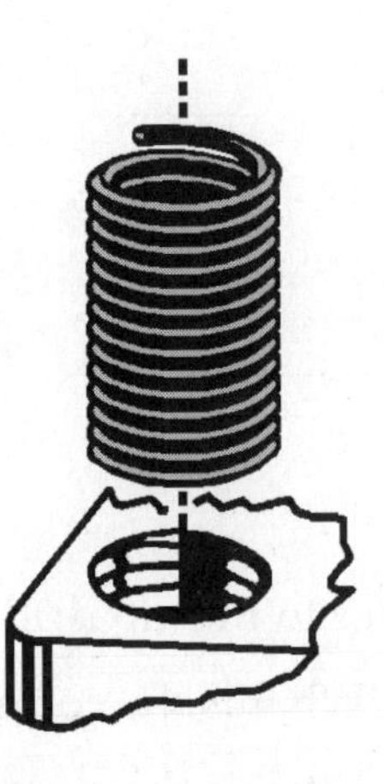

____

STOP

Instructor OK ____________ Score ____________

Activity Sheet #8

HAND TOOLS

Name______________________ Class ______________

Directions: Match the words on the left to the descriptions on the right. Write the letter for the correct word on the line provided. For the terms that you are not certain of, use the glossary in your textbook.

A. Flare nut wrench	____	Hacksaw blade that is the best choice for cutting thick steel
B. Stubby	____	Type of socket used with air tools
C. Impact driver	____	Used with a puller to pull a bearing from a shaft
D. Speed handle	____	Can be used on a 6-point fastener head
E. Dead blow hammer	____	Measures the turning effort applied to a fastener
F. Slide hammer	____	Commonly called Allen wrenches
G. Bearing separator	____	Used to move around under vehicles
H. Breaker bar	____	What a very short screwdriver is commonly called
I. Coarse tooth	____	Best socket to use on rusty fasteners
J. 8-point socket	____	Never used as a prybar
K. 12-point socket	____	Fits between a socket and a ratchet
L. 6-point socket	____	Tool used to loosen and tighten fuel line fittings
M. Fine tooth	____	Wrench used with a ratchet
N. Impact wrench	____	Always remove mushroom edge before using
O. Impact socket	____	Hand tool that is used to loosen fasteners that are very tight
P. Vise grips	____	Puller that uses a heavy weight that is slid against its handle
Q. Creeper	____	Worn whenever working in the shop
R. Torque wrench	____	Box that drill bits are stored in
S. Extension	____	Used to protect the vehicle's finish
T. Safety goggles	____	Screwdriver that is pounded on with a hammer to loosen a screw
U. Screwdriver	____	Used on square drive fastener heads
V. Chisel	____	Air-powered tool used to remove fasteners
W. Hex wrench	____	Pliers that can be locked to a part
X. Crowfoot wrench	____	Hacksaw blade that is the best choice for cutting sheetmetal
Y. Drill index	____	A soft-faced hammer that has metal shot in its head
Z. Fender cover	____	Hand tool used for quickness when assembling parts

STOP

Instructor OK ______________ Score ____________

Activity Sheet #9 Part 1

IDENTIFY SHOP TOOLS AND EQUIPMENT

Name______________________________ Class ____________________

Directions: Identify the shop tools and equipment. Place the identifying letter next to the name of the item on page 22.

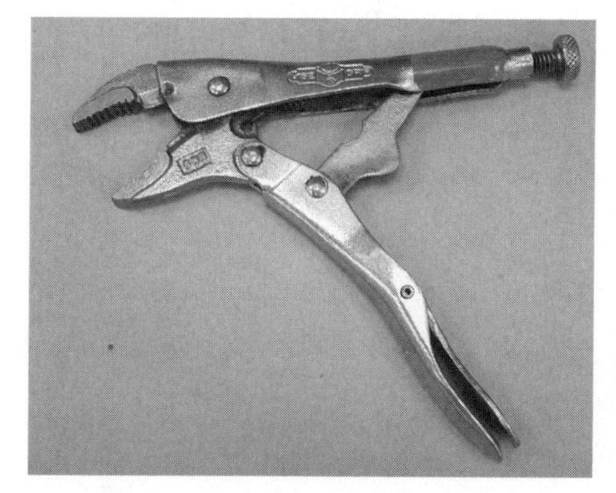

A

B

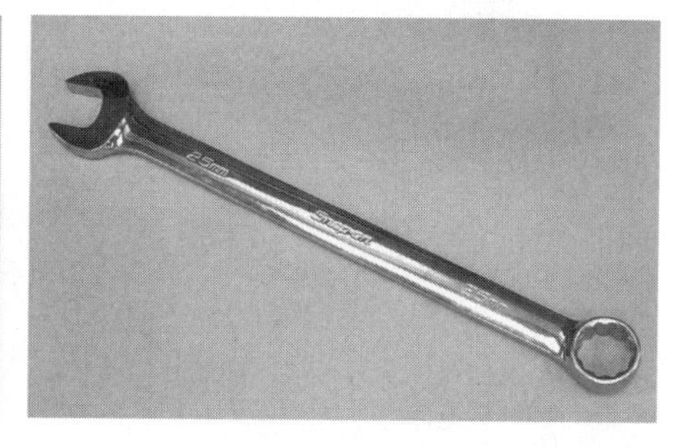

C

D

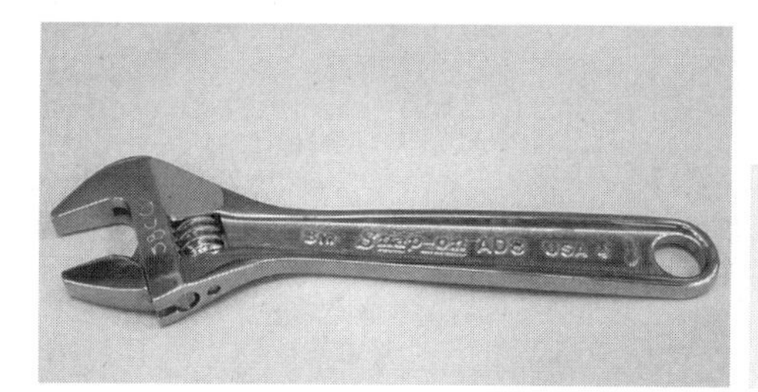

E

F

TURN

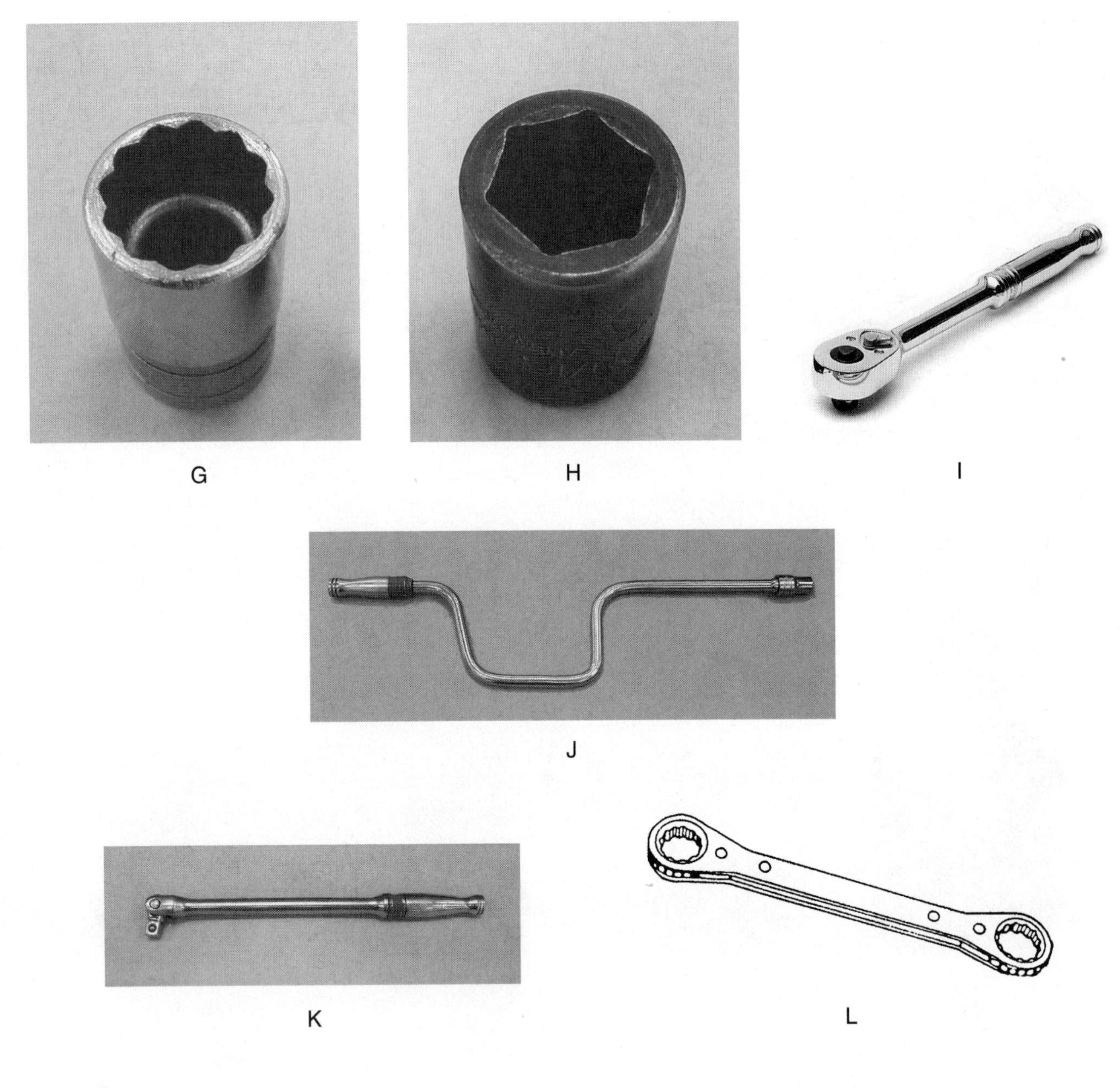

Chisel	___	Vise Grip	___	Adjustable End Wrench	___
Ratcheting Box Wrench	___	Regular Socket	___	Impact Socket	___
Ratchet	___	Socket Adapter	___	Speed Handle	___
Combination Wrench	___	Breaker Bar	___	Socket Extension	___

STOP

Instructor OK ______________ Score __________

Activity Sheet #9 Part 2

IDENTIFY SHOP TOOLS AND EQUIPMENT

Name______________________________ Class ____________________

Directions: Identify the shop tools and equipment. Place the identifying letter next to the name of the item on page 24.

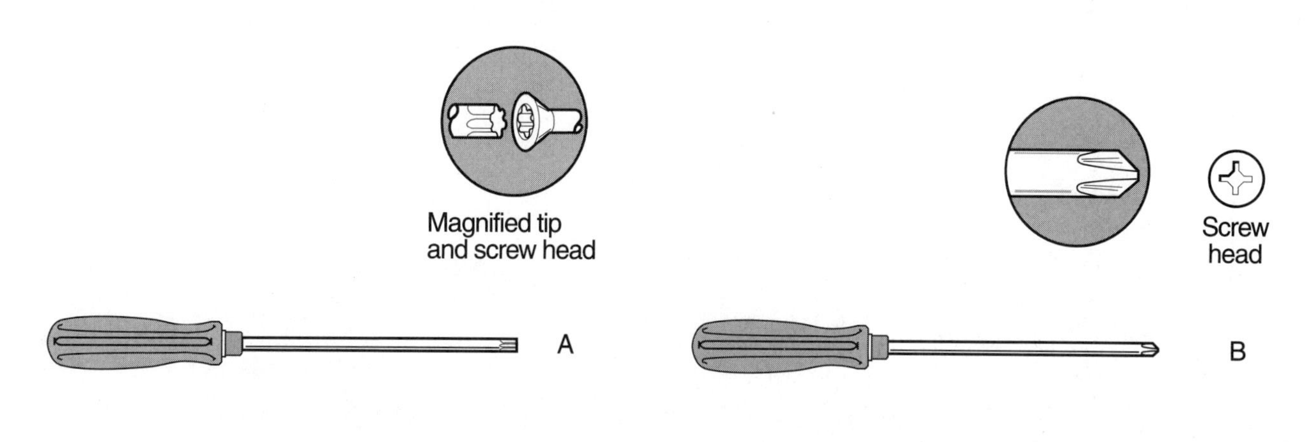

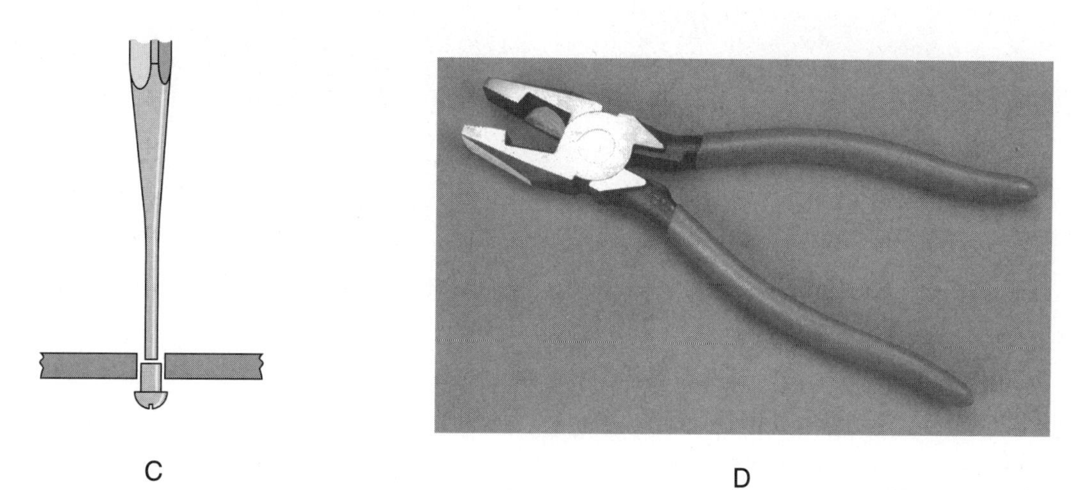

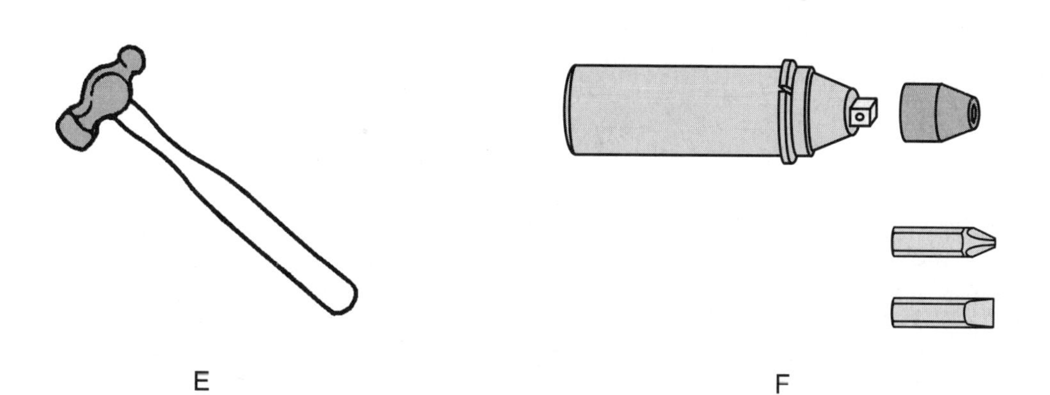

TURN

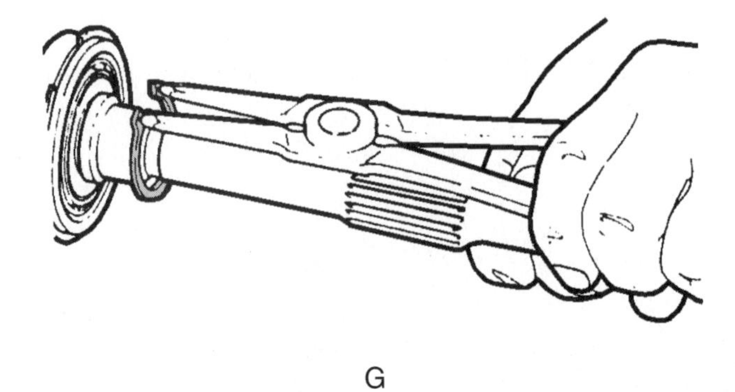

G

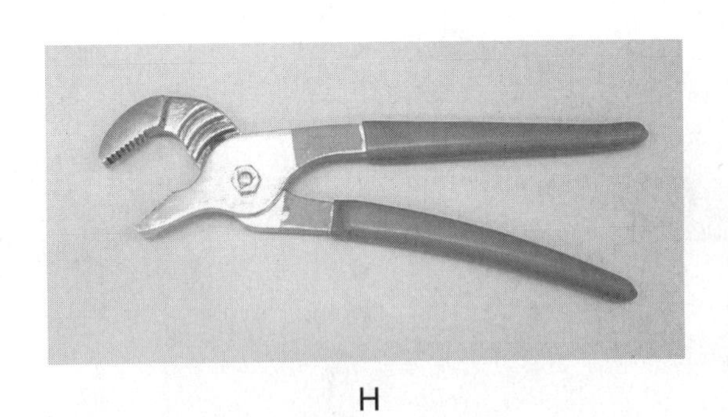

H

I

Cavity filled with shot

J

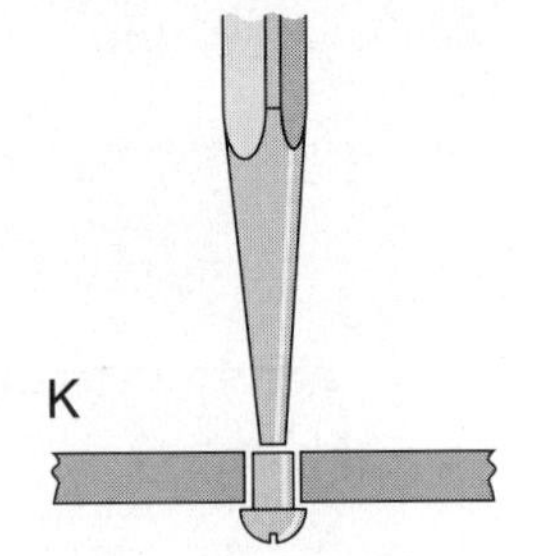

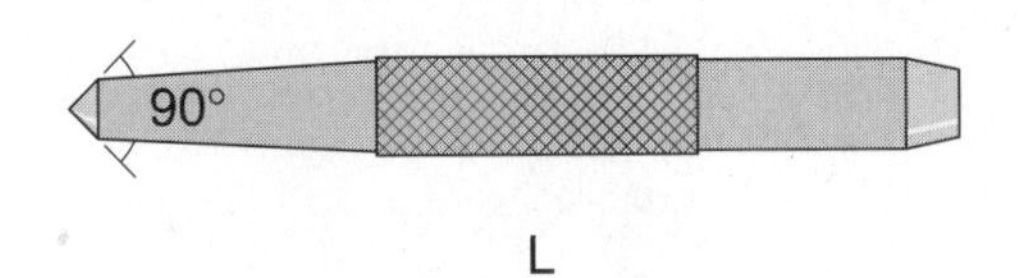

L

Rib Joint Pliers	___	Lineman's Pliers	___	High Leverage Pliers	___
Pin Punch	___	Snap Ring Pliers	___	Center Punch	___
Starting Punch	___	Impact Screwdriver	___	Torx Screwdriver	___
Ball Peen Hammer	___	Dead Blow Hammer	___	Phillips Screwdriver	___

STOP

Activity Sheet #9 Part 3

IDENTIFY SHOP TOOLS AND EQUIPMENT

Name_____________________________ Class ____________________

Directions: Identify the shop tools and equipment. Place the identifying letter next to the name of the item on page 26.

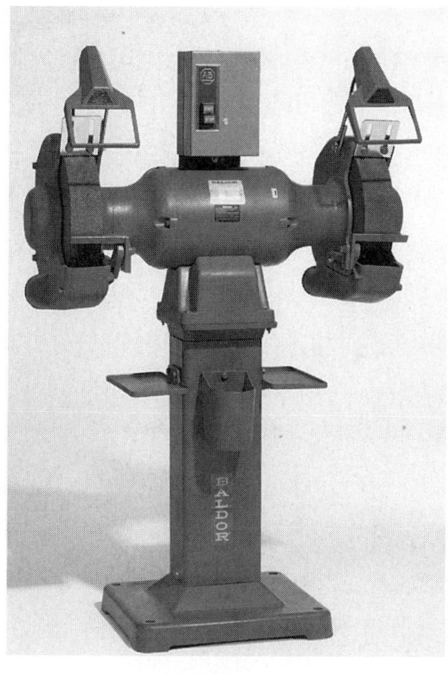

A

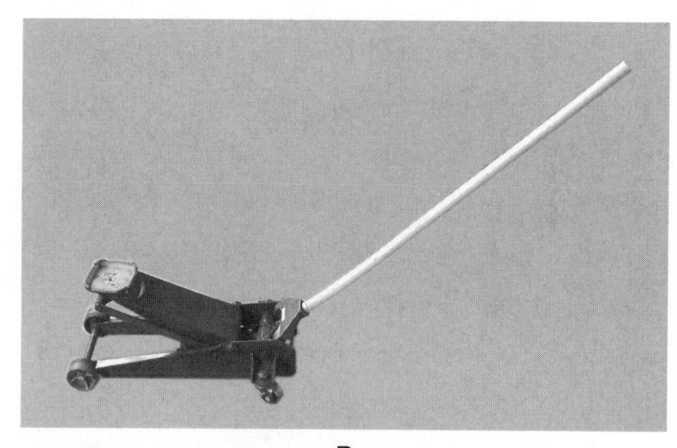

B

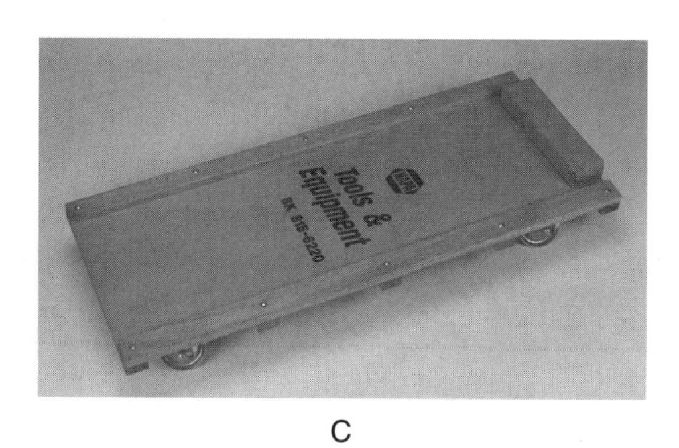

C

D

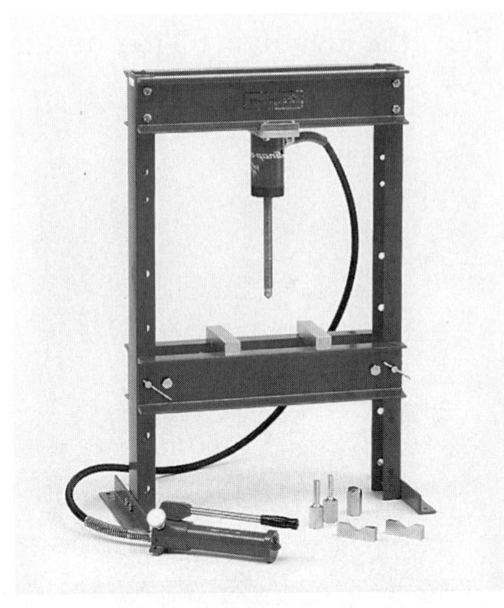

E

F

G

Jack	___	Jack Stand	___	Creeper	___
Press	___	Grinder	___	Battery Charger	___
Tire Changer	___				

STOP

Instructor OK ______________ Score __________

Activity Sheet #9 Part 4
IDENTIFY SHOP TOOLS AND EQUIPMENT

Name______________________________ Class ______________

Directions: Identify the shop tools and equipment. Place the identifying letter next to the name of the item at the top of page 28.

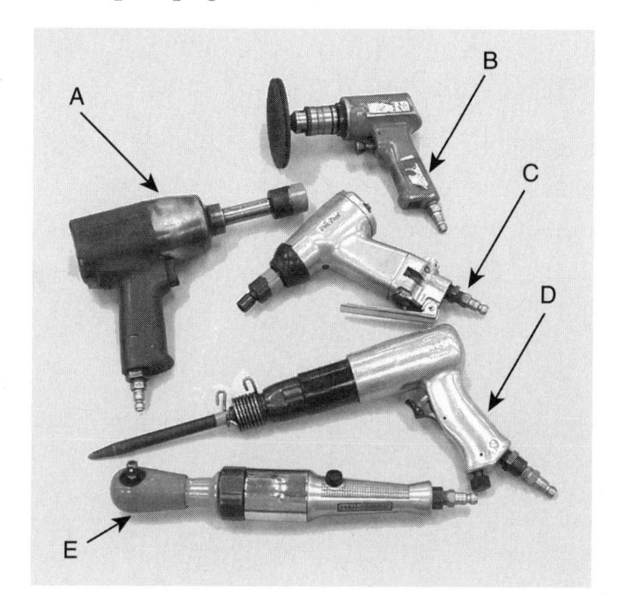

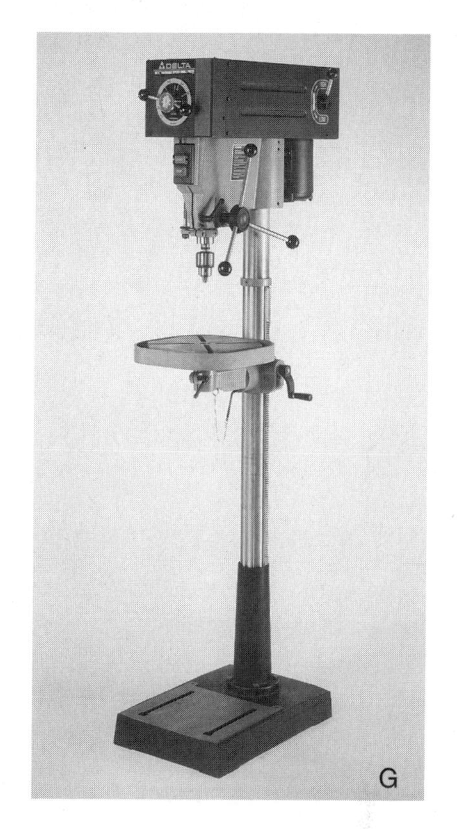

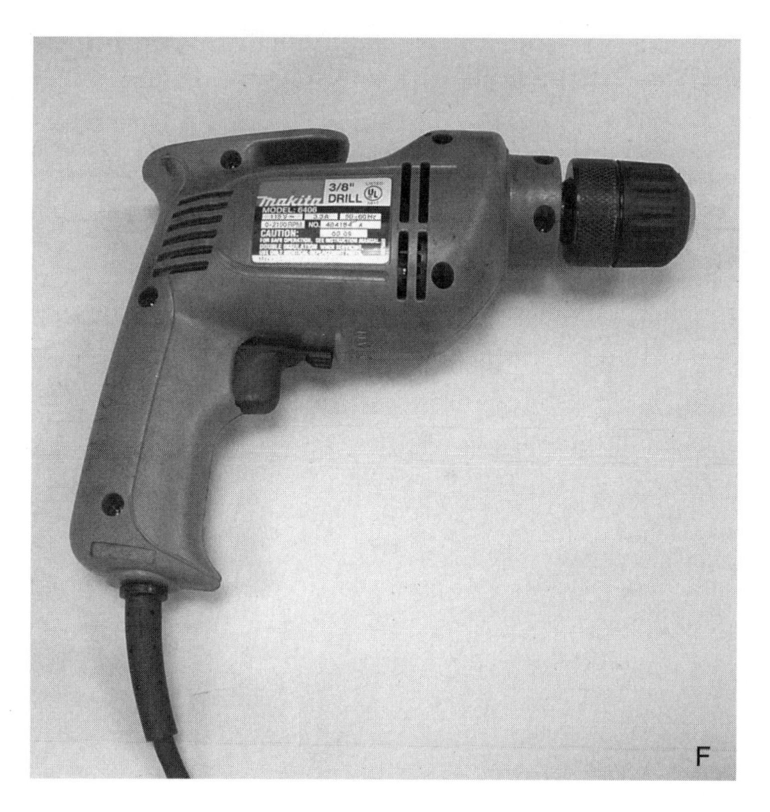

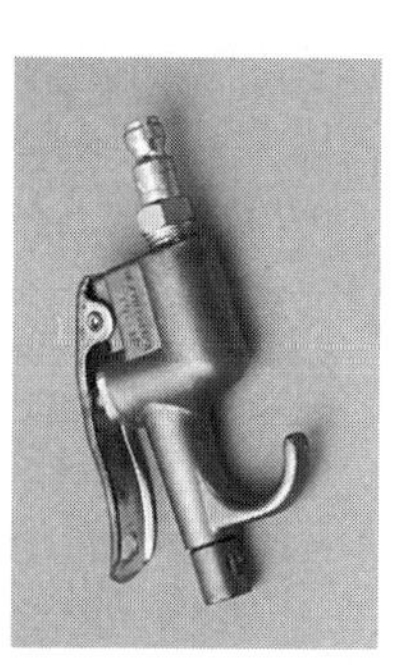

H

TURN

Electric Drill Motor ___

Air Drill ___

Drill Press ___

1/2″ Impact Wrench ___

Air Blowgun ___

Air Ratchet ___

3/8″ Impact Wrench ___

Air Hammer ___

STOP

Part I

Activity Sheets

Section Three

Vehicle Inspection (Lubrication/Safety Check)

Instructor OK ______________ Score ____________

Activity Sheet #10

FIRE EXTINGUISHERS

Name______________________________ Class ____________________

	Class of Fire	Typical Fuel Involved	Type of Extinguisher
Class **A** Fires (green)	**For Ordinary Combustibles** Put out a Class A fire by lowering its temperature or by coating the burning combustibles.	Wood Paper Cloth Rubber Plastics Rubbish Upholstery	Water*[1] Foam* Multipurpose dry chemical[4]
Class **B** Fires (red)	**For Flammable Liquids** Put out a Class B fire by smothering it. Use an extinguisher that gives a blanketing flame-interrupting effect; cover whole flaming liquid surface.	Gasoline Oil Grease Paint Lighter fluid	Foam* Carbon dioxide[5] Halogenated agent[6] Standard dry chemical[2] Purple K dry chemical[3] Multipurpose dry chemical[4]
Class **C** Fires (blue)	**For Electrical Equipment** Put out a Class C fire by shutting off power as quickly as possible and by always using a nonconducting extinguishing agent to prevent electric shock.	Motors Appliances Wiring Fuse boxes Switchboards	Carbon dioxide[5] Halogenated agent[6] Standard dry chemical[2] Purple K dry chemical[3] Multipurpose dry chemical[4]
Class **D** Fires (yellow)	**For Combustible Metals** Put out a Class D fire of metal chips, turnings, or shavings by smothering or coating with a specially designed extinguishing agent.	Aluminum Magnesium Potassium Sodium Titanium Zirconium	Dry powder extinguishers and agents only

Directions: Use the chart to answer the following questions.

1. What kind of fire extinguisher can be used to put out an oil, gasoline, or grease fire?

____

2. What kind of fire extinguisher is used to put out electrical fires?

____

3. What kind of fire extinguisher is used to put out a wood or paper fire?

____

STOP

Instructor OK ____________ Score ____________

Activity Sheet #11

ENGINE OIL

Name____________________ Class ____________

Directions: Fill in the information in the spaces provided.

1. What does SAE represent? ______________________________
2. What does API represent? ______________________________
3. What does ASTM represent? ______________________________
4. What does the "W" in the SAE rating mean? ______________________________
5. What symbol identifies that the motor oil meets tough CMA standards?

   ____

6. Which oil is thicker when hot?

   SAE 20W–40 _____

   SAE 40 _____

7. Which multiviscosity oil is recommended for normal driving at each of the following temperatures?

   20°F (–5°C) ______________________________

   40°F (5°C) ______________________________

   105°F (40°C) ______________________________

   0°F (–13°C) ______________________________

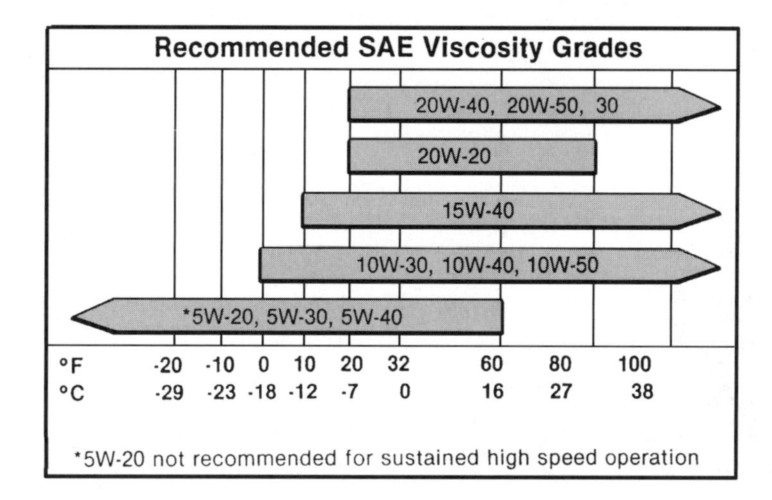

8. What single-viscosity oil is recommended for each of the following temperatures?

40°F (5°C) ______________________

105°F (40°C) ______________________

0°F (–13°C) ______________________

9. Write the following terms in their correct positions in the API donut below.

- ❑ API service SH/CE
- ❑ SAE 10W–30
- ❑ Energy conserving II

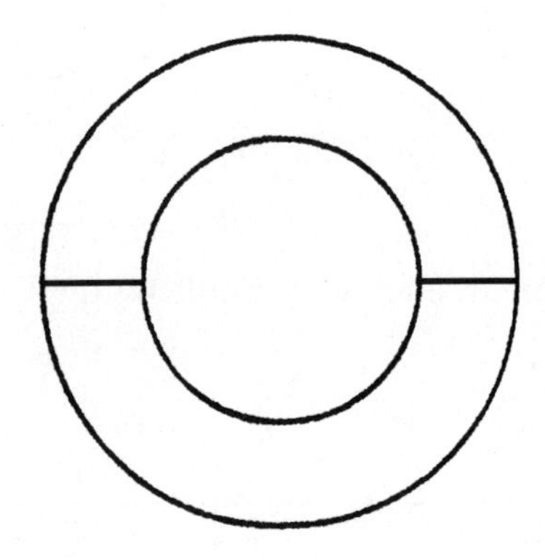

10. Place an X on the line(s) that best identifies the symbol below.

Multiviscosity oil _____

Single-viscosity oil _____

CMA standards _____

STOP

Activity Sheet #12

IDENTIFY PARTS OF THE OIL FILTER

Name______________________________ Class ____________________

Directions: Identify the parts of the oil filter in the drawing. Place the identifying letter next to the name of the part.

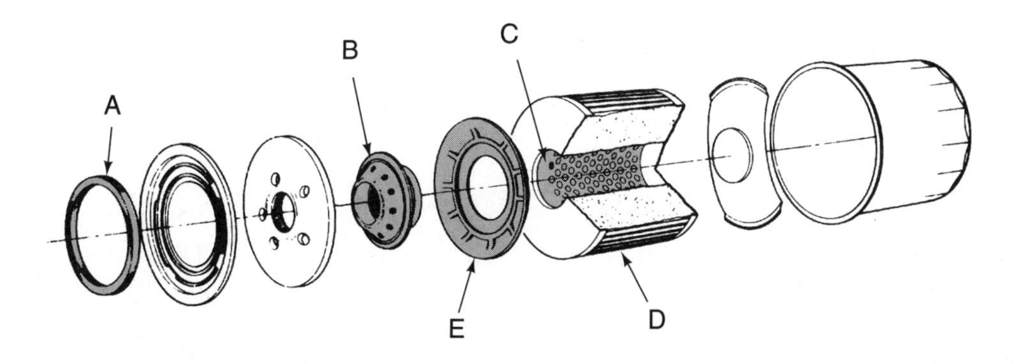

Anti-Drainback Valve ____ By-pass Valve ____ Metal Center ____

Paper Element ____ Gasket ____

1. Which way does the oil flow through the filter element?

   outside to inside ____

   inside to outside ____

2. Draw an arrow(s) to show where the oil enters the filter pictured above.

STOP

Activity Sheet #13

IDENTIFY MAJOR UNDERCAR SERVICE COMPONENTS

Name_____________________________ Class ___________________

Directions: Identify the major undercar service components of an automobile in the drawing. Place the identifying letter next to the name of the component listed below.

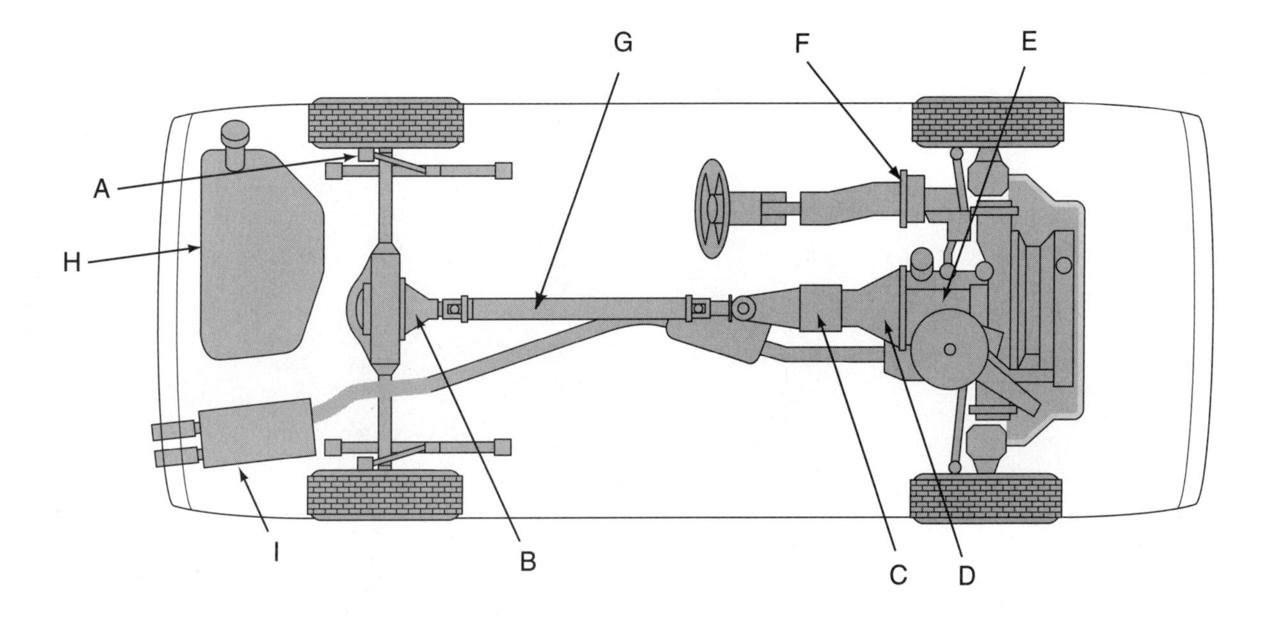

Brakes	____	Steering	____	Transmission	____
Drive Shaft	____	Differential	____	Fuel Tank	____
Clutch	____	Engine	____	Muffler	____

STOP

Part I

Activity Sheets

Section Four

Engine Operation and Service

Instructor OK ______________ Score __________

Activity Sheet #14
IDENTIFY FOUR-STROKE CYCLE

Name______________________________ Class ____________________

Directions: Identify which stroke is occurring in each sketch. Place the identifying letter next to the name of the stroke.

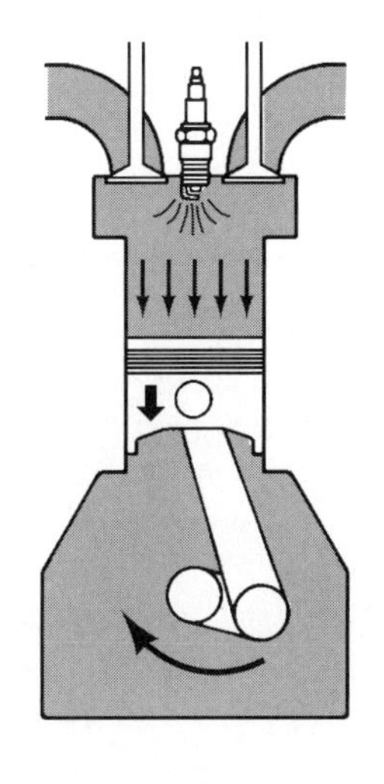

A

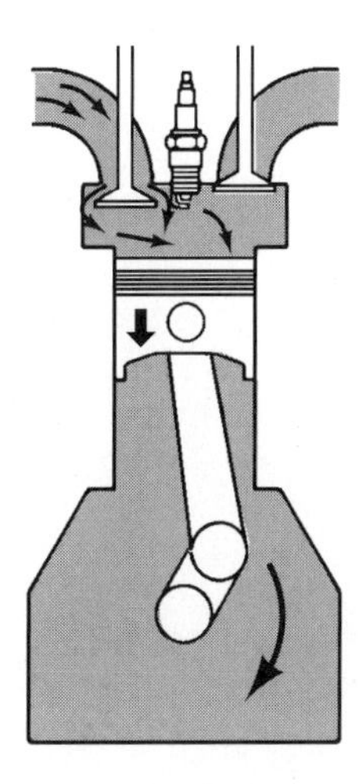

B

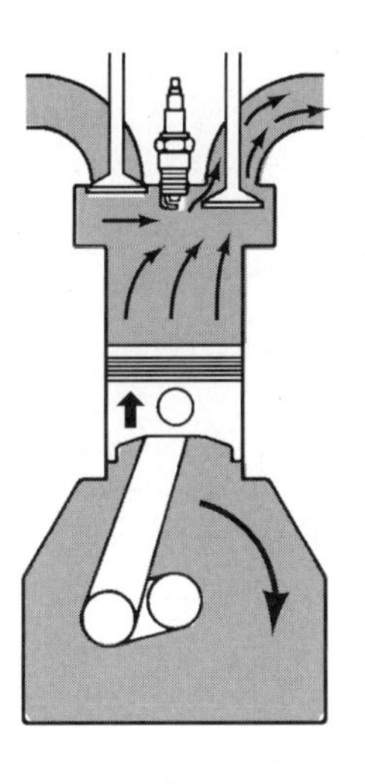

C

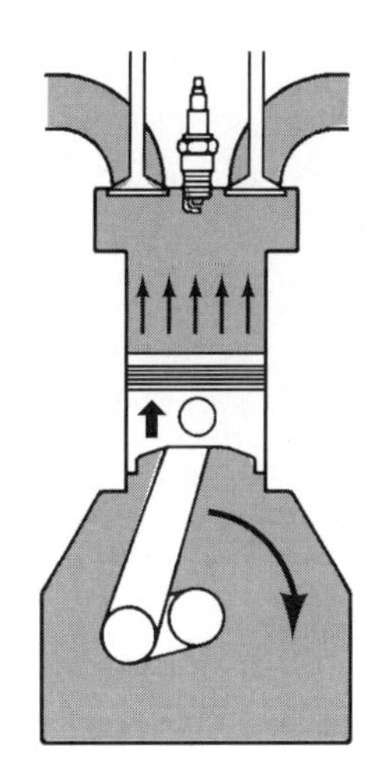

D

Intake ______ Power ______

Compression ______ Exhaust ______

STOP

Instructor OK ______________ Score __________

Activity Sheet #15

IDENTIFY ENGINE OPERATION

Name______________________________ Class ____________________

Directions: Identify the missing information in the drawing. Place the identifying letter next to the best answer.

A (full cycle); 360° | B; 180° | C | 180° | 180°

	FIRST STROKE	SECOND STROKE	THIRD STROKE	FOURTH STROKE
	180°	C	180°	180°
FIRST CYLINDER	POWER	D	INTAKE	COMPRESSION
SECOND CYLINDER	COMPRESSION	E	EXHAUST	INTAKE
THIRD CYLINDER	F	COMPRESSION	POWER	EXHAUST
FOURTH CYLINDER	EXHAUST	INTAKE	G	POWER

Power Stroke	____	Intake Stroke	____	720°	____
Compression Stroke	____	360°	____	Exhaust Stroke	____
180°	____				

Related Four–Stroke Cycle Questions

1. A revolution is ______ ° of crankshaft rotation.
2. The camshaft rotates ______ ° during one four–stroke cycle.
3. The crankshaft rotates ______ ° during one four–stroke cycle.
4. How many revolutions does the crankshaft make per second at 3000 rpm?

   ____

5. At 3000 rpm, how many times does each intake valve open in one minute?

   ____

6. At 3000 rpm, how many times does each intake valve open in one second?

   ____

7. At 3000 rpm, how many times does each exhaust valve open in one minute?

   ____

8. At 3000 rpm, how many times does each exhaust valve open in one second?

   ____

9. At 3000 rpm, how many four–stroke cycles occur each minute in a cylinder?

   ____

10. At 3000 rpm, how many four–stroke cycles occur each second in a cylinder?

    ____

Instructor OK ______________ Score __________

Activity Sheet #16

ENGINE OPERATION

Name____________________________ Class ____________________

Directions: Match the words on the left to the descriptions on the right. Write the letter for the correct word on the line provided. For the terms that you are not certain of, use the glossary in your textbook.

A. Poppet valve	____ Number of degrees the camshaft turns during one four–stroke cycle
B. Blowby	____ Number of degrees the crankshaft turns during one four–stroke cycle
C. Upper end	____ Number of times a valve opens during 720° of crankshaft rotation
D. Valvetrain	____ Forces the piston ring against the cylinder wall
E. Pushrod engine	____ Exhaust valve is open during part of this stroke
F. Overhead cam engine	____ Job of top two piston rings
G. Long block	____ Space between a bearing and shaft
H. Short block	____ Intake valve is open during part of this stroke
I. Crankcase	____ May be larger than the exhaust valve
J. End play	____ Number of cam lobes for a typical pushrod V8 camshaft
K. Bearing clearance	____ May be driven by the camshaft
L. Piston slap	____ Another name for the harmonic balancer
M. Once	____ Includes the cylinder head(s) and valvetrain
N. OHC	____ Leakage of gases past the rings
O. Exhaust	____ Engine with the cam above the cylinder head
P. 50	____ Camshaft turns ______ as fast as the crankshaft
Q. Vibration damper	____ Noise when there is excessive piston-to-cylinder wall clearance
R. Combustion	____ The parts that open and close the valves
S. Seal compression	____ Non-magnetic valve
T. 720°	____ Overhead cam
U. 360°	____ The style of valve used by four-stroke cycle internal combustion engines
V. 16	____ Back and forth clearance

TURN

W. Half	____	Number of times each valve opens every second at 6000 rpm
X. Oil pump	____	Rebuilt engine including heads
Y. Intake valve	____	Area surrounding the crankshaft
Z. Compression stroke	____	Rebuilt engine without heads
AA. Intake stroke	____	Engine with the cam in the block

STOP

Instructor OK ______________ Score ___________

Activity Sheet #17

IDENTIFY ENGINE PARTS

Name______________________________ Class ___________________

Directions: Identify the engine parts in the drawing. Place the identifying letter next to the name of the part.

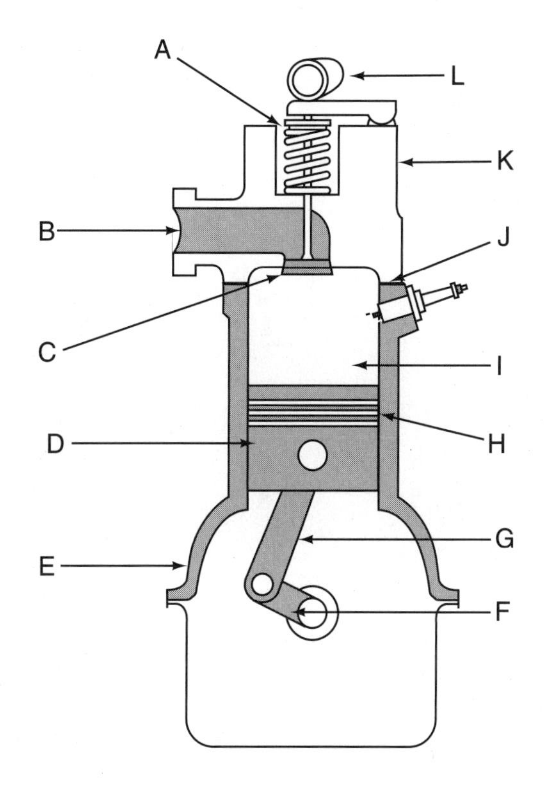

Connecting Rod	____	Cam Lobe	____	Port	____
Crankshaft	____	Block	____	Head Gasket	____
Piston	____	Valve	____	Head	____
Valve Spring	____	Cylinder	____	Piston Rings	____

STOP

Instructor OK ______________ Score __________

Activity Sheet #18

IDENTIFY CYLINDER BLOCK ASSEMBLY

Name______________________________ Class ____________________

Directions: Identify the cylinder block assembly in the drawing. Place the identifying letter next to the name of the component.

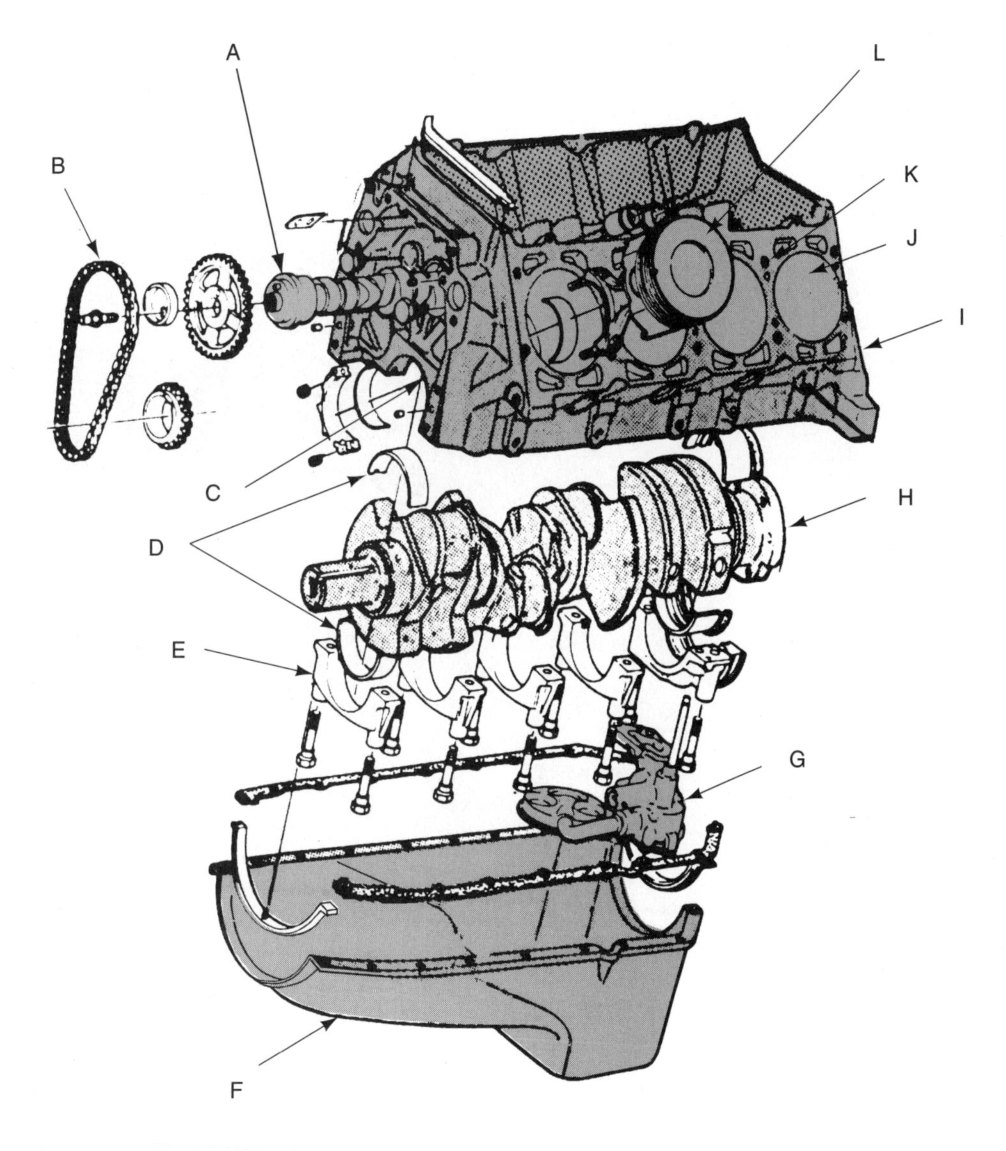

Crankshaft	____	Block	____	Bore	____
Deck	____	Piston	____	Camshaft	____
Main Caps	____	Oil Pan	____	Oil Pump	____
Main Bore	____	Main Bearing	____	Timing Chain	____

Instructor OK ______________ Score __________

Activity Sheet #19

IDENTIFY CYLINDER HEAD CLASSIFICATIONS

Name______________________________ Class __________________

Directions: Identify the cylinder head classifications in the drawing. Place the identifying letter next to the correct name at the bottom of the page.

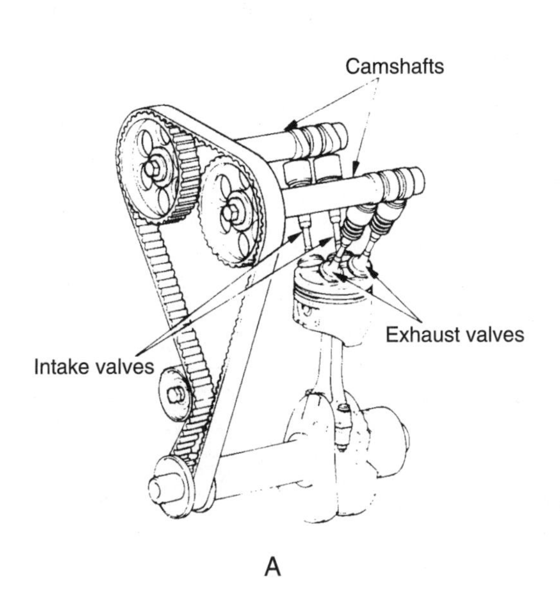

A

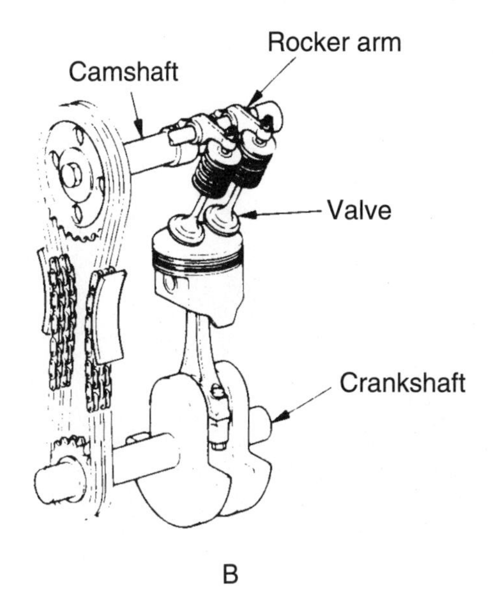

B

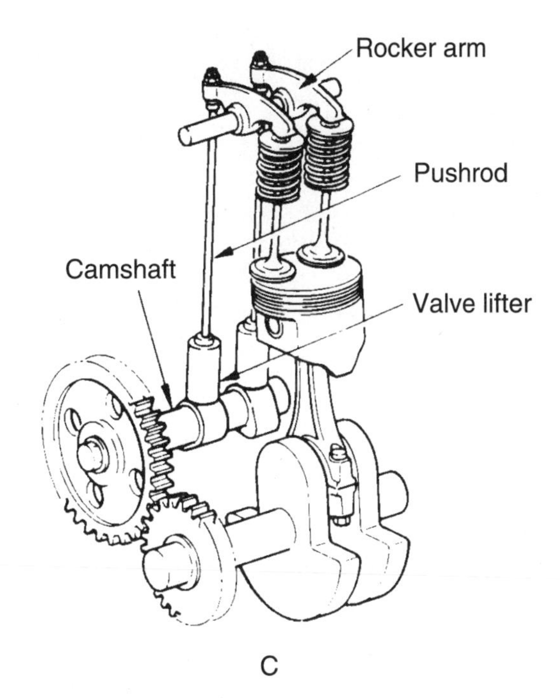

C

OHC _____

DOHC _____

Cam-in-Block _____

STOP

Instructor OK ______________ Score __________

Activity Sheet #20

IDENTIFY ENGINE BLOCK CONFIGURATIONS

Name______________________________ Class ____________________

Directions: Identify the engine block configurations in the drawing. Place the identifying letter next to the name of the design listed at the bottom of the page.

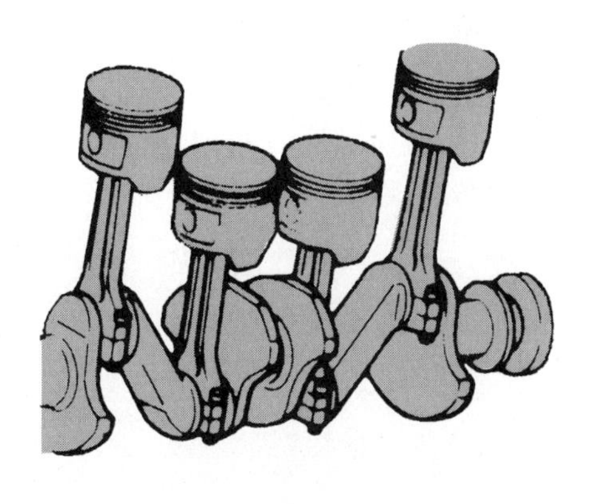

A

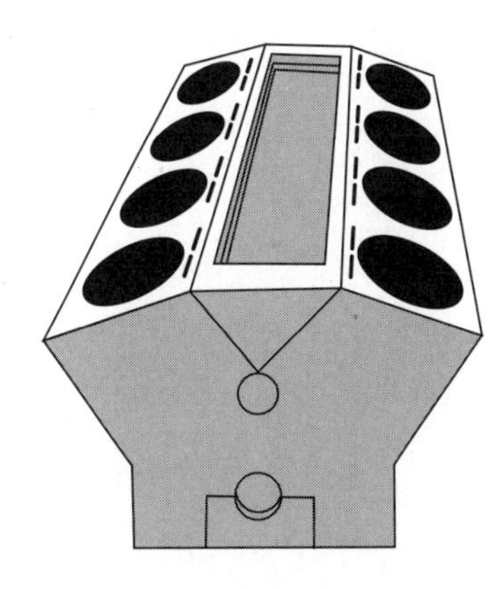

C

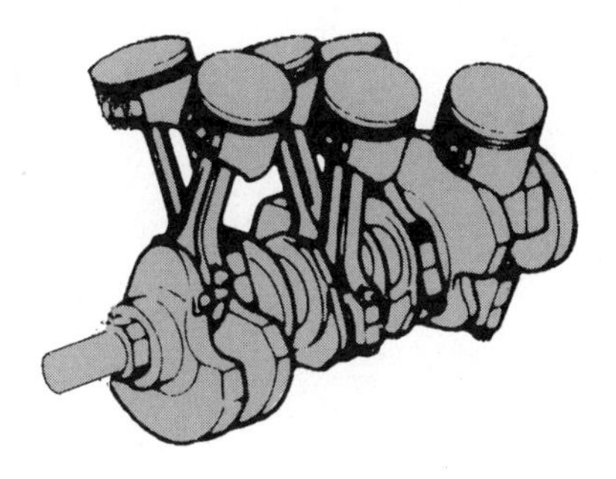

B

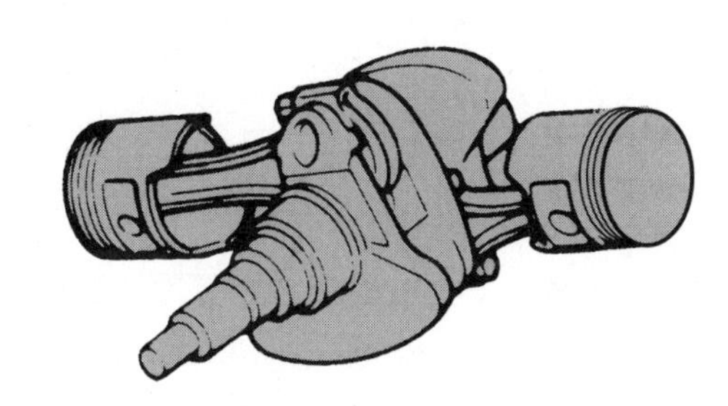

D

Opposed ____ V6 ____

In-line ____ V8 ____

STOP

Instructor OK ______________ Score __________

Activity Sheet #21
ENGINE CLASSIFICATIONS

Name____________________________ Class ____________________

Directions: Match the words on the left to the descriptions on the right. Write the letter for the correct word on the line provided. For the terms that you are not certain of, use the glossary in your textbook.

A.	Cylinder bank	____	Cam is located in the cylinder block
B.	Firing order	____	In-line, V, opposed
C.	Companion cylinders	____	Engine has four camshafts
D.	L-head	____	Turbulent combustion chamber design
E.	I-Head	____	Volume difference between TDC and BDC
F.	Pushrod engine	____	Rich air-fuel mixture starts lean mixture burning
G.	OHC	____	Four–stroke spark ignition engine
H.	SOHC	____	Four–cylinder engine firing order
I.	DOHC	____	Zero emission vehicle
J.	Cross-flow head	____	Compression ignition engine
K.	Hemi-head	____	A row of cylinders
L.	Wedge head	____	Rows of cylinders in a V-type block
M.	Stratified charge	____	Rotary engine
N.	Otto-cycle	____	Cam is located in the cylinder head
O.	S.I. engine	____	The order in which the spark plugs fire
P.	Diesel-cycle	____	Dual overhead cam
Q.	Diesel engine	____	Area between the heads on a V–block
R.	Compression ratio	____	Valve placement is in the cylinder head
S.	Wankel engine	____	Intake and exhaust ports on opposite sides of the engine
T.	Two-stroke engine	____	Ignites its fuel mixture with a spark
U.	Z.E.V.	____	Single overhead cam
V.	Hybrid vehicle	____	Engine used in chain saws
W.	Cylinder arrangements	____	Completes cycle in one revolution

TURN

X.	Bank	____	Hemispherical combustion chamber design
Y.	Valley	____	Rotary engine design
Z.	1–3–4–2	____	Ignites fuel using heat from compression
AA.	Wankel	____	The valves are in the cylinder block
AB.	Two-stroke	____	Uses more than one type of energy
AC.	Four-stroke engine	____	Two revolutions to fire all cylinders
AD.	V8 DOHC	____	Pistons come to TDC and BDC together

STOP

Instructor OK ______________ Score __________

Activity Sheet #22

ENGINE SIZES AND MEASUREMENTS

Name______________________________ Class ____________________

Objective: After completing this assignment, you should be able to identify and calculate the displacement of an engine. This task will help prepare you to pass the ASE certification examination in engine repair.

Directions: Identify the cylinder displacement terms in the drawing below. Place the identifying letter next to the name of the component.

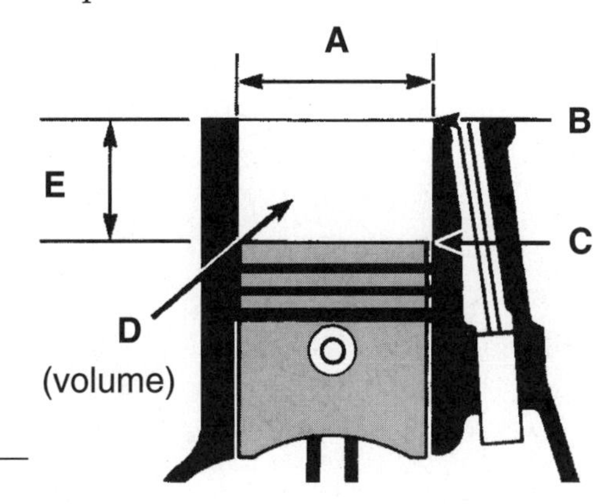

TDC __________

Bore __________

BDC __________

Stroke __________

CI, L, or CCs __________

1. Calculate the displacement of a cylinder that has a 3.5–inch bore and a 3.5–inch stroke. Show your work in the box below.

Example: Formula for displacement of a cylinder is
B × B × S × .785

Bore 4 inches
× Bore 4 inches
16 square inches
× Stroke 4 inches
64 cubic inches
× .785
50.240 cubic inches (displacement for the cylinder)

50.240 cubic inches

TURN

2. Calculate the displacement of an eight–cylinder engine that has a 4–inch bore and a 4–inch stroke. Show your work in the box.

   Note: The formula for engine displacement is:

   $B \times B \times S \times .785 \times$ number of cylinders

3. Calculate the displacement of a six–cylinder engine that has a 4–inch bore and a 4–inch stroke. Show your work in the box.

4. Calculate the displacement of a four–cylinder engine that has a 3 1/2–inch bore and a 3 9/16–inch stroke. Show your work in the box.

5. What is the displacement in cubic centimeters (cc) of the following engines?

   1.7 liters __________ cc 2 liters __________ cc 5.7 liters __________ cc

6. What is the displacement in cubic inches (ci) of the following engines?

   1.7 liters __________ ci 2 liters __________ ci 5.7 liters __________ ci

STOP

Instructor OK ______________ Score ___________

Activity Sheet #23

ENGINE MEASUREMENTS

Name______________________________ Class ____________________

Directions: Match the words on the left to the descriptions on the right. Write the letter for the correct word on the line provided. For the terms that you are not certain of, use the glossary in your textbook.

A. Bore	____	Measurement of work in which 1 pound is moved for a distance of 1 foot
B. Stroke	____	Volume displaced by the piston
C. Oversquare	____	Cylinder volume at BDC compared to volume at TDC
D. Cylinder displacement	____	Measurement comparing the volume of airflow actually entering the engine with the maximum that theoretically could enter
E. Engine displacement	____	Ability to do work
F. Compression ratio	____	The tendency of a body to keep its state of rest or motion
G. Compression pressure	____	The diameter of the cylinder
H. Force	____	Usable crankshaft horsepower
I. Work	____	Metric horsepower equivalent
J. Foot–pound	____	The measurement of an engine's ability to perform work
K. Energy	____	Cylinder displacement times number of cylinders
L. Inertia	____	Typical gasoline engine compression pressure
M. Momentum	____	Cylinder bore larger than stroke
N. Power	____	Any action that changes, or tends to change, the position of something
O. Torque	____	The turning force exerted by the crankshaft
P. Btu	____	Pressure in cylinder as piston moves up when the valves are closed
Q. One Btu	____	When an object is moved against a resistance or opposing force
R. Horsepower	____	Name of equipment used to test an engine's power

TURN

S.	Watts	____	Heat required to heat a pound of water by 1°F
T.	Brake horsepower	____	Piston travel from TDC to BDC
U.	Road horsepower	____	How fast work is done
V.	Volumetric efficiency	____	British thermal units
W.	125–175 psi	____	Body going in a straight line will keep going the same direction at the same speed if no other forces act on it
X.	Dynamometer	____	Horsepower available at the car's drive wheels

STOP

Instructor OK ______________ Score __________

Activity Sheet #24

IDENTIFY VALVE PARTS

Name______________________________ Class ____________________

Directions: Identify the valvetrain components in the drawing. Place the identifying letter next to the name of the component.

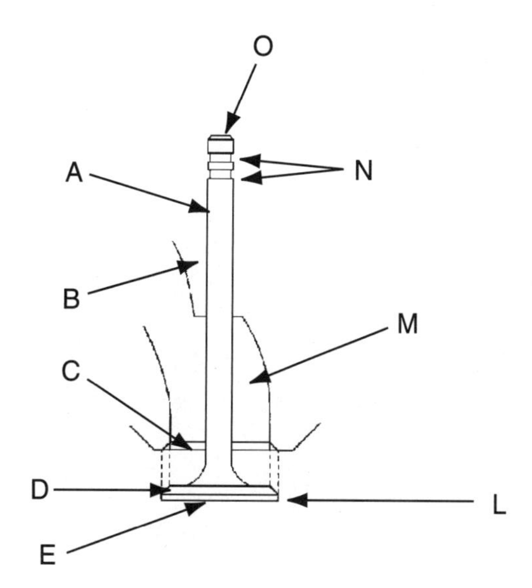

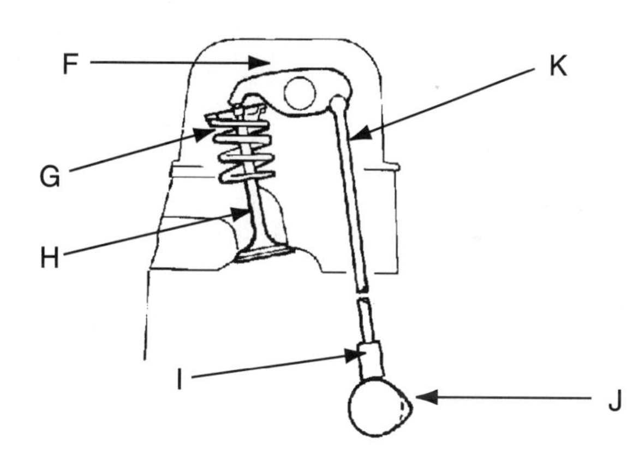

Valve Port	____	Guide	____	Lifter	____
Valve Head	____	Stem	____	Spring	____
Valve Seat	____	Keeper Groove	____	Valve	____
Margin	____	Stem Tip	____	Rocker Arm	____
Face	____	Pushrod	____	Cam Lobe	____

STOP

Instructor OK ______________ Score ____________

Activity Sheet #25

IDENTIFY OHC VALVETRAIN PARTS

Name______________________________ Class ____________________

Directions: Identify the OHC valvetrain parts in the drawing. Place the identifying letter next to the correct name.

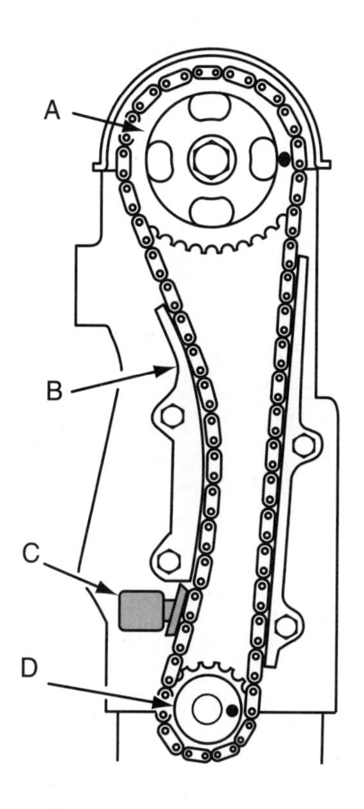

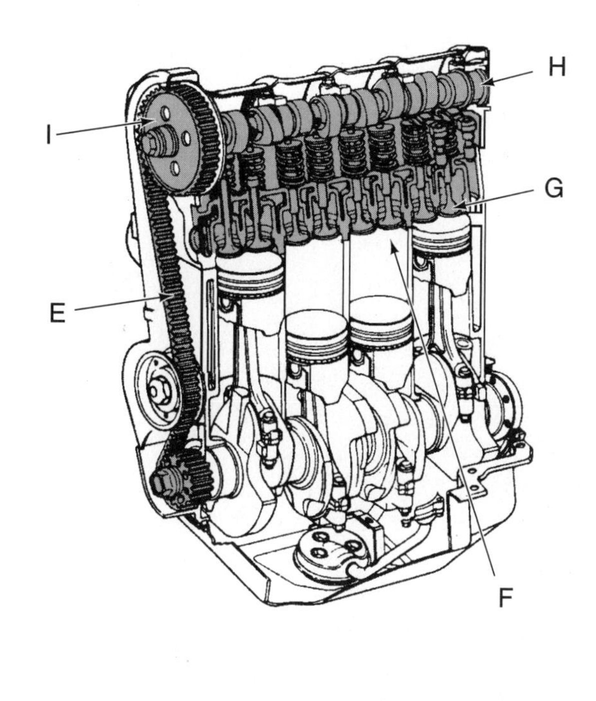

Pump Eccentric	____	Timing Belt	____	Camshaft	____
Crank Sprocket	____	Chain Tensioner	____	Valve	____
Chain Guide	____	Cam Sprocket	____	Combustion Chamber	____

STOP

Instructor OK ______________ Score ____________

Activity Sheet #26

IDENTIFY PUSHROD ENGINE COMPONENTS

Name______________________________ Class ____________________

Directions: Identify the pushrod engine components in the drawing. Place the identifying letter next to the name of the component.

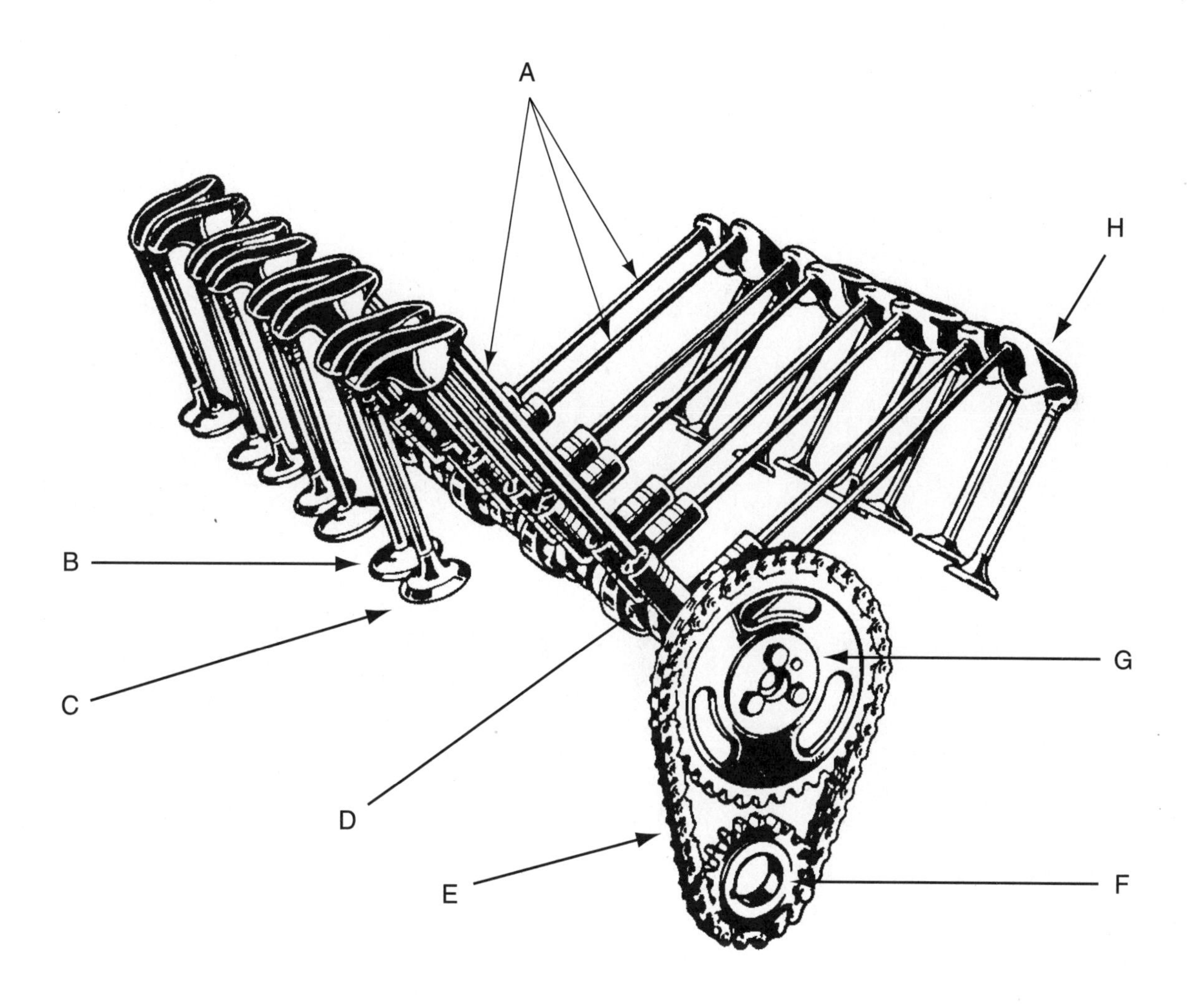

Rocket Arm	____	Pushrods	____	Timing Chain	____
Exhaust Valve	____	Crank Sprocket	____	Lifter	____
Intake Valve	____	Cam Sprocket	____		

STOP

Instructor OK ______________ Score ___________

Activity Sheet #27

CYLINDER HEAD

Name______________________________ Class ____________________

Directions: Match the words on the left to the descriptions on the right. Write the letter for the correct word on the line provided. For the terms that you are not certain of, use the glossary in your textbook.

A. Cylinder head	____ Engine produced at the factory
B. Integral guide	____ Two types of valve guides
C. Induction hardened	____ Types of valve guide seals
D. Poppet valves	____ Moves up and down with the valve stem
E. Production engine	____ Attached to the top of the valve guide
F. Stock	____ Valve guides that are made as part of the head
G. Iron or aluminum	____ Type of valve guide used in aluminum heads
H. O-ring, positive and umbrella	____ Style of valve used in cylinder heads
I. Core	____ Original equipment
J. Insert guide	____ Cylinder heads can be made of two materials
K. Umbrella seal	____ How integral valve seats are hardened
L. Positive valve seal	____ Integral seats are integral with the _____
M. Integral or insert	____ Used part that is returned

STOP

Instructor OK ______________ Score ____________

Activity Sheet #28

IDENTIFY FOUR–STROKE CYCLE EVENTS

Name______________________________ Class ____________________

Directions: Identify the events in the four–stroke cycle on the sketch. Place the identifying letter next to its name.

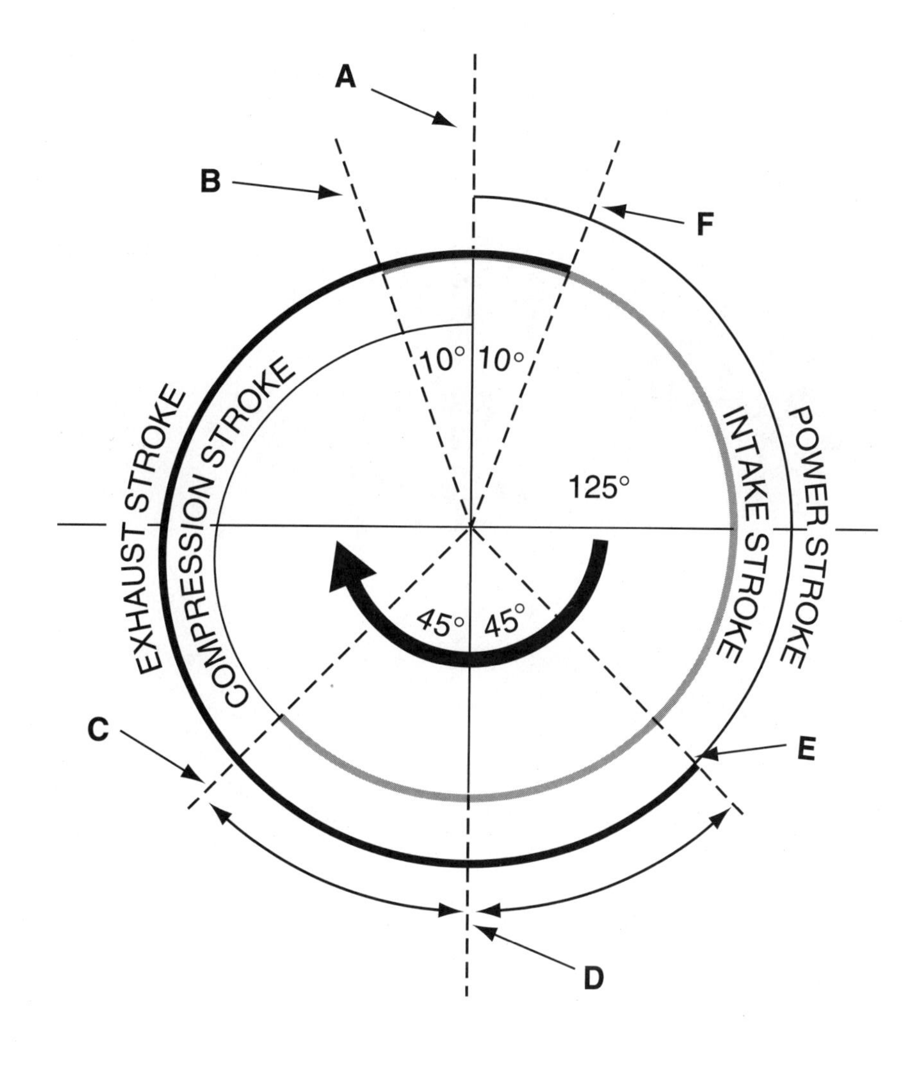

Intake Valve Opens	____	Exhaust Valve Opens	____	BDC	____
Intake Valve Closes	____	TDC	____	Exhaust Valve Closes	____

STOP

Instructor OK ______________ Score ____________

Activity Sheet #29

CAM AND LIFTERS

Name______________________________ Class ____________________

Directions: Match the words on the left to the descriptions on the right. Write the letter for the correct word on the line provided. For the terms that you are not certain of, use the glossary in your textbook.

A. Positive stop	_____ Ways that a camshaft can be driven
B. Base circle	_____ When there is no clearance
C. Lift	_____ Automatic adjustment feature used with hydraulic valve lifters
D. Duration	_____ Engine that relies on atmospheric pressure
E. Valve overlap	_____ Parts that the camshaft can drive
F. Zero-lash	_____ Lifter that must be held from turning
G. Interference engine	_____ Height the cam lobe raises the lifter
H. Freewheeling engine	_____ What would be left if the cam lobe were removed?
I. Naturally aspirated	_____ The number of degrees of crankshaft travel when the valve is open
J. Roller lifter	_____ An engine that will *not* experience piston-to-valve interference if the timing chain or belt skips or breaks
K. Chain, belt, gears	_____ An engine that will experience piston-to-valve contact if the timing chain or belt skips or breaks
L. Fuel pump, oil pump, and distributor	_____ The time that both the intake exhaust valves are open at the same time

STOP

Instructor OK ______________ Score ________

Activity Sheet #30

IDENTIFY LUBRICATION SYSTEM COMPONENTS

Name_________________________ Class ________________

Directions: Identify the lubrication system components in the drawing. Place the identifying letter next to the name of the component. Color the oil galleries with a highlighter or a colored pen.

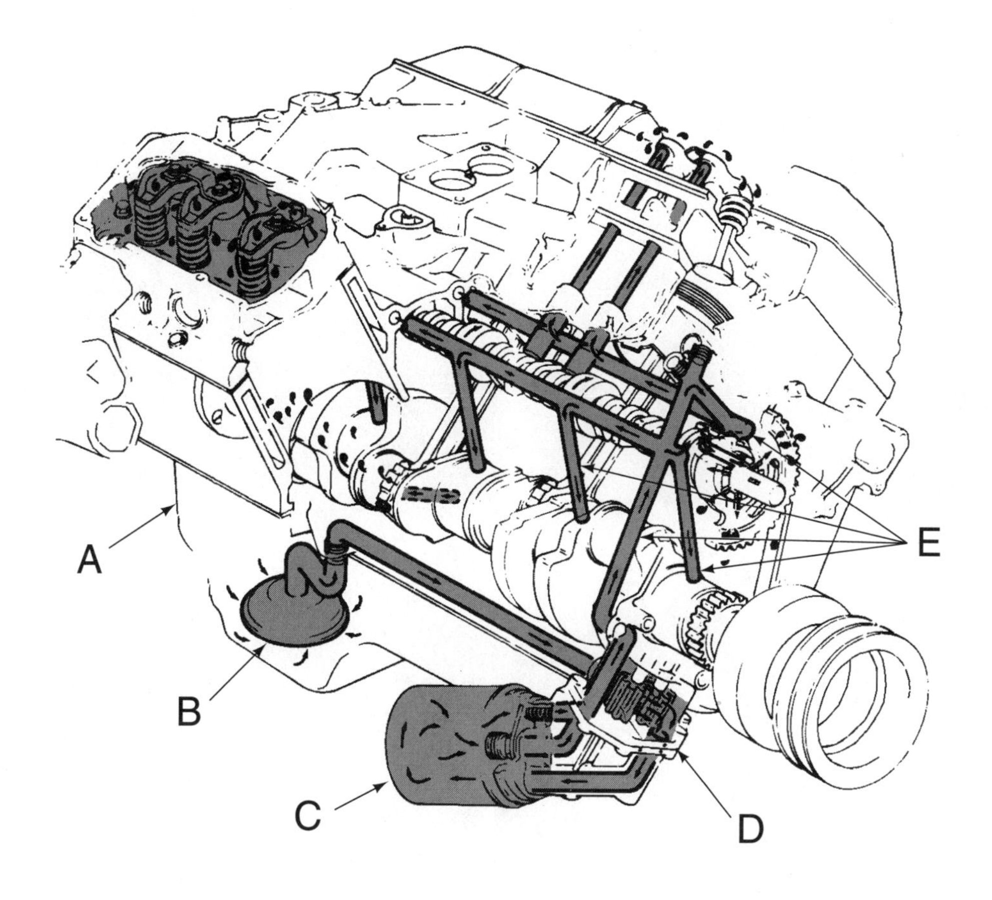

Oil Pump	____	Oil Pan	____	Oil Galleries	____
Pick-up Screen	____	Oil Filter	____		

STOP

Instructor OK ______________ Score ________

Activity Sheet #31

IDENTIFY OIL PUMP COMPONENTS

Name______________________________ Class ____________________

Directions: Identify the oil pump components in the drawing. Place the identifying letter next to the name of the component listed at the bottom of the page.

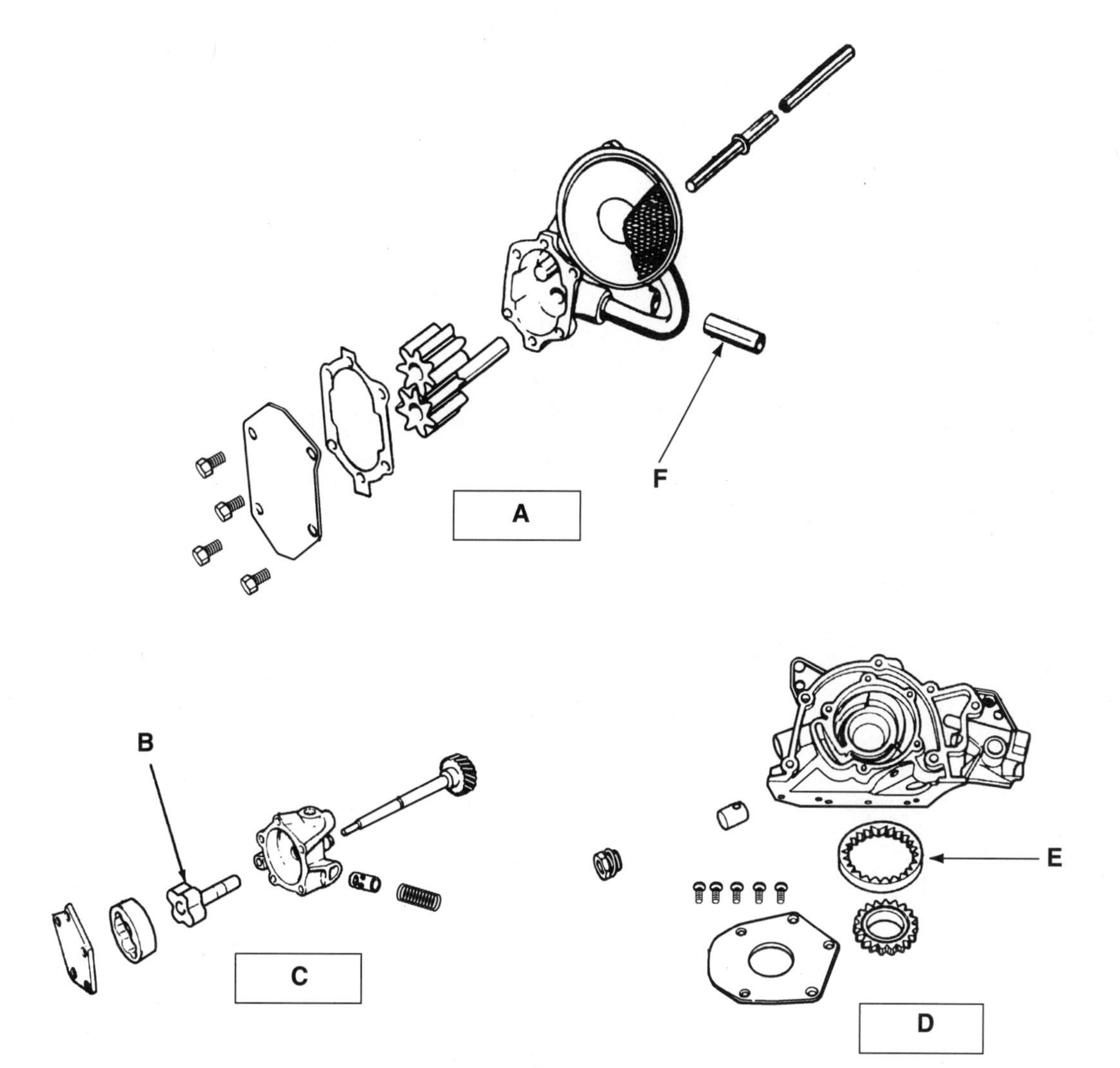

Relief Valve	____	Gear Pump	____	Rotor	____
Internal Gear	____	Internal and External Gear Pump	____	Rotor Pump	____

Instructor OK ______________ Score __________

Activity Sheet #32

IDENTIFY PISTON COMPONENTS

Name______________________________ Class ______________________

Directions: Identify the piston components in the drawing. Place the identifying letter next to the name of the component.

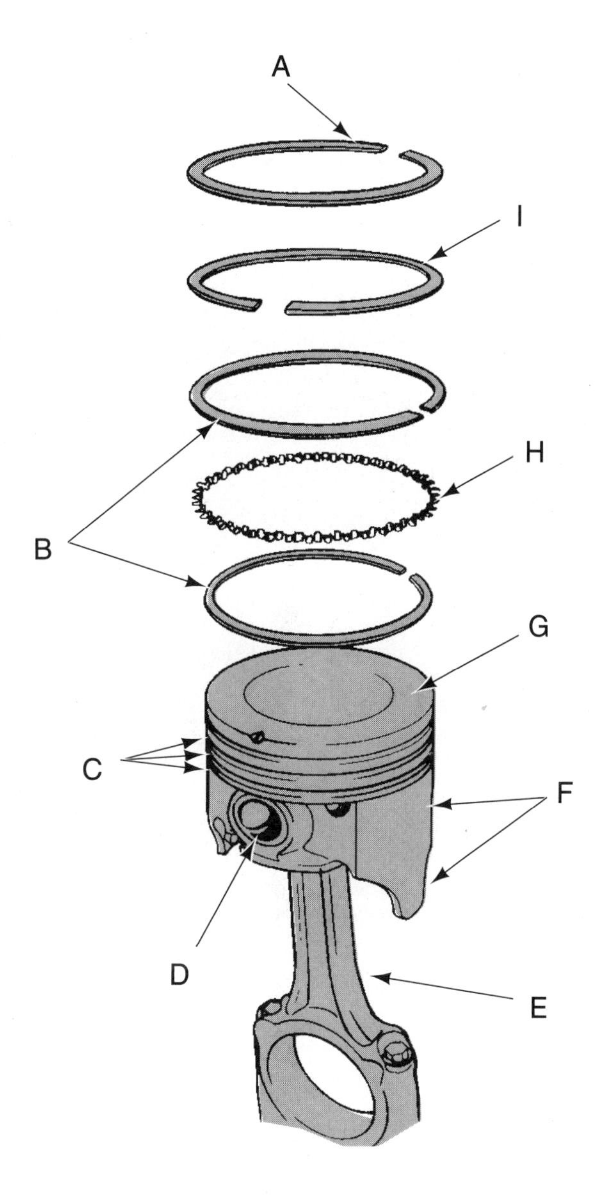

Second Ring	____	Expander Spacer	____	Piston Head	____
Rails	____	Top Ring	____	Skirt	____
Connecting Rod	____	Piston Pin	____	Ring Grooves	____

Instructor OK ______________ Score __________

Activity Sheet #33

ENGINE BLOCK

Name______________________________ Class ____________________

Directions: Match the words on the left to the descriptions on the right. Write the letter for the correct word on the line provided. For the terms that you are not certain of, use the glossary in your textbook.

A. Lower end	____	A part installed to correct a damaged cylinder
B. Cylinder taper	____	Prevents the bearing from turning in its bore
C. End thrust	____	The number of times the piston must start and stop in one second at 3000 rpm
D. Torsional vibration	____	Two types of ring facings
E. Bearing spread	____	Front or back force against a shaft
F. Bearing crush	____	Wear occurring in the top inch of a cylinder wall
G. Cam ground	____	Result of tapered wear
H. Sleeve	____	The name that describes all of the parts of a short block
I. One hundred	____	Oil passages
J. Moly, chrome	____	Space between a bearing and journal
K. Cylinder ridge	____	Piston's cold shape
L. Bearing clearance	____	Occurs when force on the pistons is imparted to the crankshaft of a V-type engine
M. Galleries	____	Holds bearing insert in place during engine assembly

STOP

Part I

Activity Sheets

Section Five

Cooling and Fuel System Theory and Service

Instructor OK ______________ Score __________

Activity Sheet #34

IDENTIFY COOLING SYSTEM COMPONENTS

Name______________________________ Class ____________________

Directions: Identify the cooling system components in the drawing. Place the identifying letter next to the name of the component.

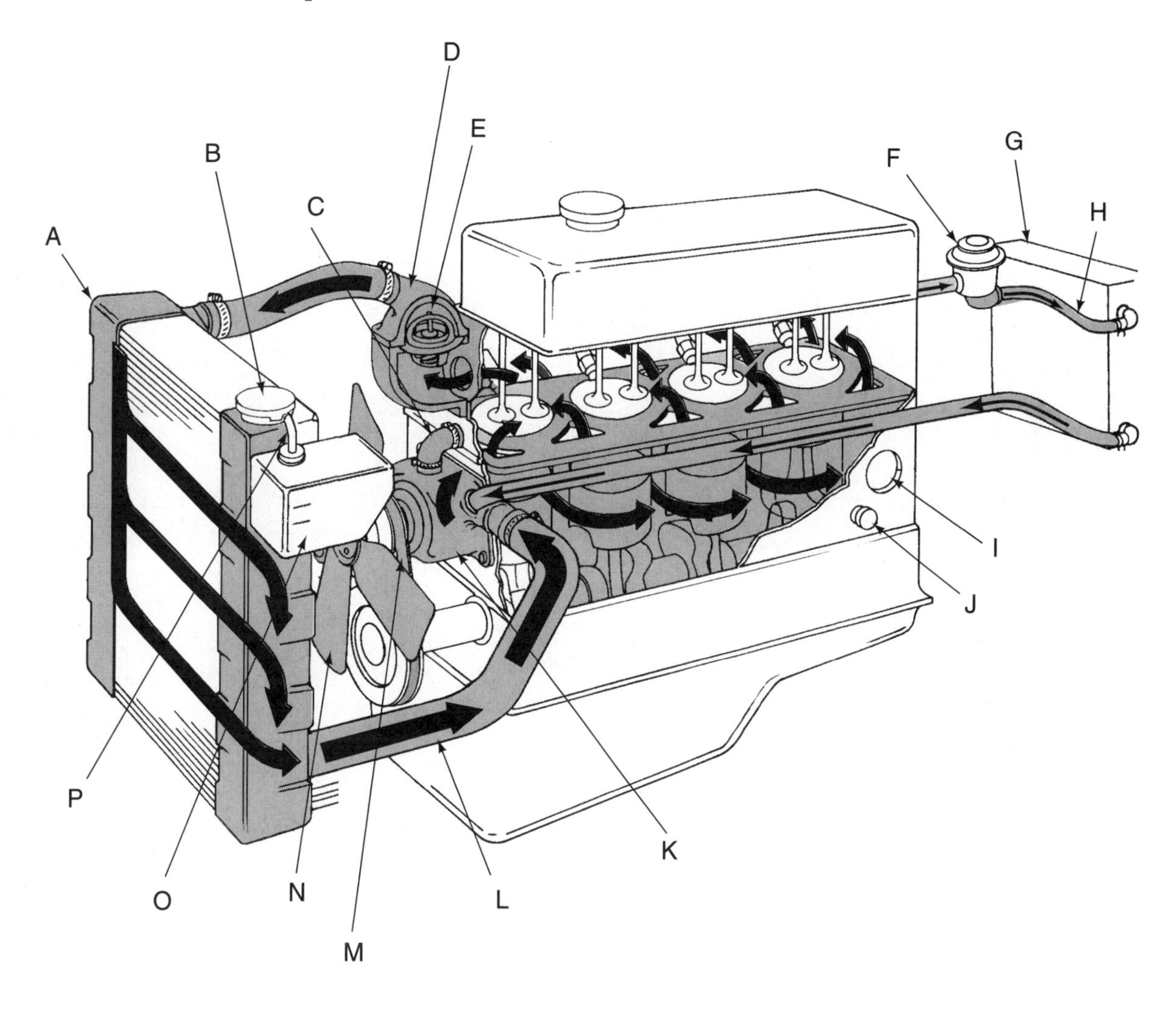

V-Belt	____	Water Pump	____	Thermostat	____
Fan	____	Heater Supply Hose	____	Overflow Tube	____
Drain Plug	____	Thermostat Housing	____	Radiator	____
Coolant Recovery Tank	____	Heater Core	____	Pressure Cap	____
Core Plug	____	By-pass Hose	____	Radiator Hose	____
Heater Control Valve	____				

Instructor OK ______________ Score __________

Activity Sheet #35

IDENTIFY RADIATOR CAP COMPONENTS

Name______________________________ Class ____________________

Directions: Identify the radiator cap components in the drawing. Place the identifying letter next to the name of the component.

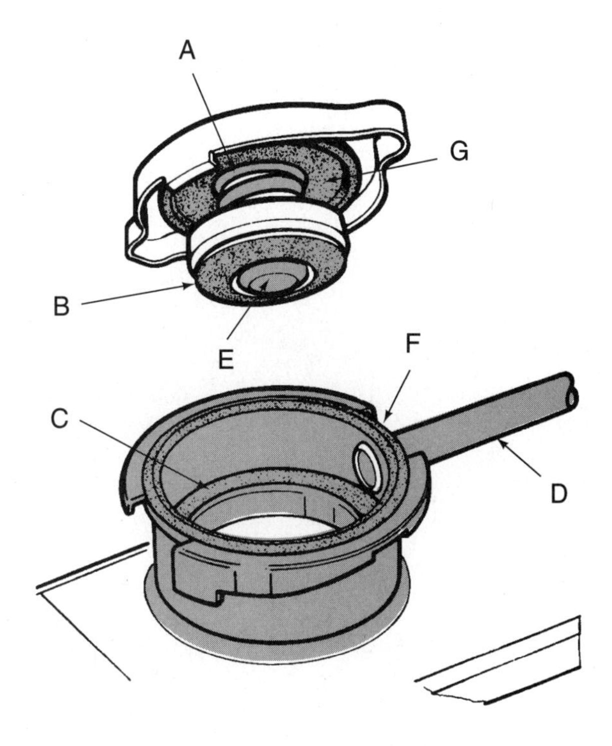

Overflow Hose	____	Vacuum Valve	____	Pressure Relief Spring	____
Upper Sealing Surface	____	Pressure Seal	____		
Lower Sealing Surface	____	Upper Sealing Gasket	____		

STOP

Instructor OK ______________ Score __________

Activity Sheet #36

ENGINE COOLING

Name______________________________ Class ____________________

Directions: Match the words on the left to the descriptions on the right. Write the letter for the correct word on the line provided. For the terms that you are not certain of, use the glossary in your textbook.

A. Down-flow radiator
B. Cross-flow radiator
C. Heat exchanger
D. Oil cooler
E. Thermostat bypass
F. Sending unit
G. Fan clutch
H. Bimetal coil spring
I. Heater core
J. Ethylene glycol
K. Cylinder blocks
L. Electrolysis
M. Silicate
N. 180–212°F
O. 3°F
P. Vacuum valve

____ Temperature or torque sensitive clutch attached to a belt-driven cooling fan

____ Amount boiling point of coolant increases under 1 PSI of pressure

____ Thermostatic coil consisting of two types of metal wound together

____ Result of two dissimilar metals in a liquid

____ Controls engine temperature

____ Normal operating temperature of an engine

____ Automotive coolant

____ Found in the radiator to cool the transmission

____ Pulls air through the radiator when the engine is warm

____ Radiator design where coolant flows from top to bottom

____ Small radiator for passenger heat

____ Another name for a heat exchanger

____ Expands to open the thermostat

____ Sensing device for gauges

____ Radiator design where coolant flows from side to side

____ Allows coolant to circulate when the thermostat is closed

TURN

Q.	Wax	____	Made of iron or aluminum
R.	Fan	____	Coolant additive that protects aluminum
S.	Thermostat	____	Small valve in the center of a radiator pressure cap

STOP

Instructor OK ______________ Score __________

Activity Sheet #37

IDENTIFY BELT-DRIVEN ACCESSORIES

Name_________________________ Class ________________

Directions: Identify the belt-driven accessories in the drawing. Place the identifying letter next to the name of the component listed below.

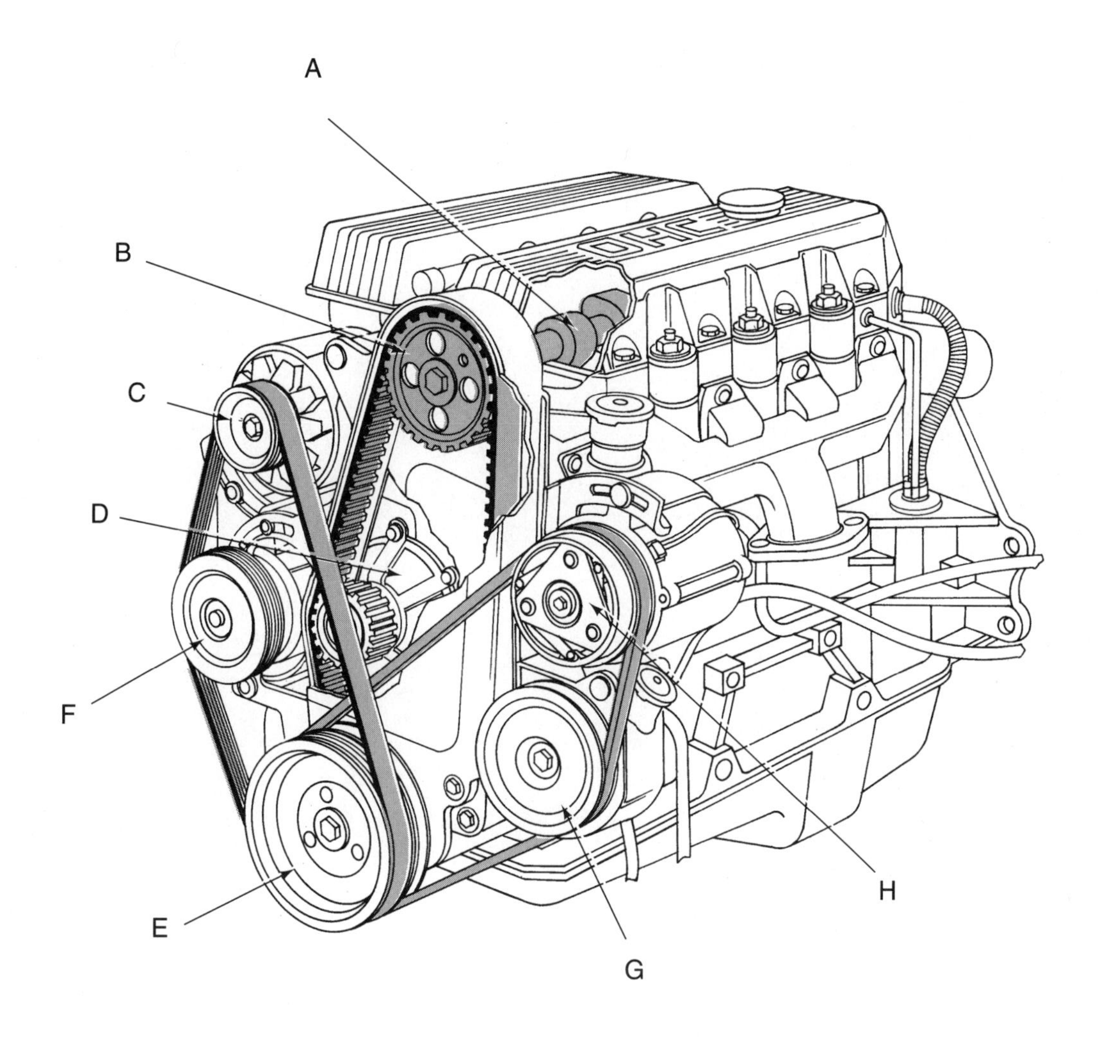

Power Steering Pump	____	Cam Sprocket	____	Air-Conditioning Compressor	____
Air Pump	____	Water Pump	____	Alternator	____
Overhead Cam	____	Crankshaft Pulley	____		

Instructor OK ______________ Score __________

Activity Sheet #38

BELTS

Name______________________ Class ______________

Directions: Match the words on the left to the descriptions on the right. Write the letter for the correct word on the line provided. For the terms that you are not certain of, use the glossary in your textbook.

A. Tensile cords	____ Belt that has multiple ribs on one side and is flat on the other side
B. Neoprene	____ Used to apply leverage with a tool when tightening a drive belt
C. High cordline belt	____ Screw for adjusting belt tension
D. V-ribbed belt	____ Provide strength to the belts
E. Serpentine belt	____ Used for checking V-ribbed belt and timing belt tension
F. Jackscrew	____ Higher quality V-belt with the tensile cord above center
G. V-belt and V-ribbed	____ Oil-resistant artificial rubber
H. Square bracket hole	____ Types of accessory drive belts
I. Click-type gauge	____ Belt that follows a snake-like path

STOP

Instructor OK ______________ Score __________

Activity Sheet #39

IDENTIFY CARBURETOR CIRCUITS

Name______________________________ Class ____________________

Directions: Identify the carburetor circuits in the drawing. Place the identifying letter next to the name of the circuit.

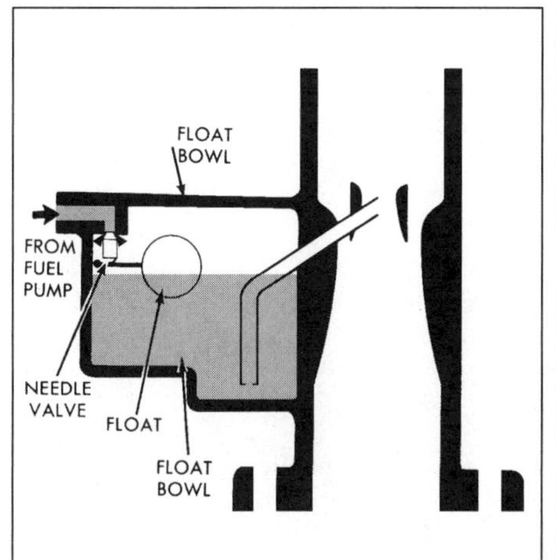

A

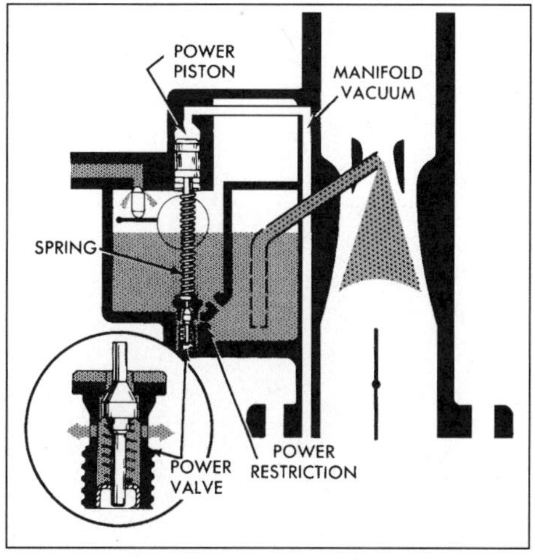

B

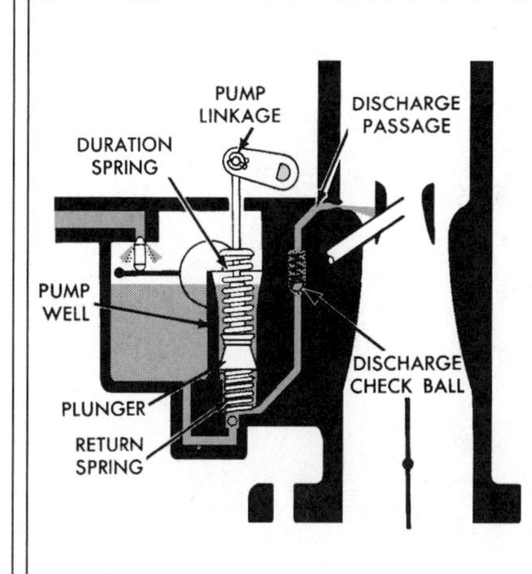

C

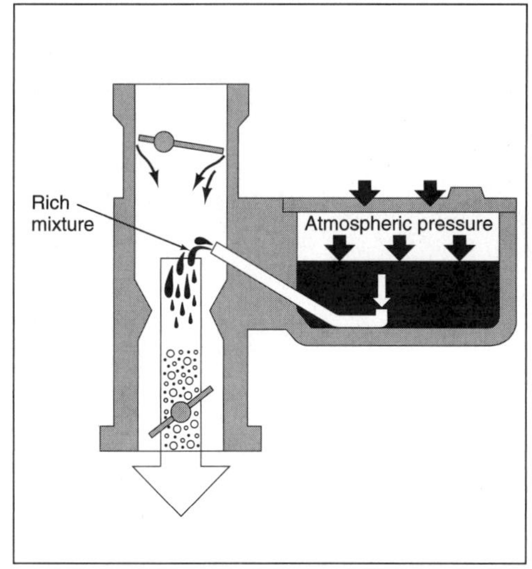

D

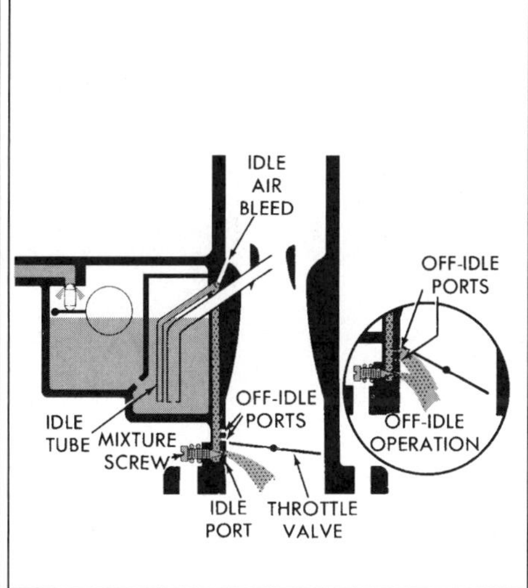

E

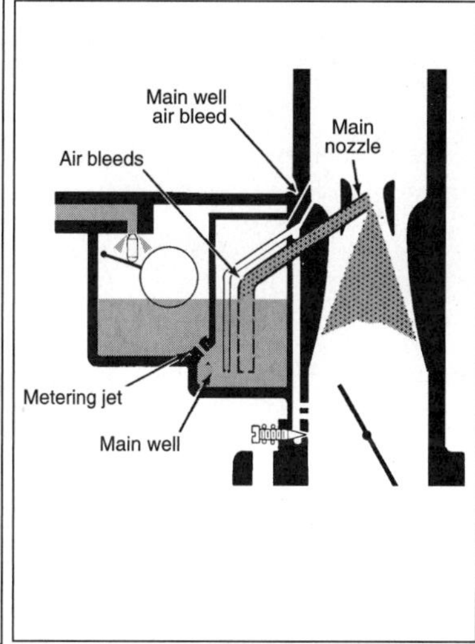

F

Float Circuit	___	Acceleration Circuit	___	Power Circuit	___
Idle Circuit	___	Main Metering Circuit	___	Choke Circuit	___

STOP

Instructor OK ______________ Score __________

Activity Sheet #40

IDENTIFY FUEL INJECTION COMPONENTS

Name______________________________ Class ____________________

Directions: Identify the fuel injection components in the drawing. Place the identifying letter next to the name of the component.

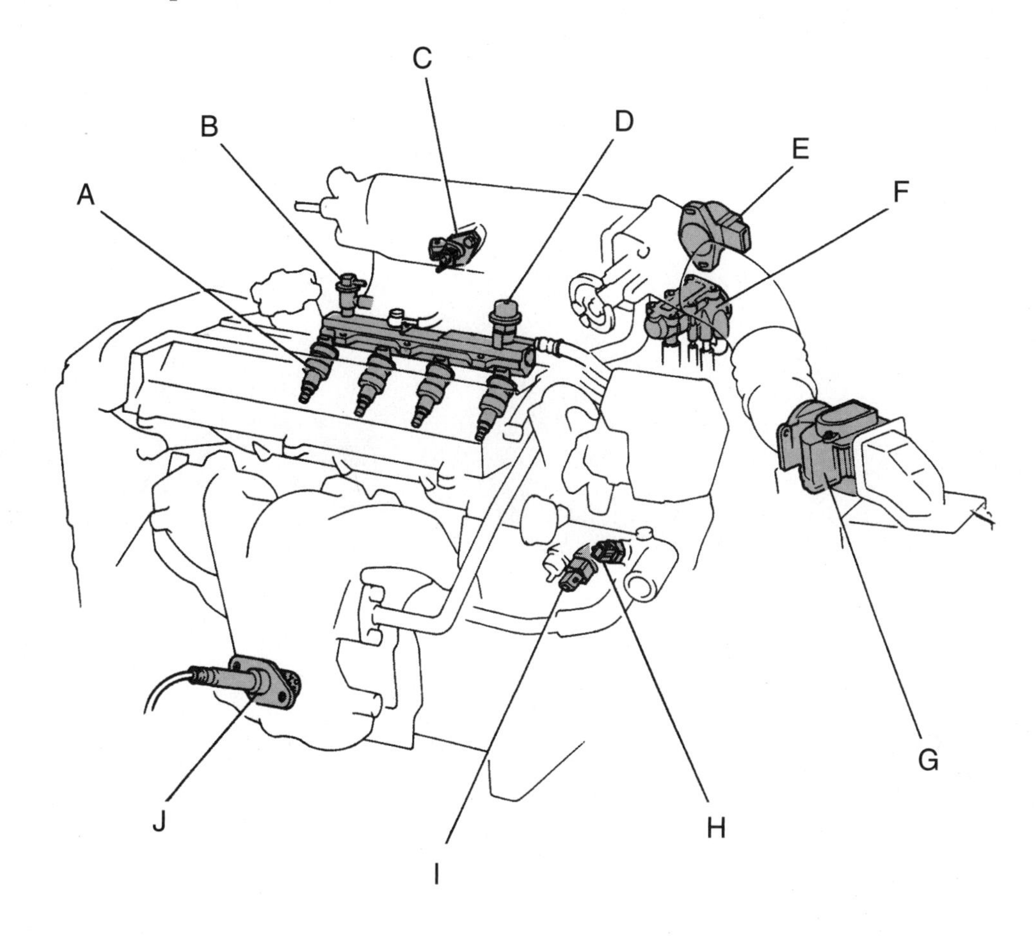

Thermo-time Switch	____	Engine Temperature Sensor	____	Fuel Pulsation Damper	____
Airflow Meter	____	Fuel Pressure Regulator	____	Idle Speed Control	____
Fuel Injector	____	Oxygen Sensor	____		
Throttle Position Sensor	____	Cold Start Injector	____		

STOP

Instructor OK ______________ Score __________

Activity Sheet #41

FUEL SYSTEMS

Name______________________________ Class ____________________

Directions: Match the words on the left to the descriptions on the right. Write the letter for the correct word on the line provided. For the terms that you are not certain of, use the glossary in your textbook.

A. Filter sock	____	Name of the process by which the fuel is suspended in the air
B. Manifold vacuum	____	When the computer uses a manifold absolute pressure (MAP) sensor and engine rpm (tach) signal to calculate the amount of air entering the engine
C. Atomization	____	Electronic fuel injection
D. Vaporization	____	Abbreviation for throttle-body injection
E. Venturi	____	Uses an intake manifold similar to a carbureted system
F. Feedback systems	____	When the computer is controlling the fuel system
G. Barrels	____	Length of time that an injector remains open
H. Pulse width	____	When the oxygen sensor does not send signals to the computer
I. EFI	____	The filter inside the gas tank
J. TBI	____	Carburetor circuit that allows more fuel under load
K. Throttle-body injection	____	Devices that relay information to the computer
L. Port injection	____	Smaller area in the carburetor that restricts airflow
M. Speed density system	____	The oxygen sensor signal for rich air-fuel mixture
N. Air density systems	____	When atomized fuel turns into a gas
O. Sensors	____	Name of the butterfly valve that controls the amount of air entering the engine
P. Actuators	____	Computer fuel systems that monitor the oxygen content in the exhaust
Q. Closed loop	____	When the carburetor _______ is too high the air-fuel mixture will be richer

R. Open loop	____	Term for the number of throttle passages found in a carburetor
S. Throttle plate	____	Fuel injection system that uses individual fuel injectors at each intake port
T. Power enrichment	____	When an airflow sensor measures the volume of air entering the engine
U. Float level	____	Devices that carry out an assigned change from the computer
V. Higher than 0.5 volt	____	Pressure inside the intake manifold when the engine is running

STOP

Instructor OK ______________ Score __________

Activity Sheet #42

IDENTIFY EXHAUST SYSTEM COMPONENTS

Name______________________________ Class ____________________

Directions: Identify the exhaust system components in the drawing. Place the identifying letter next to the name of the component listed below.

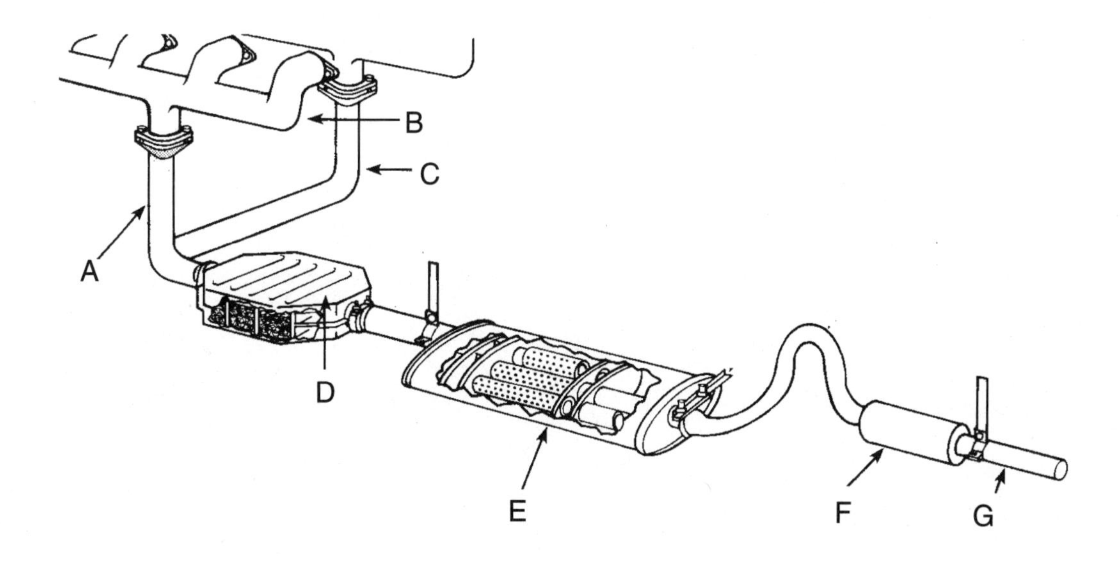

Exhaust Pipe	____	Tailpipe	____	Resonator	____
Exhaust Manifold	____	Catalytic Converter	____	Crossover Pipe	____
Muffler	____				

STOP

Instructor OK ______________ Score __________

Activity Sheet #43

INTAKE AND EXHAUST SYSTEMS

Name______________________________ Class ____________________

Directions: Match the words on the left to the descriptions on the right. Write the letter for the correct word on the line provided. For the terms that you are not certain of, use the glossary in your textbook.

A. Heat riser	____ When each barrel of the carburetor supplies half of an engine's cylinders
B. Runners	____ Butterfly valve that fits between the exhaust manifold and the exhaust pipe
C. Cross-flow head	____ A second muffler in line with the main muffler
D. Siamese runners	____ Intake fan driven by the exhaust flow
E. Dual-plane manifold	____ Exhaust manifolds made of tube steel
F. Headers	____ When each barrel of the carburetor serves all of an engine's cylinders
G. Sound	____ When intake and exhaust manifolds are on opposite sides of an in-line engine
H. Resonator	____ When one runner feeds two neighboring cylinders
I. Catalytic converter	____ Passages in the intake manifold
J. Turbocharger	____ Cleans up engine emissions before they leave the tailpipe
K. Backpressure	____ Caused by restriction in the exhaust system
L. Supercharger	____ Vibration in the air
M. Single-plane manifold	____ Belt-driven intake air pump

STOP

Part I
Activity Sheets

Section Six

Electrical System Theory and Service

Instructor OK ______________ Score __________

Activity Sheet #44
ELECTRICAL THEORY

Name______________________ Class ______________

Directions: Match the words on the left to the descriptions on the right. Write the letter for the correct word on the line provided. For the terms that you are not certain of, use the glossary in your textbook.

A. Atoms	____ Also called a capacitor
B. Electricity	____ Electrical flow
C. Circuit	____ Composed of protons, neutrons, and electrons
D. Switch	____ Unit of measurement for electrical resistance
E. Fuse	____ Stores electricity
F. Volt	____ An obstruction to electrical flow
G. Current	____ Varies the voltage in a circuit
H. Amp	____ Path for electrical flow
I. Capacitor	____ Type of meter that has a dial
J. Resistance	____ Magnetically controlled switch
K. Ohm's law	____ Circuit protection device
L. Ohm	____ Solid state material that can act as either an insulator or a conductor
M. Current draw	____ Unit of measurement for electrical pressure
N. Rheostat	____ Unit of measurement for electrical current flow
O. Potentiometer	____ Loss of voltage caused by current flow through a resistance
P. Series circuit	____ Flow of electrons between atoms
Q. Parallel circuit	____ The law governing the relationship between volts, ohms, and amps
R. Relay	____ Oscillating electrical current
S. Alternating current	____ Amount of current required to operate a load
T. Condenser	____ A meter used only on a circuit that has no electrical power
U. Semiconductor	____ Digital multimeter
V. Diode	____ Used to turn a circuit on or off

TURN

W. Transistor	____	Varies current flow through the circuit
X. Analog meter	____	As resistance increases, current ____
Y. DMM	____	Electrical pressure
Z. Voltage drop	____	An electrical circuit where current must flow through all parts
AA. Open circuit	____	Circuit with different branches that current can flow through
AB. Grounded circuit	____	Name for a complete path provided for electrical flow
AC. Closed circuit	____	When current goes directly to ground
AD. Decreases	____	When there is a break in the path of electrical flow in a circuit
AE. Ohmmeter	____	An electronic relay
AF. Voltage	____	An electronic one-way check valve

STOP

Instructor OK ______________ Score ____________

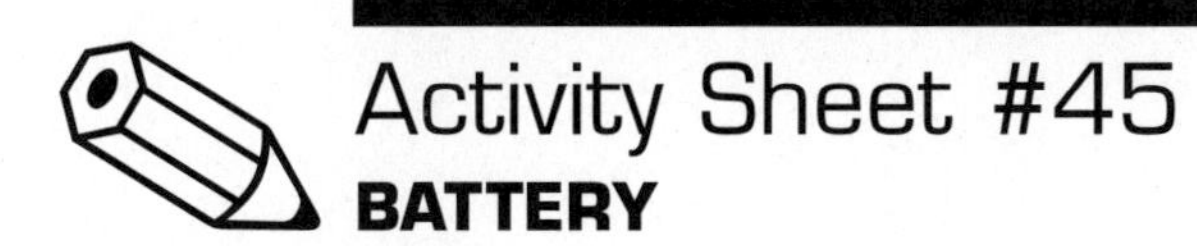

Activity Sheet #45

BATTERY

Name______________________________ Class ____________________

Directions: Match the words on the left to the descriptions on the right. Write the letter for the correct word on the line provided. For the terms that you are not certain of, use the glossary in your textbook.

A Electrolyte	____ Unit of measurement for electrical power
B. Element	____ Low-maintenance battery
C. Battery terminal	____ When the battery is allowed to run almost completely dead, and then is recharged
D. BCI	____ Maintenance-free battery
E. CCA	____ Largest vehicle load on a car battery
F. Reserve capacity	____ Fully charged cell voltage
G. Deep-cycle	____ Gas given off as a battery is charged
H. Watt	____ Measurement of the battery's ability to provide current when there is no electricity from the charging system
I. Cannot add water	____ Battery Council International
J. May need water	____ Voltage of a fully charged battery
K. Starter draw	____ Cold cranking amps
L. DC	____ Largest battery post
M. 2.1 volts	____ Mixture of sulfuric acid and water used in automotive batteries
N. Hydrogen	____ Connection on the side or top of a battery
O. 12.6 volts	____ Group of battery plates connected in parallel
P. Positive	____ Type of electrical current produced by a battery

STOP

Instructor OK ______________ Score __________

Activity Sheet #46

JUMP STARTING

Name______________________ Class ______________

Directions: Draw the jumper cables. Place a number next to each cable clamp to show the order in which each connection should be made.

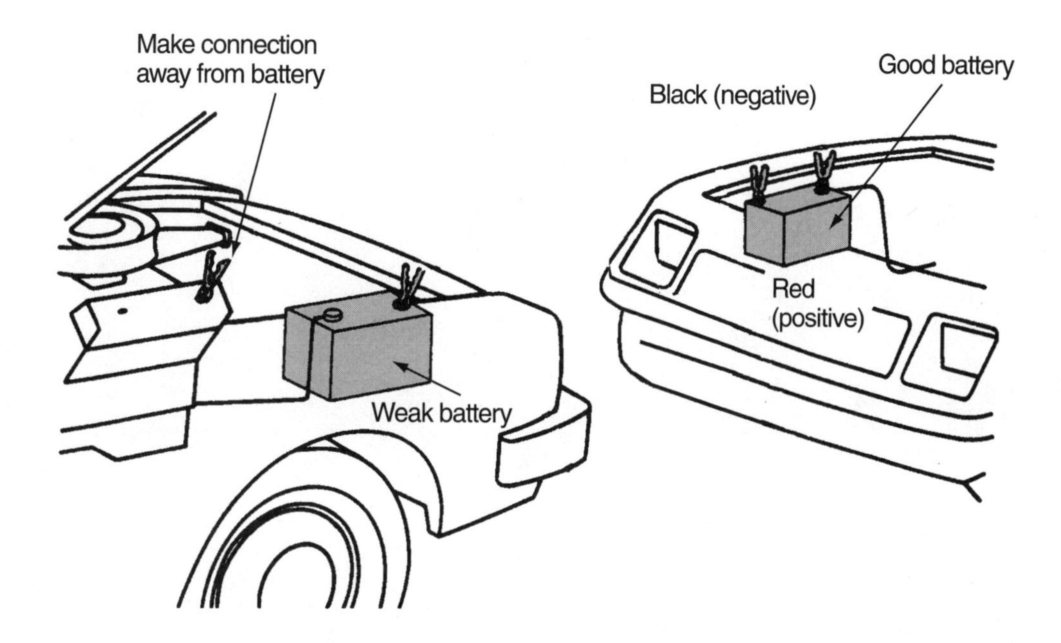

Note: The last connection should be lower than the battery and more than 12 inches away from the battery.

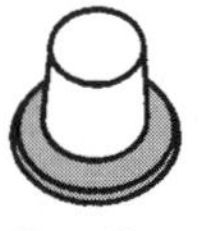

Smaller Larger

Top-terminal batteries

Write a (+) and a (–) on the correct battery terminal.

Instructor OK ______________ Score ____________

Activity Sheet #47

IDENTIFY STARTING SYSTEM COMPONENTS

Name______________________________ Class ____________________

Directions: Identify the starting system components in the drawing. Place the identifying letter next to the name of the component.

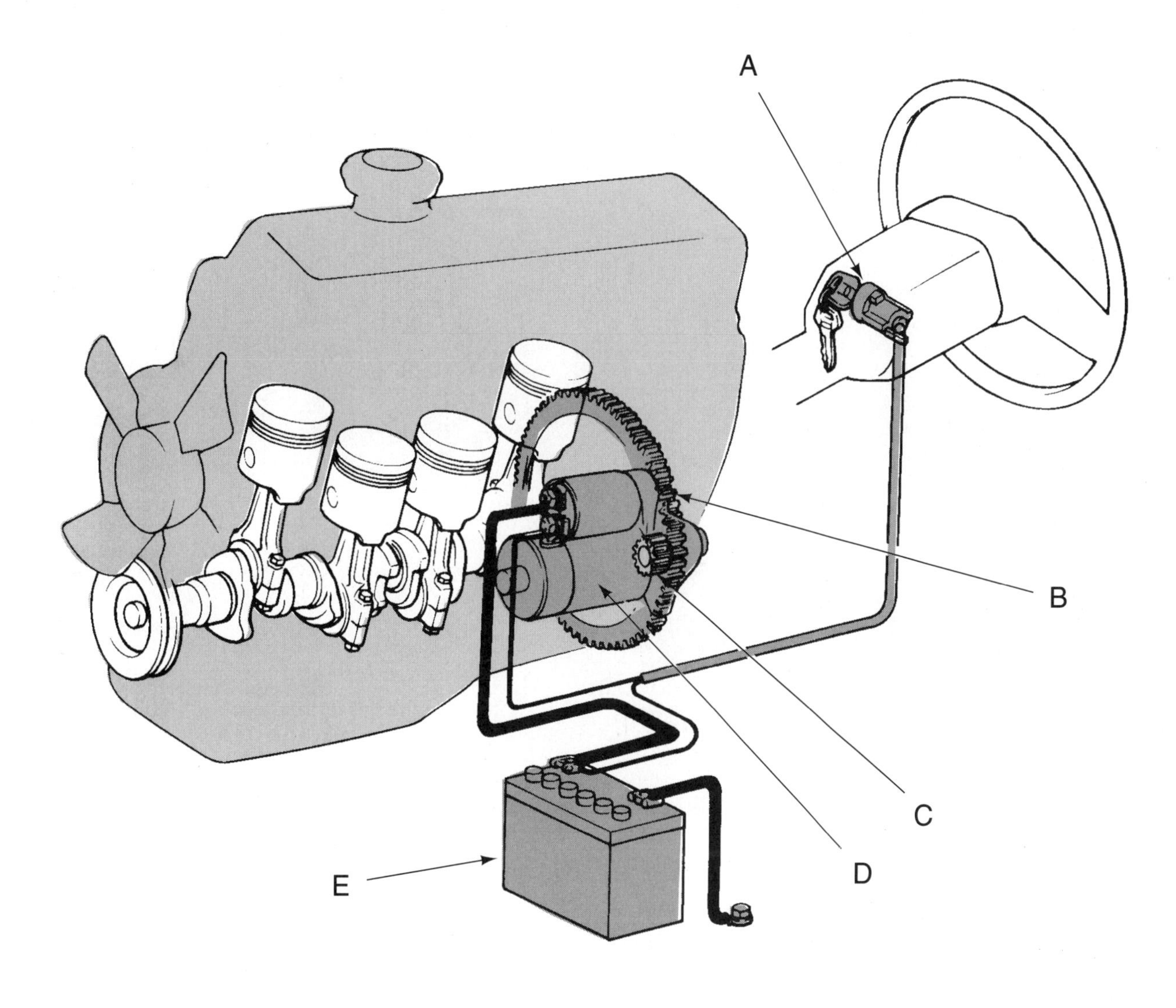

Pinion Drive Gear	____	Ring Gear	____	Battery	____
Ignition Switch	____	Starter Motor	____		

STOP

Instructor OK ________________ Score ____________

Activity Sheet #48

IDENTIFY STARTER MOTOR COMPONENTS

Name______________________________ Class ____________________

Directions: Identify the starter motor components in the drawing. Place the identifying letter next to the name of the component.

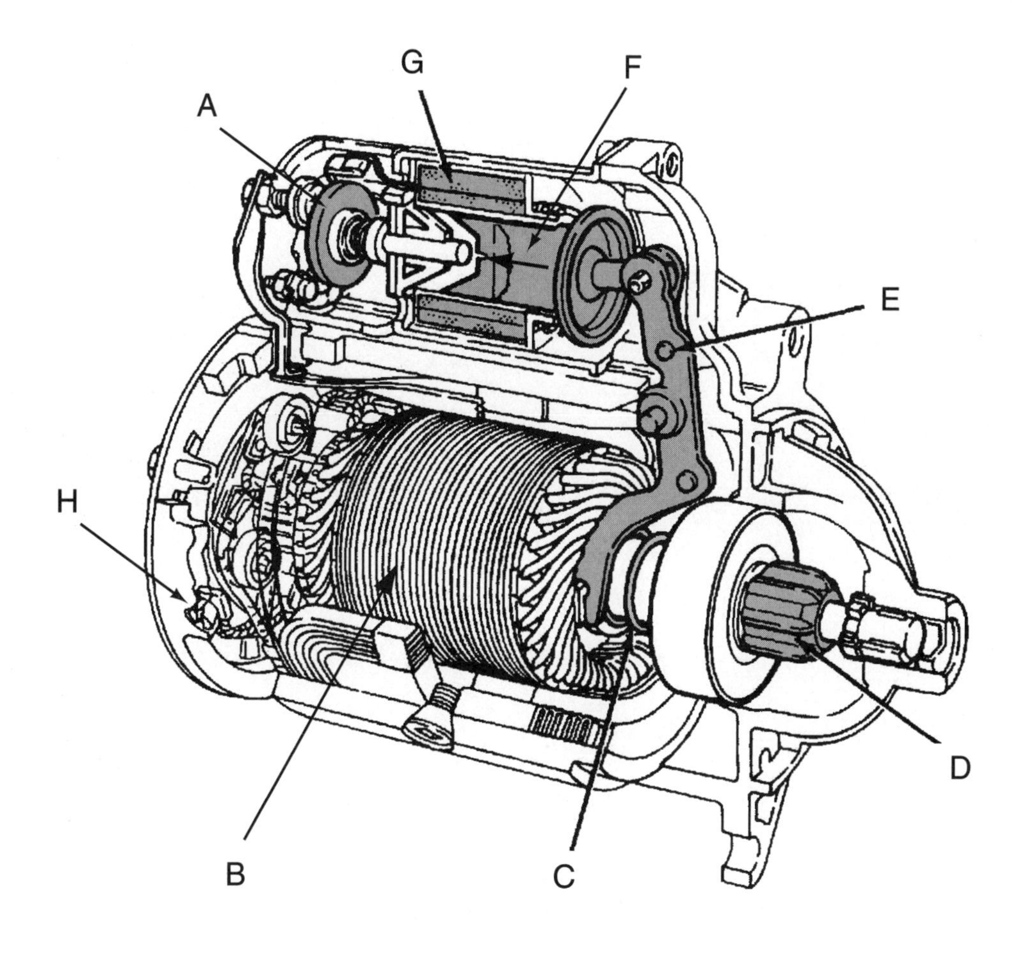

Solenoid Windings	_____	Contact Disc	_____	Drive Pinion	_____
Piston (Plunger)	_____	Return Spring	_____	Armature	_____
Pivot Fork	_____	Brushes	_____		

STOP

Instructor OK ________________ Score ____________

Activity Sheet #49

IDENTIFY CHARGING SYSTEM COMPONENTS

Name________________________________ Class ____________________

Directions: Identify the charging system components in the drawing. Place the identifying letter next to the name of the component.

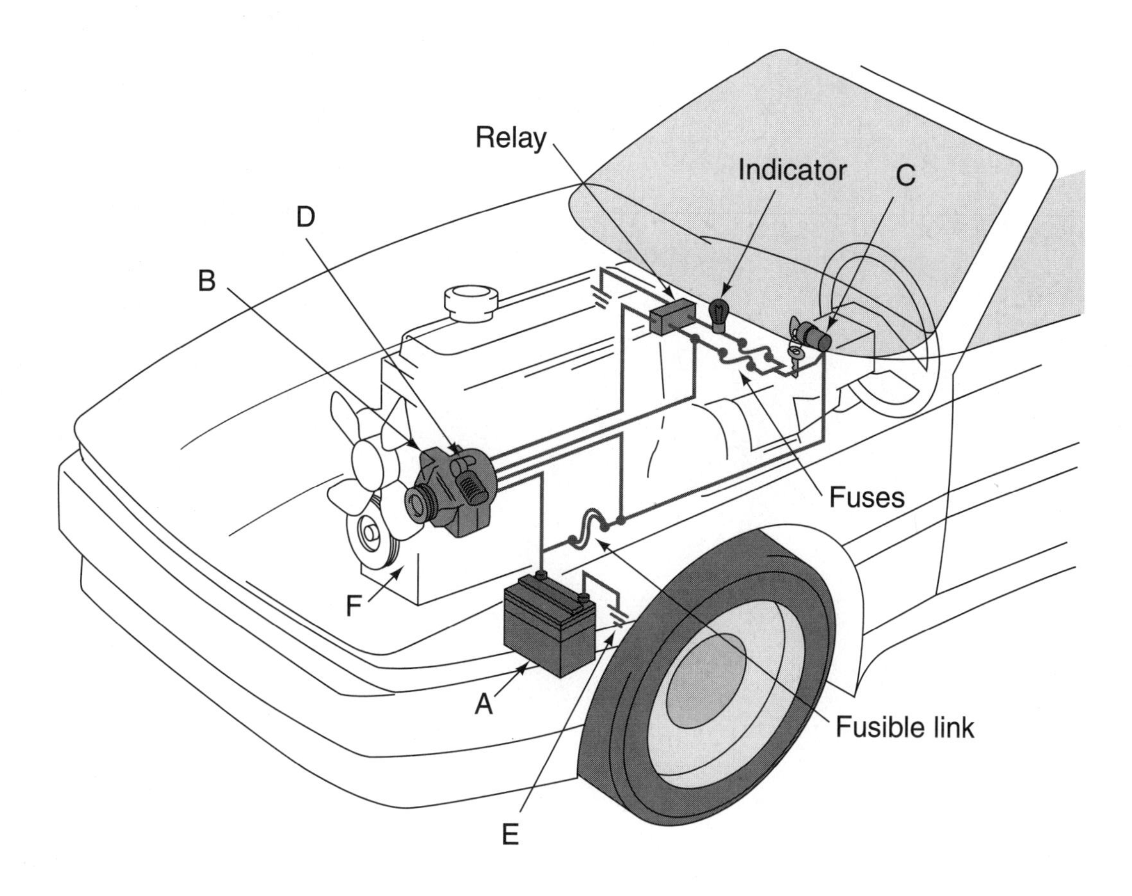

Regulator	____	Ignition Switch	____	Battery Ground	____
Alternator	____	Battery	____	Drive Belt	____

STOP

Instructor OK ______________ Score __________

Activity Sheet #50

IDENTIFY ALTERNATOR COMPONENTS

Name_____________________________ Class ____________________

Directions: Identify the alternator components in the drawing. Place the identifying letter next to the name of the component.

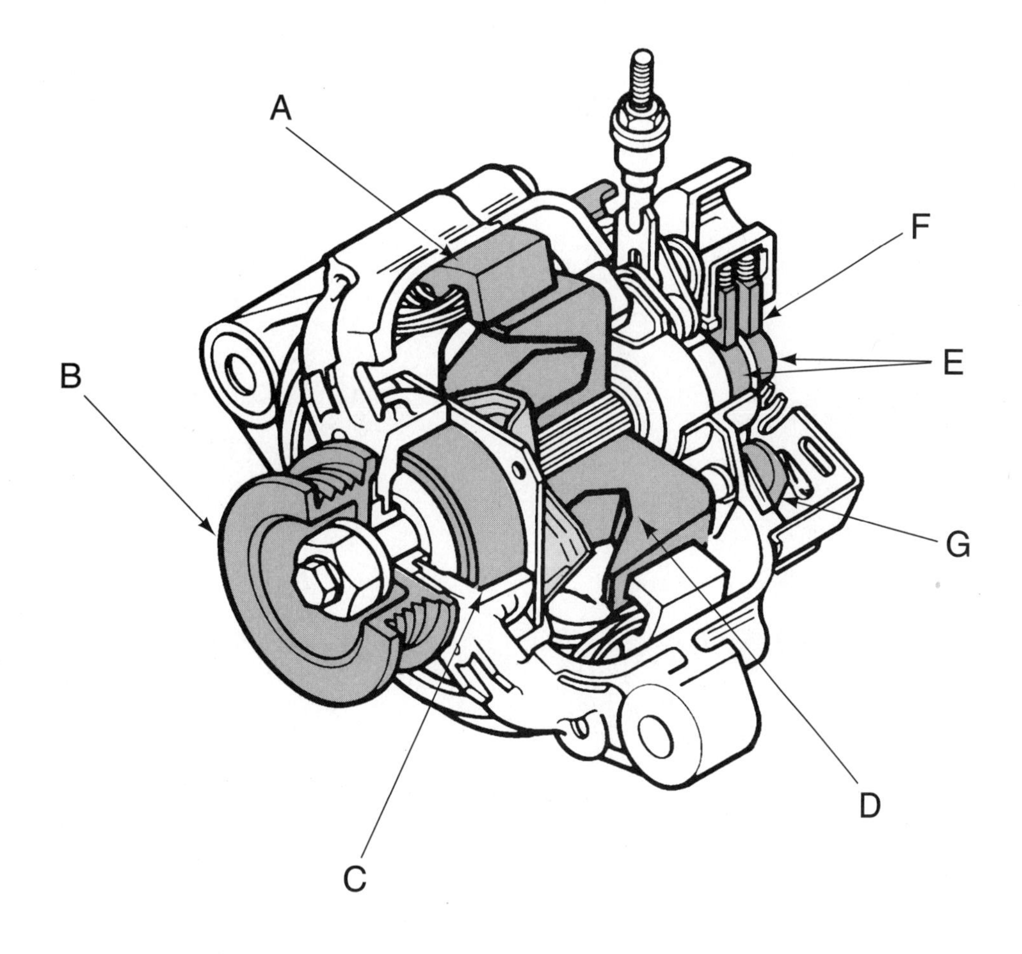

Rotor	____	Bearing	____	Pulley	____
Slip Rings	____	Brushes	____	Stator	____
Rectifier	____				

STOP

Instructor OK _______________ Score _______________

Activity Sheet #51

IDENTIFY LIGHTING AND WIRING COMPONENTS

Name_______________________________ Class _______________________

Directions: Identify the wiring and lighting components in the drawing. Place the identifying letter next to the name of the component.

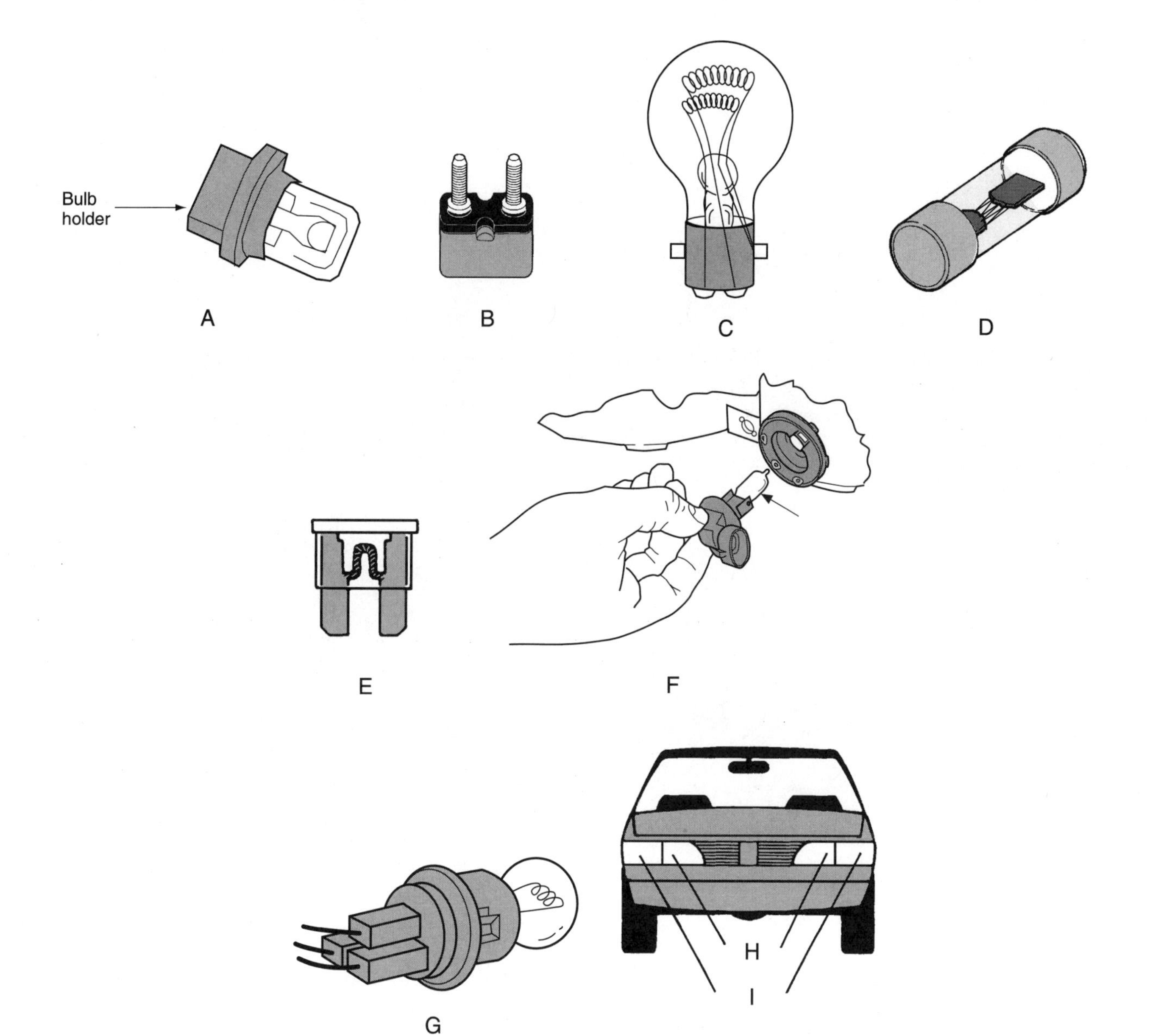

Halogen Headlamp	____	Taillight Bulb	____	Dash Light Bulb	____
Blade Fuse	____	High Beam	____	Low Beam	____
Circuit Breaker	____	Cartridge Fuse	____	Dual Filament Bulb	____

TURN

Directions: Identify the vehicle lights in the drawing. Place the identifying letter next to the name of the light.

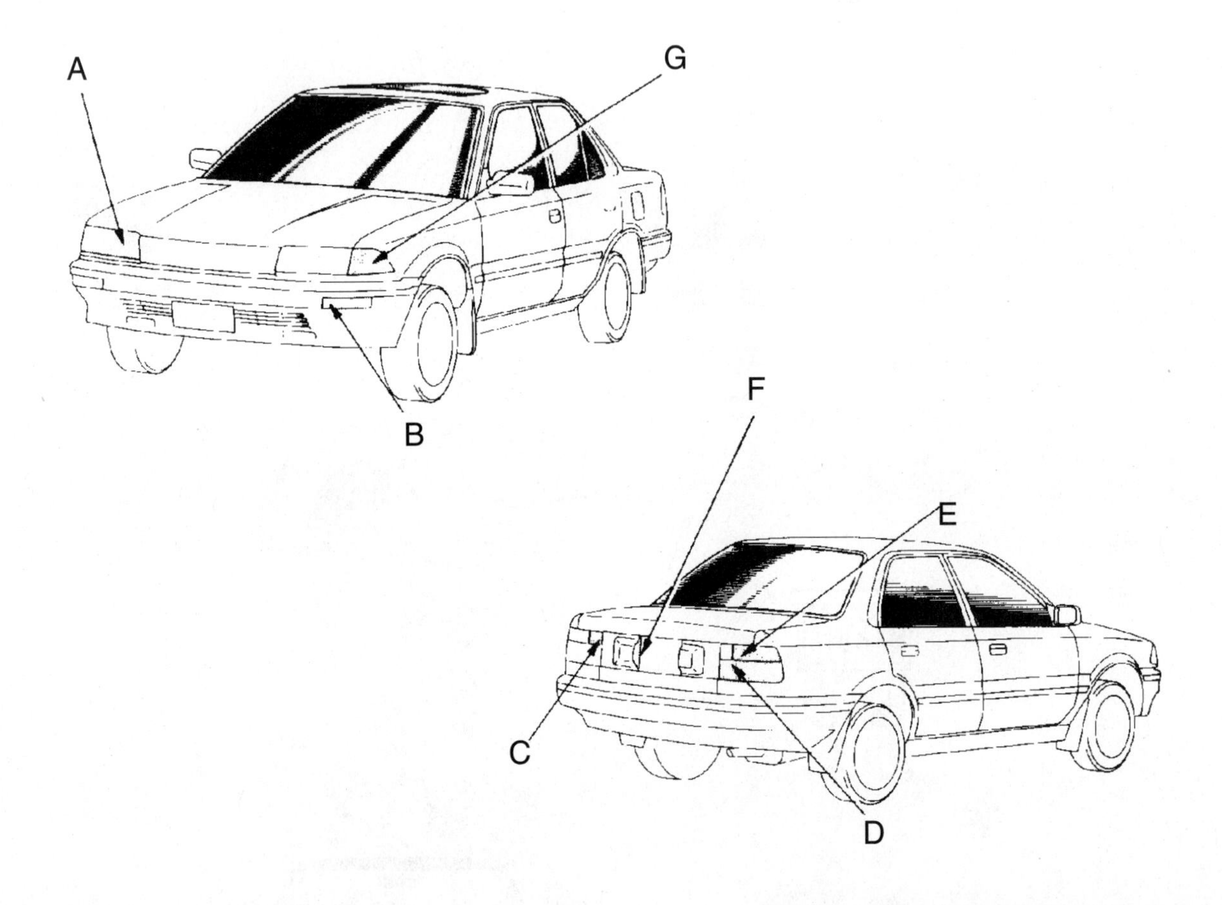

Tail and Brake Light	____	Clearance Light	____	Front Turn Signal	____
Backup Light	____	Headlight	____	Rear Turn Signal	____
License Plate Light	____				

STOP

Instructor OK ______________ Score __________

Activity Sheet #52

LIGHTING AND WIRING

Name______________________________ Class ____________________

Directions: Match the words on the left to the descriptions on the right. Write the letter for the correct word on the line provided. For the terms that you are not certain of, use the glossary in your textbook.

A. Primary wiring	____	A wire in a light bulb that provides a resistance to electron flow. When it heats up, it causes light
B. Secondary wiring	____	The number that identifies a bulb for all manufacturers
C. Cables	____	A rating for the intensity of a headlamp
D. AWG	____	Two metal strips with different expansion rates
E. Fuse	____	Has high beam only
F. SFE	____	A device activated by the heat of the electricity that causes the turn signal bulbs to flash
G. Mini-fuse	____	Larger wire means a _______ AWG number
H. Fuse link	____	American Wire Gauge
I. Circuit breaker	____	A dimmer switch and turn signal lever together
J. Bimetal strip	____	American National Standards Institute
K. Filament	____	The Society of Fuse Engineers
L. Candlepower	____	A type of fuse that allows for a heavier startup draw
M. Type I headlamp	____	A circuit protection device designed to melt when the flow of current becomes too high for the wires or loads in the circuit
N. Type II headlamp	____	Larger wires that allow more electrical current flow
O. Halogen headlamp	____	The year rectangular headlamps were introduced
P. Composite headlamp	____	The year hazard flashers were introduced
Q. ANSI	____	A circuit protection device that resets automatically or can be manually reset after it trips
R. Bulb trade number	____	Has both low and high beams
S. NA	____	A smaller type of blade fuse; it has a fuse element cast into a clear plastic outer body

TURN

T.	Signal flasher	____	A circuit protection device that is a length of wire smaller in diameter than the wire it is connected to
U.	Lower	____	A brighter headlamp used on newer cars
V.	Slow blow fuse	____	A natural amber light bulb
W.	Multifunction switch	____	Low-voltage wiring
X.	1975	____	A headlamp housing with a glass balloon that the halogen lamp fits inside
Y.	1967	____	High-voltage ignition wiring

STOP

Instructor OK ______________ Score __________

Activity Sheet #53

IDENTIFY ELECTRICAL TEST INSTRUMENTS

Name______________________________ Class ____________________

Directions: Identify the electrical test instruments in the drawing. Place the identifying letter next to the name of the instrument.

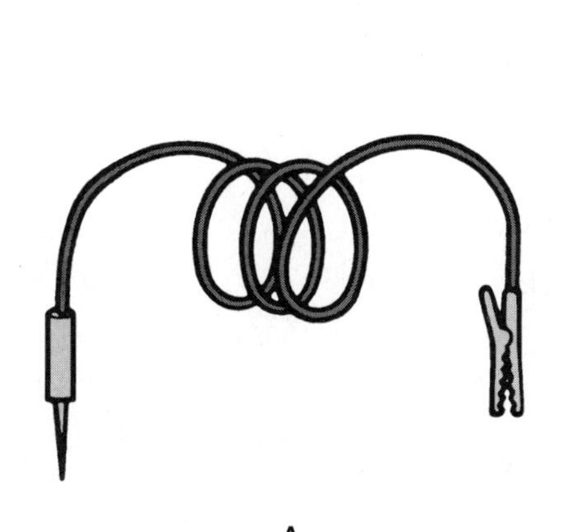

A

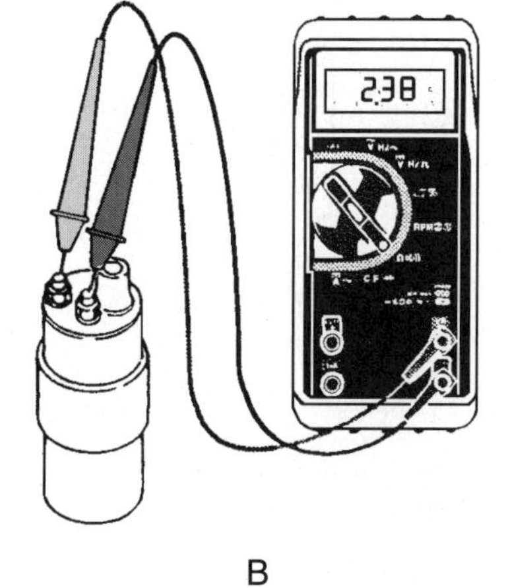

B

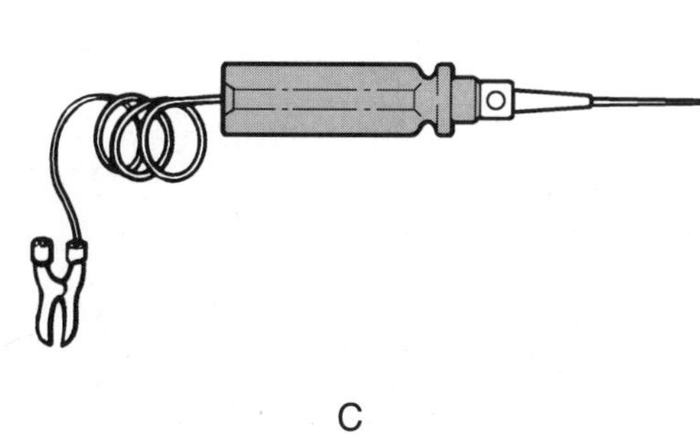

C

D

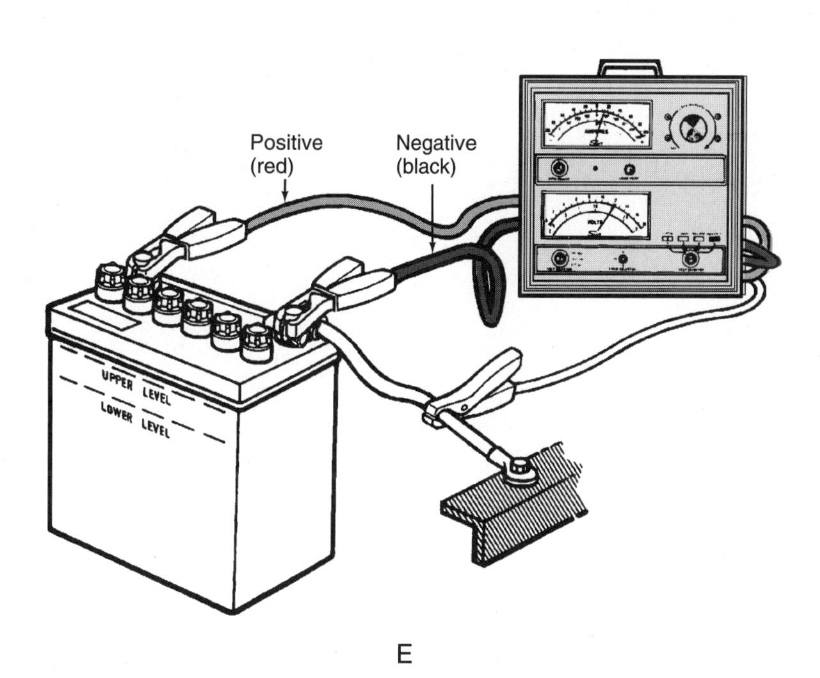

E

Analog Meter	____	Volt-Amp Tester	____	Jumper Wire	____
Digital Multimeter	____	Circuit Tester	____		

STOP

Part I

Activity Sheets

Section Seven

Engine Performance Diagnosis Theory and Service

Instructor OK ______________ Score __________

Activity Sheet #54

IDENTIFY IGNITION SYSTEM COMPONENTS

Name______________________________ Class ____________________

Directions: Identify the ignition system components in the drawing. Place the identifying letter next to the name of the component.

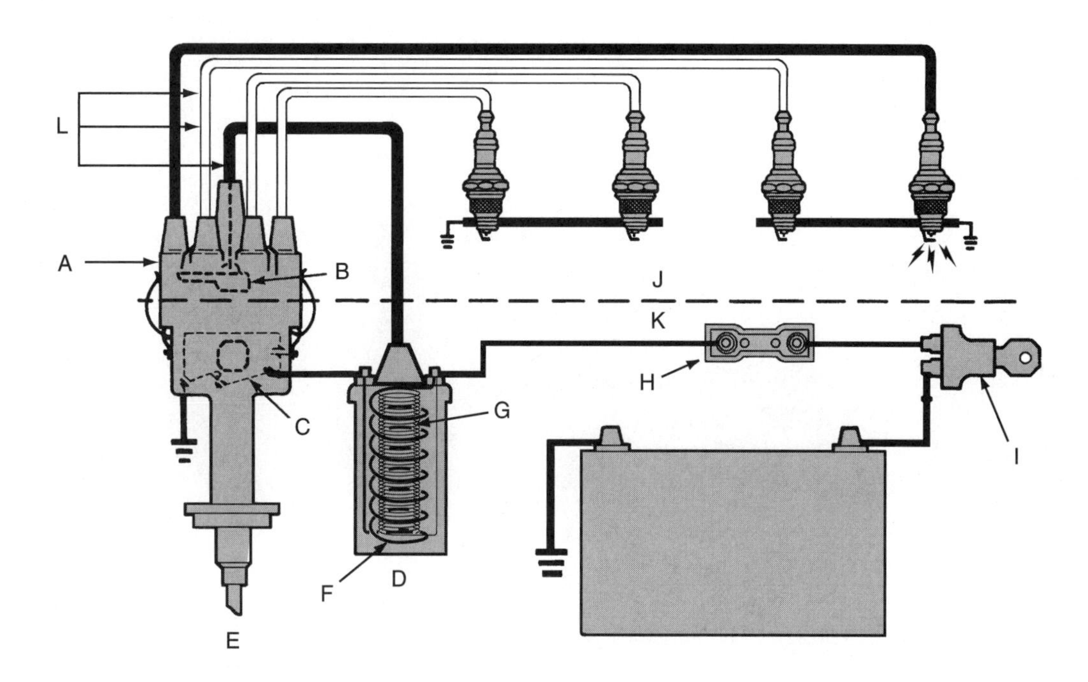

Distributor Cap	____	Secondary Winding	____	Points or Electronic Switch	____
Coil	____	Rotor	____	Resistor	____
Secondary Circuit	____	Primary Winding	____	Ignition Switch	____
Distributor	____	Primary Circuit	____	Spark Plug Cables	____

STOP

Instructor OK ______________ Score __________

Activity Sheet #55

IDENTIFY IGNITION SYSTEM PARTS

Name______________________________ Class ____________________

Directions: Identify the ignition system parts in the drawing. Place the identifying letter next to the name of the part.

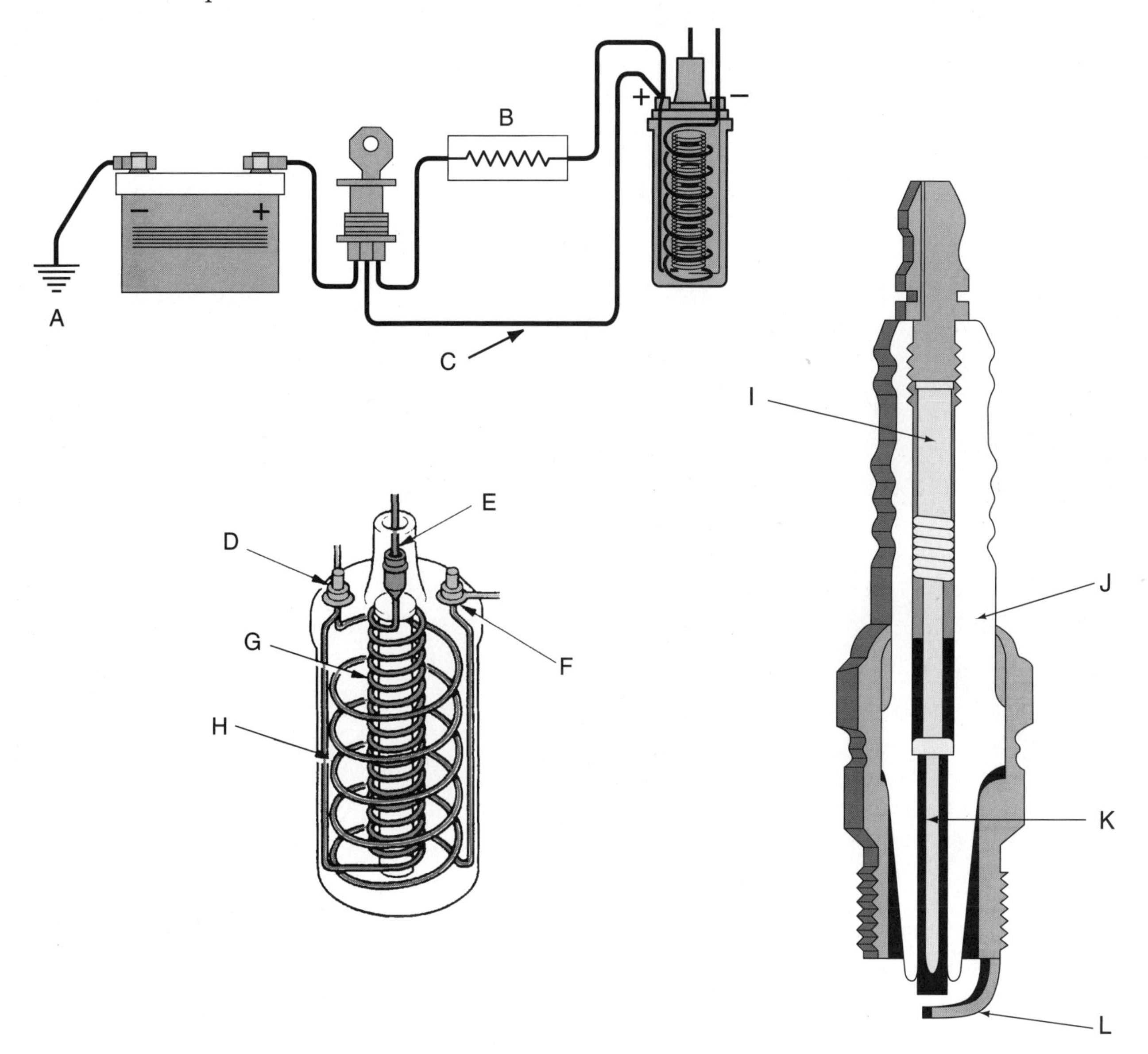

Secondary Winding	____	Coil Secondary Terminal	____	Primary Winding	____
Coil (–) Terminal	____	Resistor Bypass	____	Ignition Resistor	____
Ground	____	Coil (+) Terminal	____	Ground Electrode	____
Insulator	____	Center Electrode	____	Spark Plug Resistor	____

STOP

Instructor OK ________________ Score ____________

Activity Sheet #56

IDENTIFY IGNITION SYSTEM OPERATION

Name______________________________ Class ____________________

Directions: Identify the ignition system operation in the drawing by drawing the missing wiring.

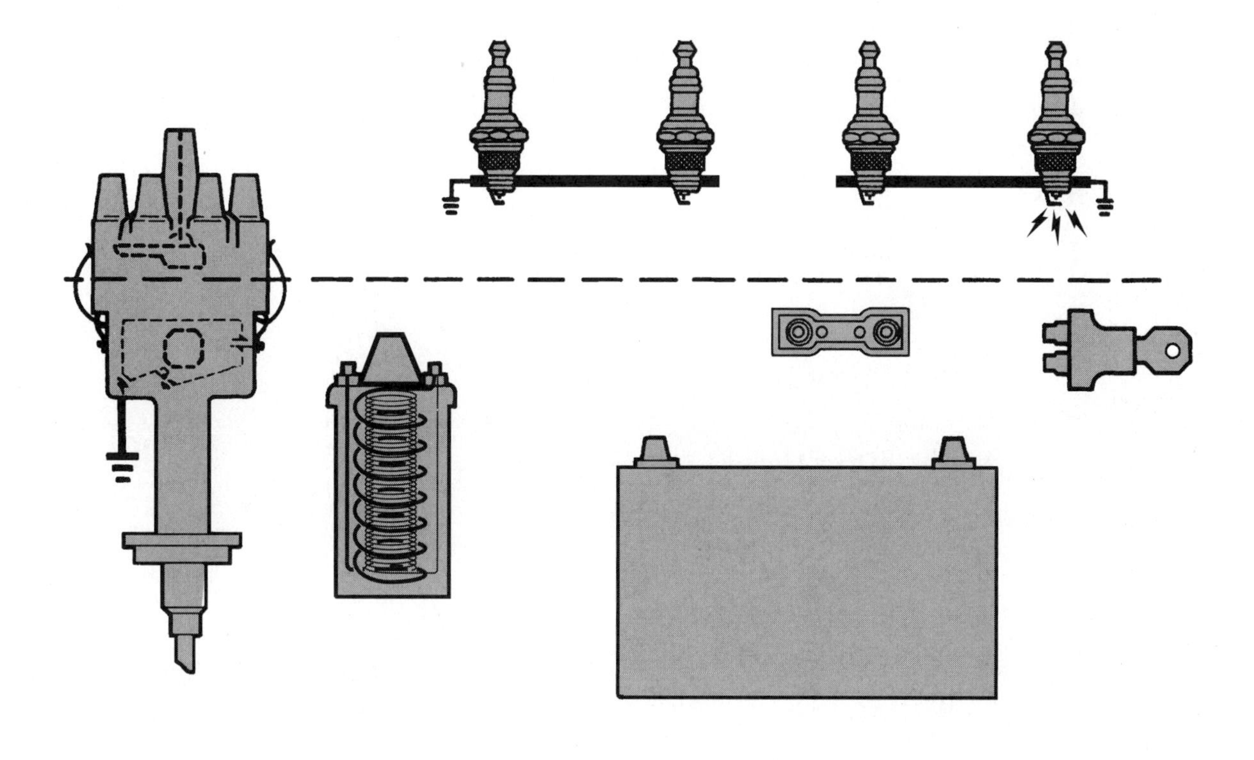

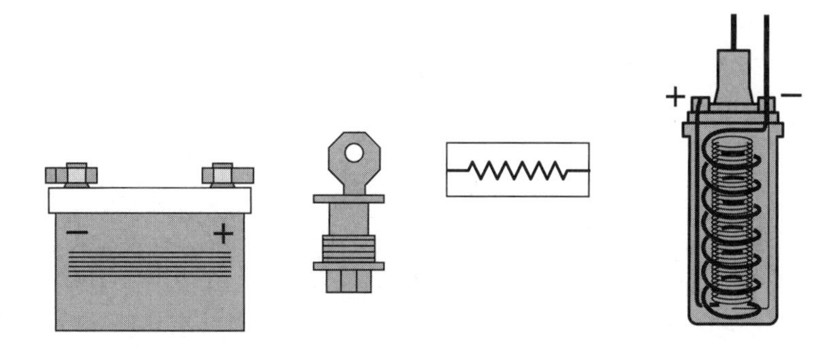

Instructor OK ______________ Score __________

Activity Sheet #57
IGNITION SYSTEMS

Name______________________________ Class ____________________

Directions: Match the words on the left to the descriptions on the right. Write the letter for the correct word on the line provided. For the terms that you are not certain of, use the glossary in your textbook.

A. Fouled spark plug	____	Also called parade pattern, it displays all of the cylinders next to each other (side by side), so that the heights of the voltage spikes can be compared
B. Crossfire induction	____	When ignition parts like a module or pickup coil are cooled or heated while watching the scope pattern
C. Carbon trail	____	Scope pattern is upside down
D. Timing light	____	Also called stacked pattern, it displays all of the cylinders vertically, one above the next.
E. Static timing	____	Combination of initial, mechanical, and vacuum advance
F. Firing line	____	Vertical movement on the oscilloscope screen
G. Spark line	____	A strobe light that is triggered by the voltage going through the number one spark plug cable
H. Display pattern	____	When one spark plug firing induces a spark in the one next to it, causing it to fire before its time
I. Raster pattern	____	Buildup of carbon that shorts out a spark plug
J. Superimposed pattern	____	A horizontal line on the scope pattern that begins at the voltage level where electrons start to flow across the spark plug gap
K. Stress test	____	Timing the distributor with the engine stopped
L. Mechanical	____	The upward line that starts the scope pattern
M. Total advance	____	A scope pattern used to compare all of the cylinders while their patterns are displayed one on top of the other
N. Voltage	____	Timing advance controlled by engine speed
O. Reversed coil polarity	____	A line of electrically conducting carbon that forms in a cracked distributor cap

STOP

Instructor OK ______________ Score __________

Activity Sheet #58
FIRING ORDER

Name______________________________ Class ____________________

Directions: Draw the cylinder numbers in their correct positions in the distributor cap spark plug wire terminal holes.

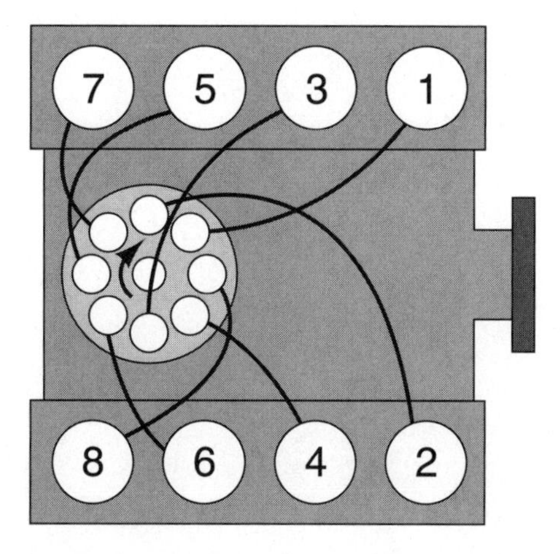

A. Firing Order: __ __ __ __ __ __ __ __

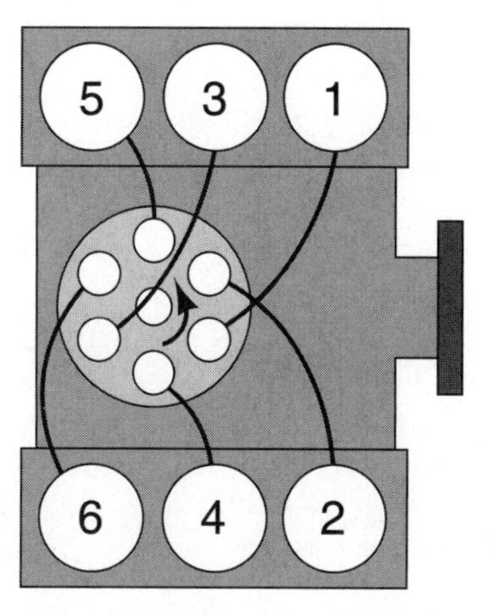

B. Firing Order: __ __ __ __ __ __

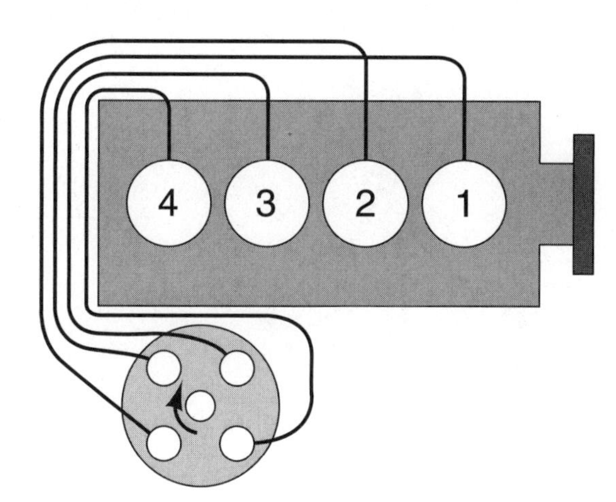

C. Firing Order: __ __ __ __

STOP

Instructor OK ________________ Score ____________

Activity Sheet #59

IDENTIFY THE DISTRIBUTORLESS IGNITION SYSTEM (DIS) COMPONENTS

Name______________________________ Class ____________________

Directions: Identify the distributorless ignition system (DIS) components in the drawing. Place the identifying letter next to the name of the component.

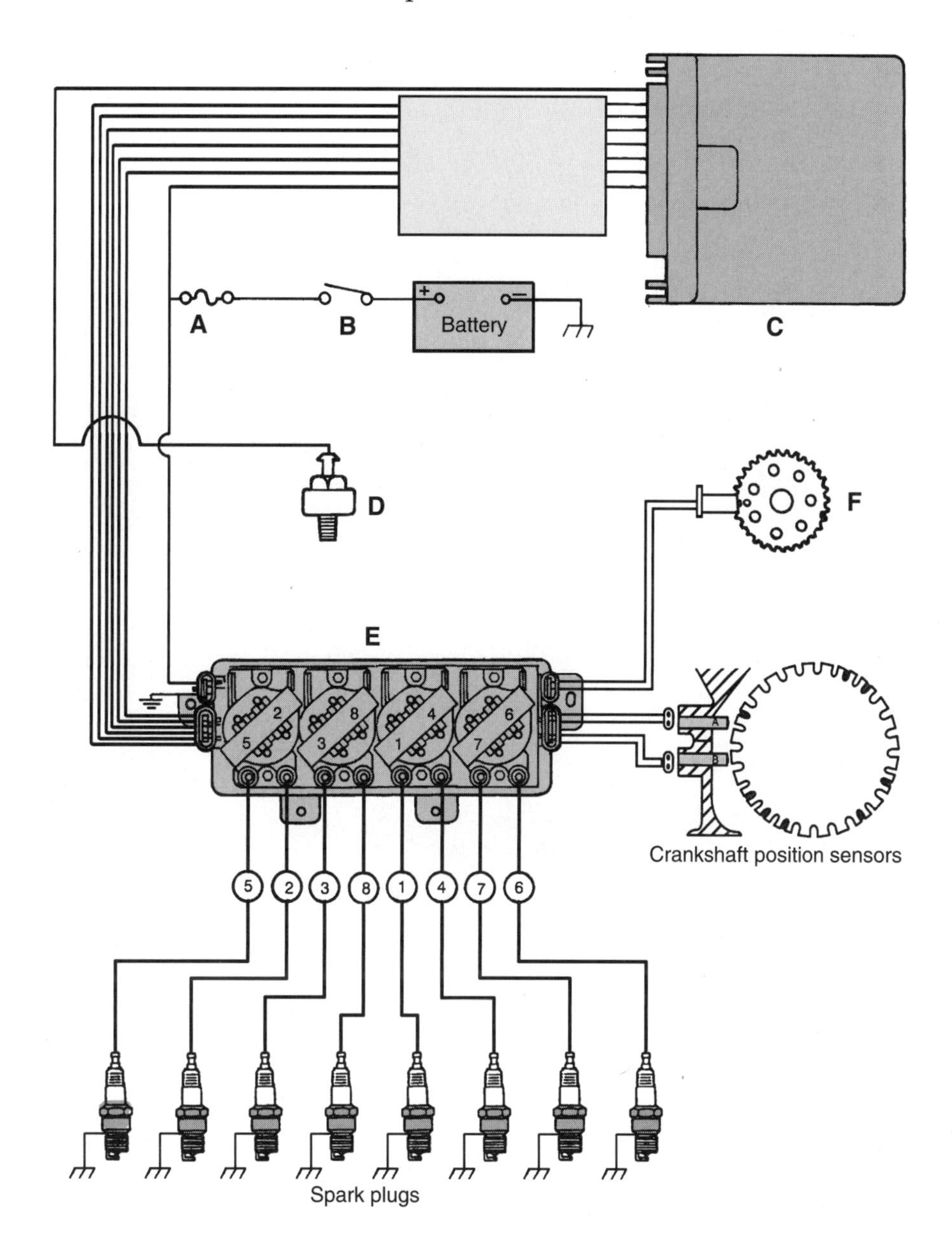

Cam Position Sensor	____	Ignition Switch	____	Knock Sensor	____
Coil Pack	____	Fuse	____	Computer	____

Instructor OK _______________ Score __________

Activity Sheet #60

IDENTIFY EMISSION CONTROL SYSTEM PARTS

Name_____________________________ Class ____________________

Directions: Identify the emission control system parts in the drawing. Place the identifying letter next to the name of the part on page 140.

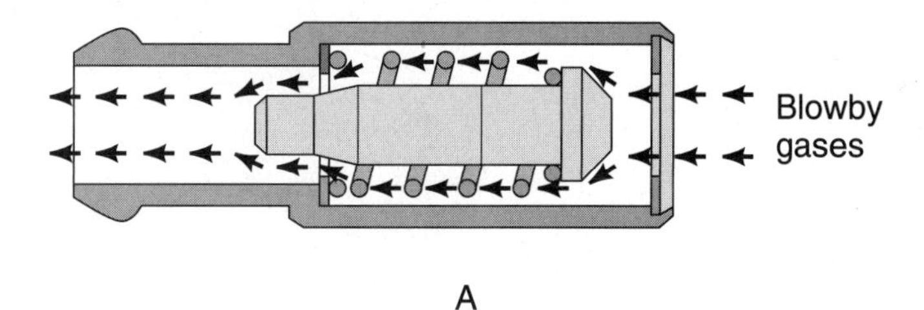

A

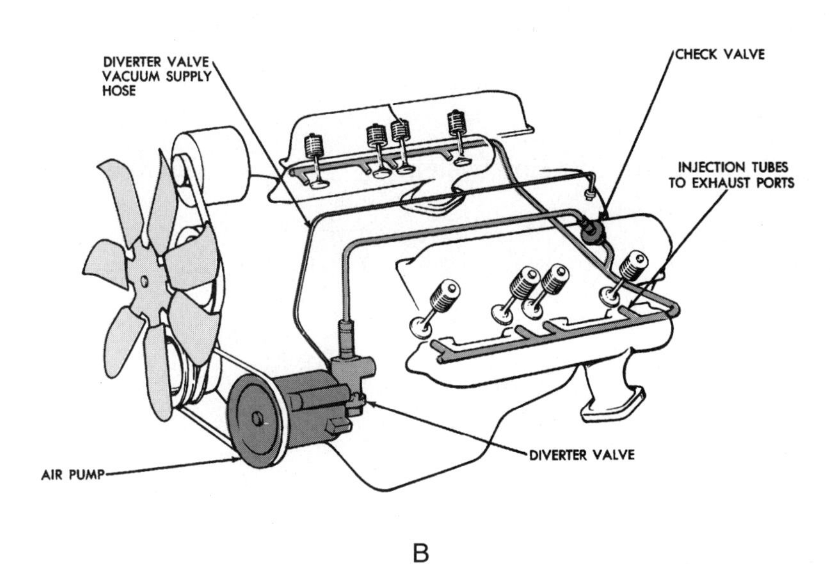

B

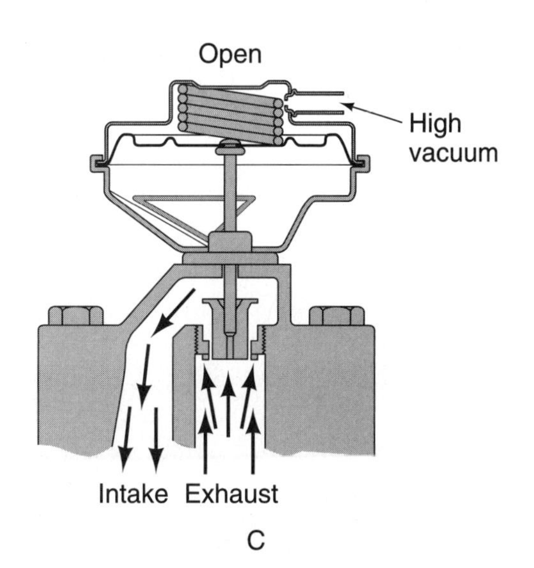

C

TURN

D

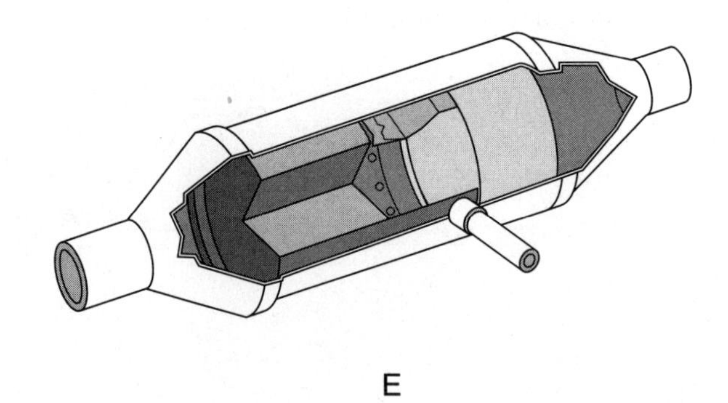

E

Charcoal Canister	____	Air Injection System	____	Catalytic Converter	____
PCV Valve	____	EGR Valve	____		

STOP

Instructor OK ______________ Score __________

Activity Sheet #61

EMISSION CONTROL SYSTEMS

Name______________________________ Class ____________________

Directions: Match the words on the left to the descriptions on the right. Write the letter for the correct word on the line provided. For the terms that you are not certain of, use the glossary in your textbook.

A. Underhood label	____ Pollutant formed under high heat and pressure in the engine
B. Oxidizing	____ Uniting fuel with oxygen during combustion
C. Light off	____ The ideal air-fuel ratio, 14.7:1 by weight, in which all of the oxygen is consumed in the burning of the fuel
D. Pyrometer	____ A good rich indicator
E. Infrared thermometer	____ Term for carbon monoxide measurement
F. Combustion	____ Thermal vacuum switch
G. Flame front	____ A label found on cars since 1972 that describes such things as the size of the engine, ignition timing specifications, idle speed, valve lash clearance adjustment, and the emission devices that are included on the engine
H. Stoichiometric	____ Deceleration ___ can be due to a faulty air injection system valve.
I. Oxides of nitrogen	____ Hydrocarbons are measured in ________.
J. Idle should drop	____ Burning of fuel
K. TVS	____ CO_2 reading from an engine
L. 500°F	____ When the catalytic converter becomes hot enough and begins to oxidize pollutants
M. PPM	____ Condition under which NO_X is tested
N. CO_2	____ When heat ignites molecules next to already burning molecules and a chain reaction takes place, which results in a flame expanding evenly across the cylinder
O. 13% to 16%	____ A good lean indicator
P. Percentage	____ A temperature measuring device that is touched against a surface to obtain a reading

TURN

Q. Air pump	____	A thermometer that takes a temperature reading when it is aimed toward a surface
R. Hydrocarbons	____	Engine idle when the PCV valve is plugged
S. O_2 reading 2-5%	____	The catalytic converter must be heated to approximately ____ °F before it starts to work.
T. Backfire	____	Disabled during emission analyzer diagnosis
U. CO	____	Analyzer oxygen reading with exhaust dilution or smog pump working
V. O_2	____	If an engine that will not start is being cranked with an emission analyzer connected, the presence of what gas tells you that there is a fuel supply?
W. Under load	____	Good indicator of engine efficiency

STOP

Instructor OK ______________ Score __________

Activity Sheet #62

IDENTIFY ENGINE PERFORMANCE TEST INSTRUMENTS

Name______________________________ Class ____________________

Directions: Identify the engine performance test instruments in the drawing. Place the identifying letter next to the name of the item on page 144.

A

B

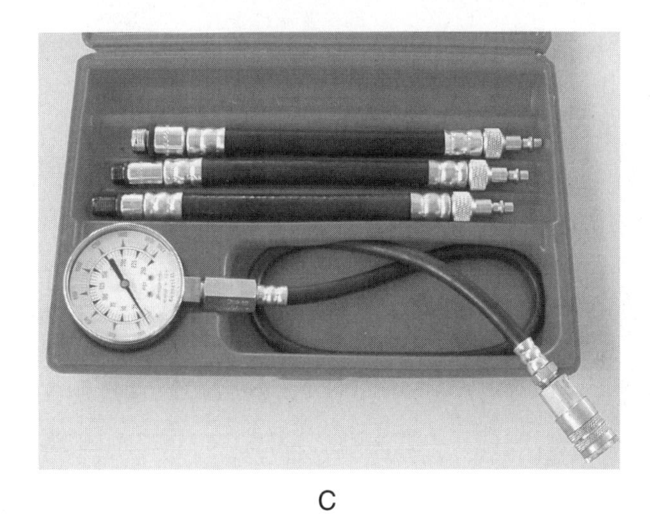

C

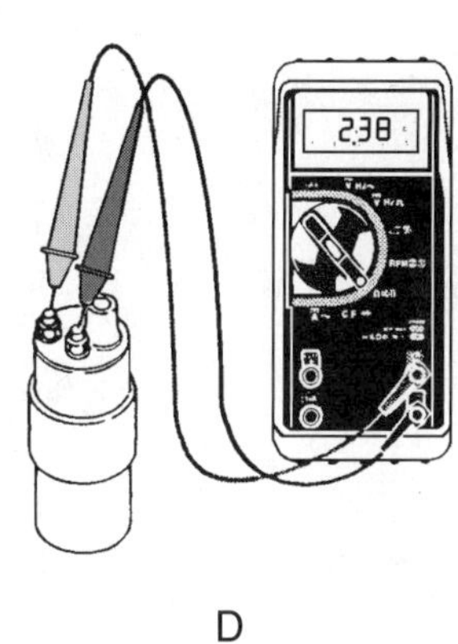

D

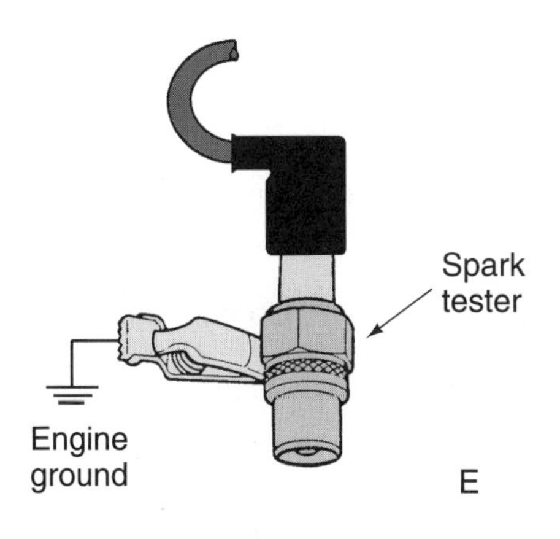

E

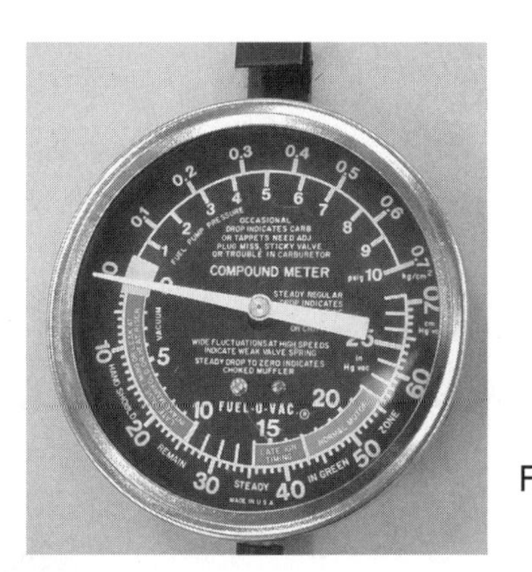

F

TURN

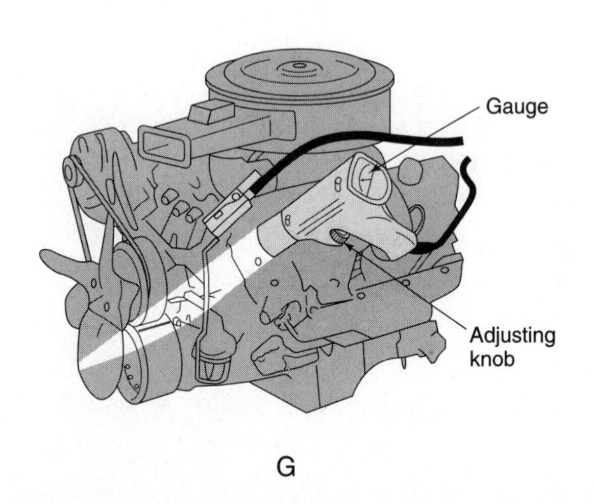

G

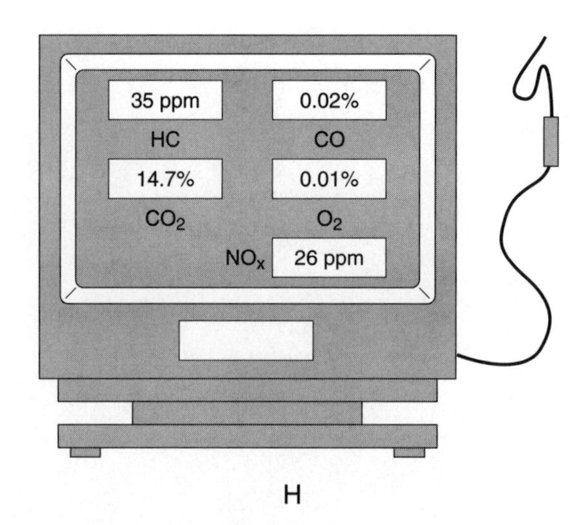

H

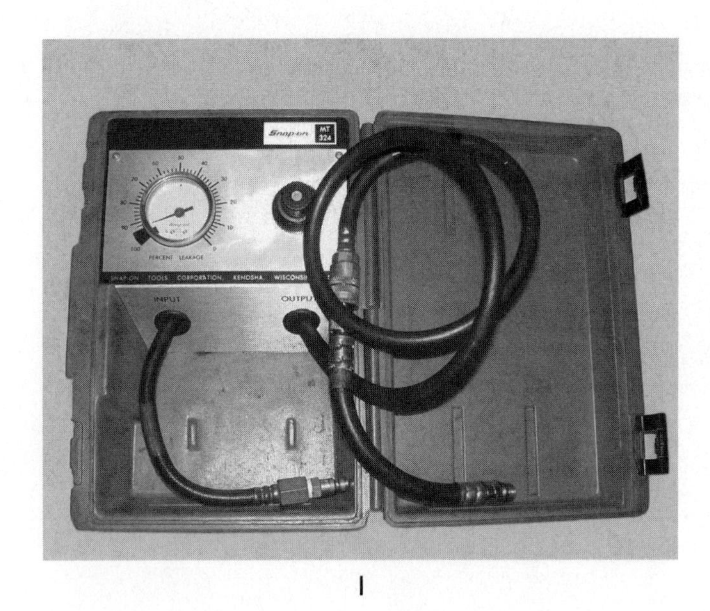

I

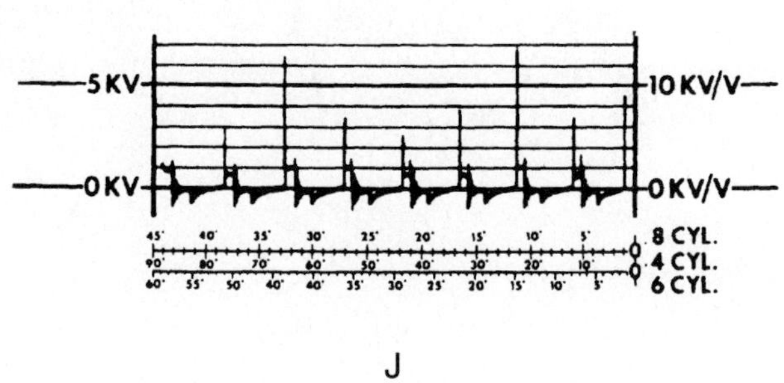

J

Scope	____	Spark Tester	____	DMM	____
Vacuum Gauge	____	Scan Tool	____	Timing Light	____
Cylinder Leakage Tester	____	Combustion Leak Tester	____	Compression Tester	____
Emission Analyzer	____				

Instructor OK ______________ Score __________

Activity Sheet #63

IDENTIFY ENGINE LEAKS

Name______________________________ Class ____________________

Directions: Identify the places where an engine's gaskets can leak. Place the identifying letter next to the correct description.

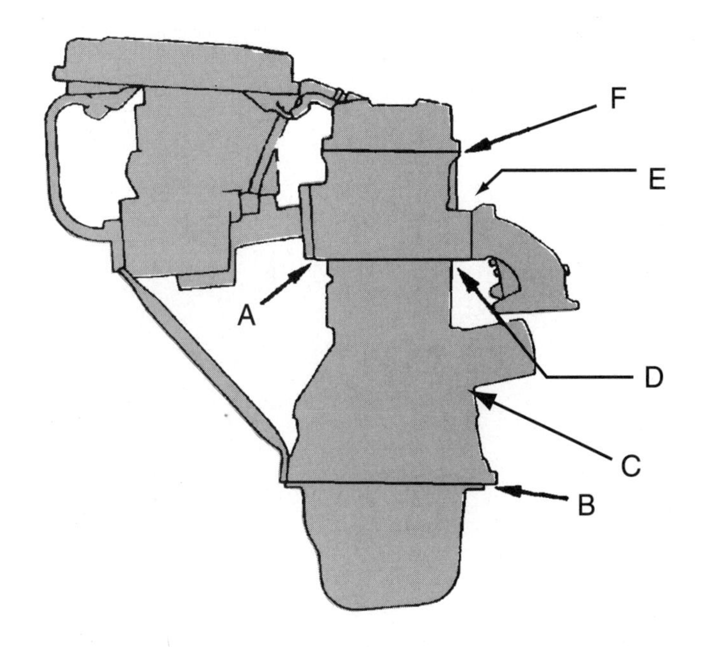

Oil Pan	____	Intake Manifold	____	Exhaust Manifold	____
Oil Filter	____	Valve Cover	____	Head Gasket	____

STOP

Part I

Activity Sheets

Section Eight

Automotive Engine Service and Repair

Instructor OK ______________ Score __________

Activity Sheet #64

ENGINE MECHANICAL PROBLEM DIAGNOSIS

Name____________________________ Class ________________

Directions: Match the words on the left to the descriptions on the right. Write the letter for the correct word on the line provided. For the terms that you are not certain of, use the glossary in your textbook.

A. Valvetrain noise	____ When a cam lobe wears out
B. Black light	____ Noises that result from excessive clearance or abnormal combustion
C. Engine knocks	____ Term used when a crankshaft will not turn
D. Piston slap	____ Can occur at 1/2 engine rpm
E. Flat cam	____ Carbon has built up in the neck area of a valve. What is the probable cause?
F. Seized engine	____ Problem with an engine that has lower oil pressure at idle speed
G. Hydrolocked engine	____ A test for oil leakage that uses fluorescent dye and an ultraviolet light source
H. Valve guide seal	____ Bad valve guide seals could cause oil smoke from the exhaust during __________ .
I. Burned spark plug	____ Noise that results from excessive clearance between the piston and cylinder
J. Worn lower main bearings	____ Damage that results when an engine runs for a long period with an excessively lean air-fuel mixture
K. Deceleration	____ When coolant or fuel in a cylinder prevents an engine from turning over

STOP

Part I

Activity Sheets

Section Nine

Miscellaneous Chassis Theory and Service

Instructor OK ______________ Score ____________

Activity Sheet #65

IDENTIFY BRAKE SYSTEM COMPONENTS

Name______________________________ Class ____________________

Directions: Identify the brake system components in the drawing. Place the identifying letter next to the name of the component.

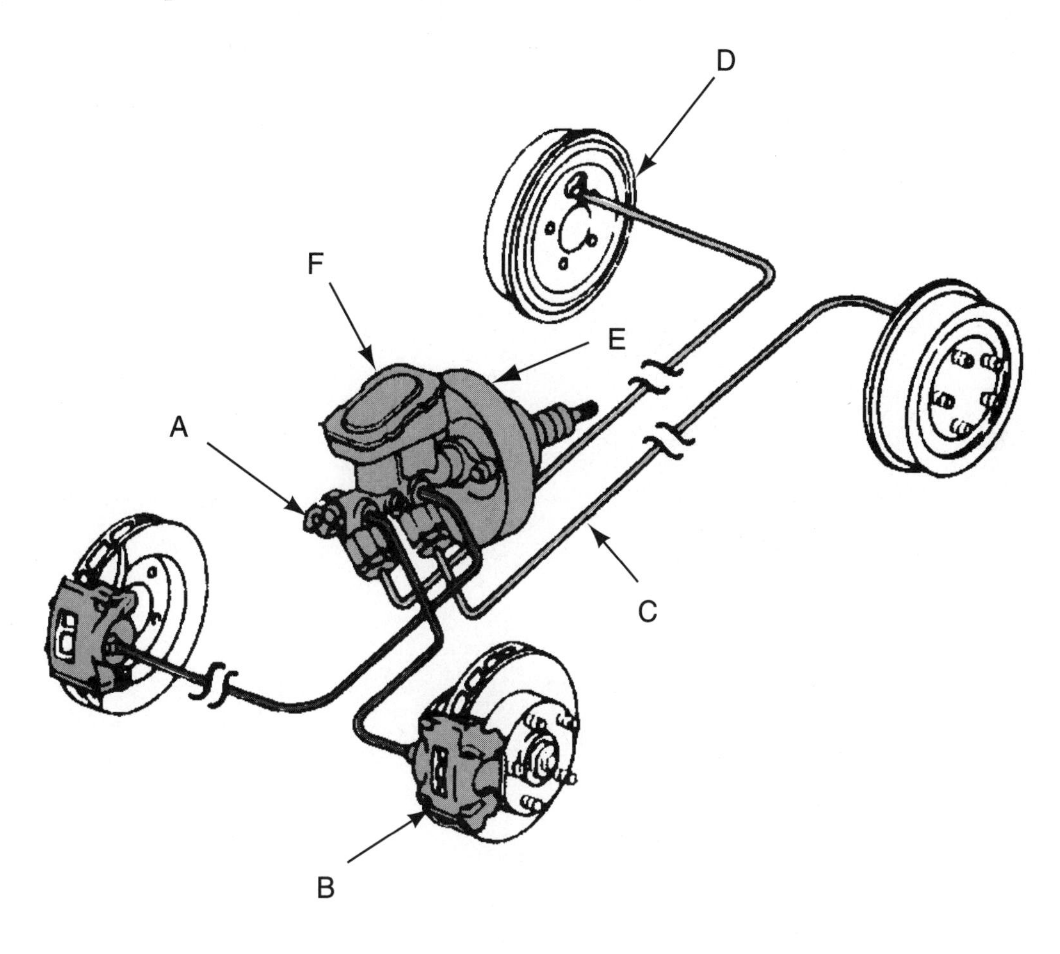

Drum Brakes	____	Brake Line	____	Master Cylinder	____
Disc Brakes	____	Brake Fluid Reservoir	____	Power Brake Booster	____

STOP

Instructor OK ________________ Score ____________

Activity Sheet #66

IDENTIFY MASTER CYLINDER COMPONENTS

Name______________________________ Class ____________________

Directions: Identify the master cylinder components in the drawing. Place the identifying letter next to the name of the component.

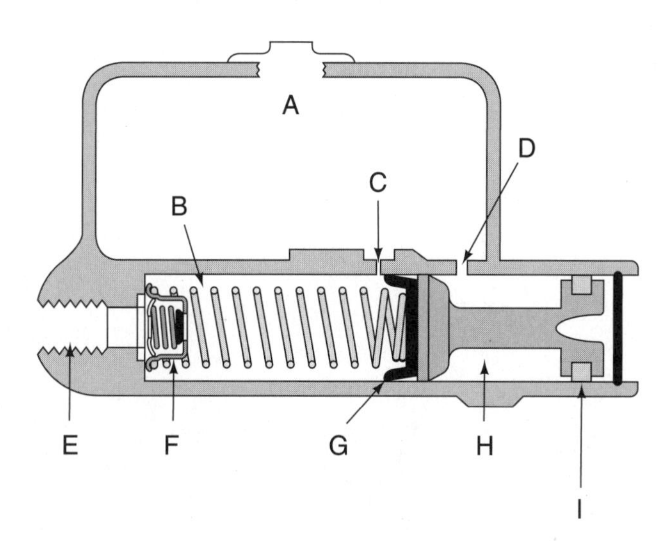

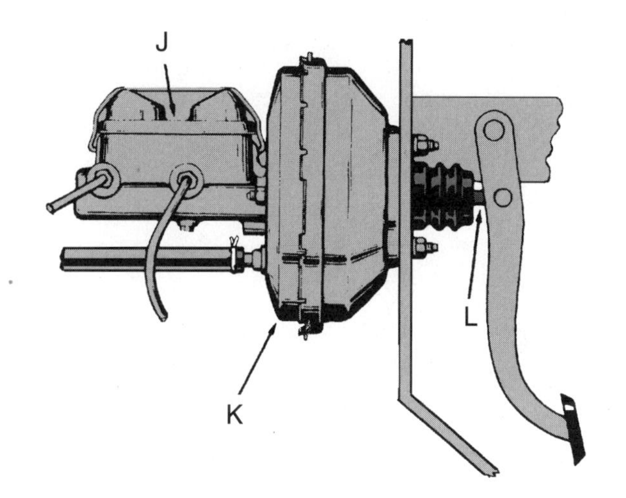

Pressure Chamber	____	Compensating Port	____	Secondary Reservoir	____
Tandem Cylinder	____	Power Brake	____	Pushrod	____
Fluid Outlet	____	Reservoir	____	Primary Cup	____
Inlet Port	____	Secondary Cup	____	Check Valve	____

Instructor OK ______________ Score __________

Activity Sheet #67

IDENTIFY WHEEL CYLINDER PARTS

Name______________________________ Class ____________________

Directions: Identify the wheel cylinder parts in the drawing. Place the identifying letter next to the name of the part.

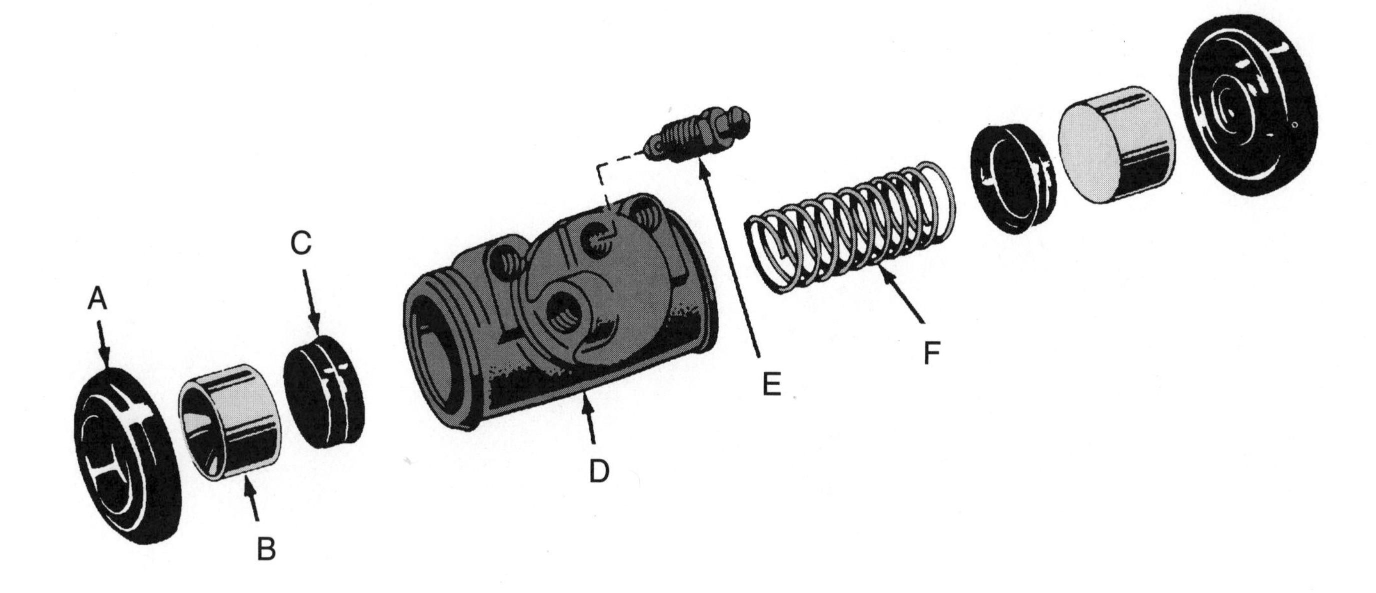

Cylinder	____	Cup	____	Piston	____
Bleed Screw	____	Boot	____	Return Spring	____

STOP

Instructor OK ______________ Score ______________

Activity Sheet #68

IDENTIFY DRUM BRAKE COMPONENTS

Name______________________________ Class ____________________

Directions: Identify the drum brake components in the drawing. Place the identifying letter next to the name of the component.

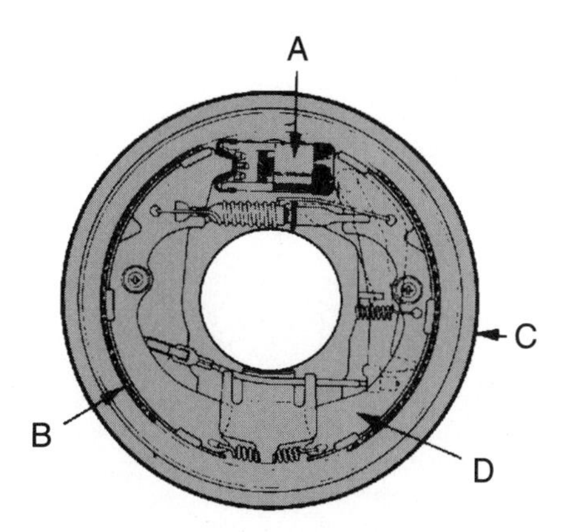

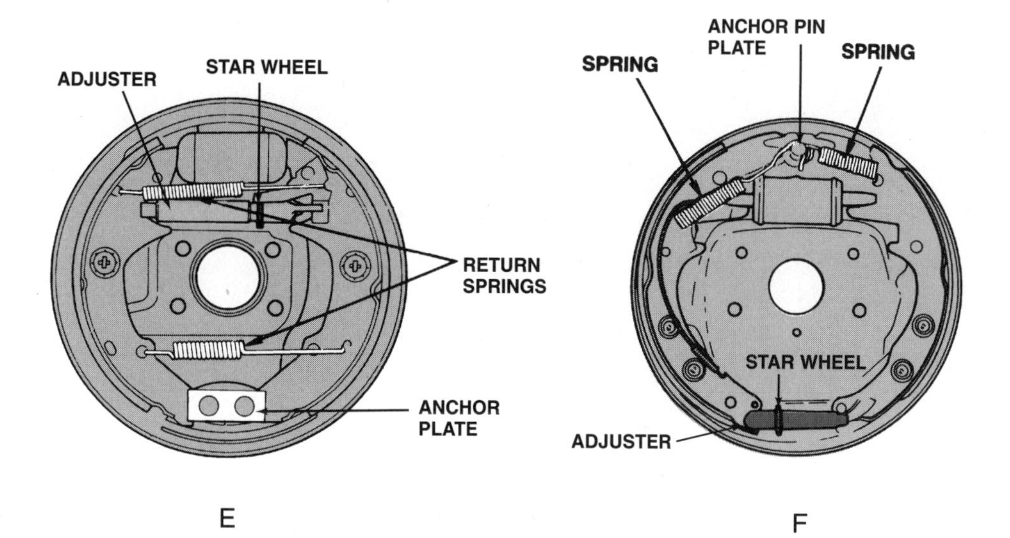

Shoe	____	Backing Plate	____	Wheel Cylinder	____
Leading/Trailing	____	Dual Servo	____	Lining	____

STOP

Instructor OK ____________ Score ____________

Activity Sheet #69

IDENTIFY DISC BRAKE COMPONENTS

Name______________________ Class ______________

Directions: Identify the disc brake components in the drawing. Place the identifying letter next to the name of the component.

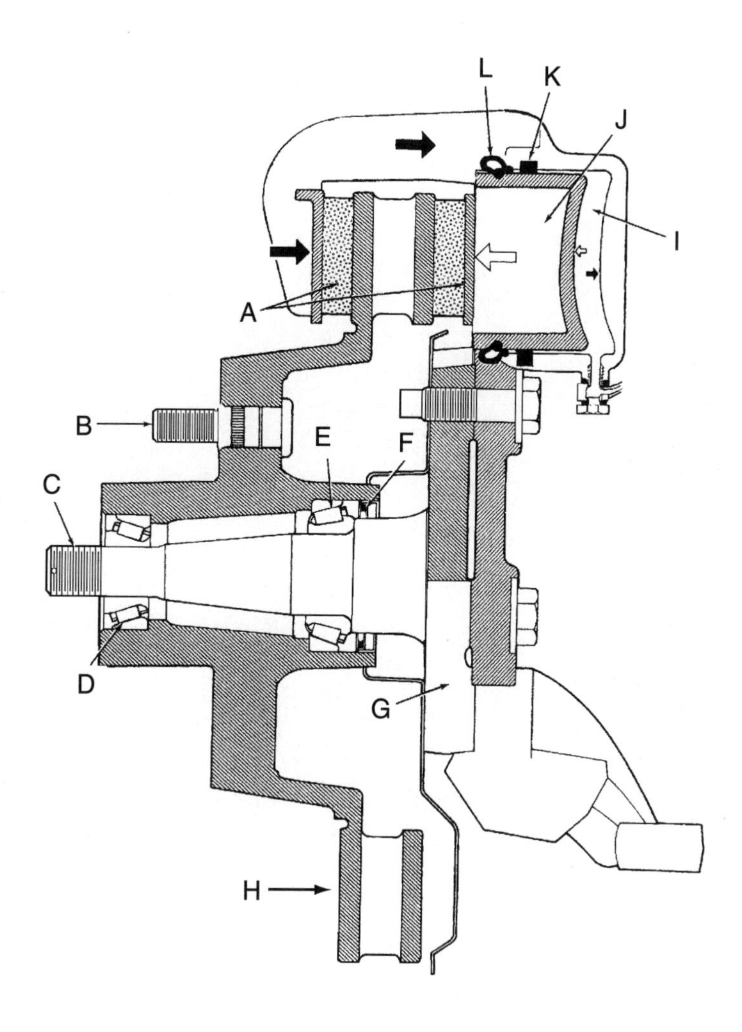

Inner Bearing	____	Piston Seal	____	Piston	____
Wheel Stud	____	Steering Knuckle	____	Outer Bearing	____
Boot	____	Seal	____	Lining	____
Disc	____	Spindle	____	Brake Fluid	____

STOP

Activity Sheet #70

IDENTIFY BRAKE HYDRAULIC COMPONENTS

Name____________________________ Class ____________________

Directions: Identify the brake hydraulic components in the drawing. Place the identifying letter next to the name of the component.

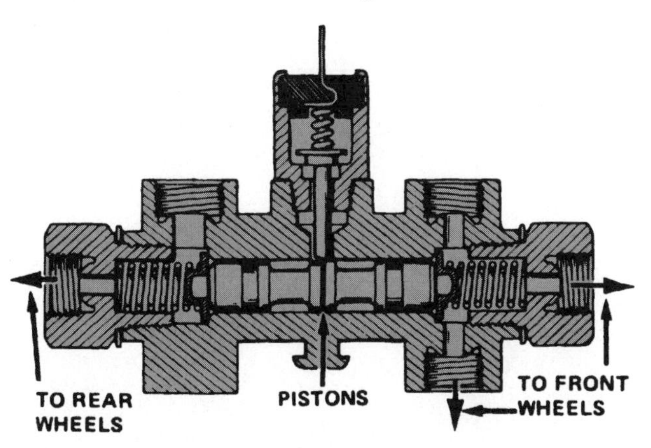

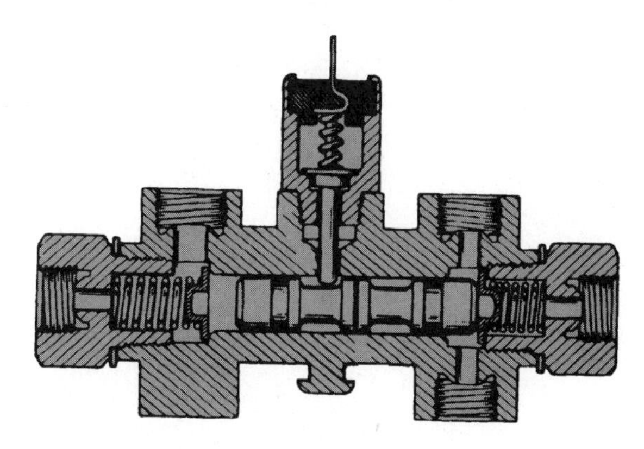

A

Related Questions:

1. Which of the above has a pressure loss?

   a. left b. right

2. Which part of the system has a pressure loss?

   a. front b. rear

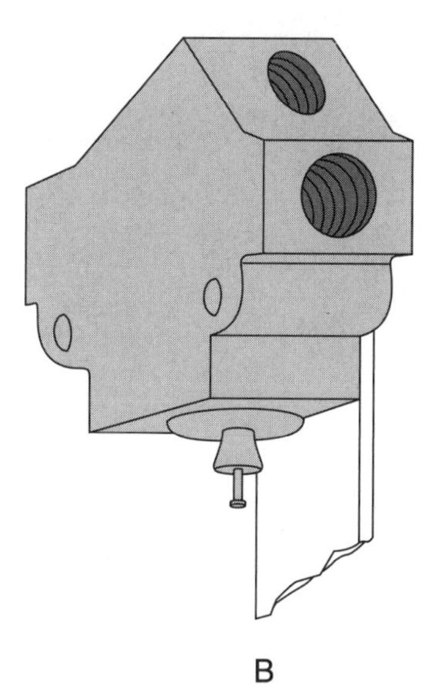

B

Pressure Differential Valve ____ Metering Valve ____

STOP

Instructor OK ______________ Score __________

Activity Sheet #71

BRAKE SYSTEMS

Name______________________________ Class ____________________

Directions: Match the words on the left to the descriptions on the right. Write the letter for the correct word on the line provided. For the terms that you are not certain of, use the glossary in your textbook.

A.	Kinetic energy	____	Linings attached to their backings with fasteners
B.	Coefficient of friction	____	When the leading shoe on a drum brake is forced into the brake drum
C.	Bonded linings	____	When liquid is used to transfer motion or apply force
D.	Riveted linings	____	Disc brake friction linings are sometimes called brake _________ .
E.	Semi-metallic linings	____	The ratio of the force required to slide one surface over another
F.	Metallic linings	____	Fluid that absorbed 2% water
G.	Brake shoes	____	Energy of motion
H.	Brake pads	____	Operates the brakes on opposite corners of the vehicle
I.	Hydraulics	____	Operates the front and rear brakes separately
J.	Pascal's law	____	Organic linings with sponge iron and steel fibers mixed into them to add strength and temperature resistance
K.	Hygroscopic	____	Linings made of metal that are used in heavy-duty conditions
L.	DOT wet specification	____	On disc brakes the friction linings are called ____________________ .
M.	DOT dry specification	____	Holds drum brake friction material
N.	Longitudinal braking	____	Loss of coefficient of friction in hot brakes
O.	Diagonal braking system	____	The weight not supported by springs
P.	Flapper valve	____	Specification for new fluid
Q.	Self-energization	____	Linings that are glued to the brake shoe
R.	Bleeding	____	The law of hydraulics
S.	Brake fade	____	DOT number for synthetic brake fluid

TURN

T.	Unsprung weight	____	Helps to prevent dangerous skids
U.	Metering valve	____	Another name for bulkhead
V.	Bulkhead	____	Allows fluid to flow one direction only
W.	Firewall	____	Disc brakes do not require this
X.	ABS	____	Removing air from a brake hydraulic system
Y.	Pads	____	Material that absorbs water
Z.	Equal	____	Separates engine and passenger compartments
AA.	DOT 5	____	Disc brake design that does not allow caliper to move
AB.	Ethylene glycol	____	Caliper that is able to slide during and after application
AC.	Trailing	____	Antilock brake systems
AD.	Engine vacuum	____	Energy source for power-assisted brakes
AE.	Return springs	____	Pressure in an enclosed system is _____ and undiminished in all directions
AF.	Fixed caliper	____	Duo servo and the leading- _____ shoe are two types of drum brake designs.
AG.	Floating caliper	____	Used to make brake fluid and automotive coolant

STOP

Instructor OK ______________ Score ________

Activity Sheet #72

IDENTIFY TAPERED WHEEL BEARING PARTS

Name______________________ Class ______________

Directions: Identify the tapered wheel bearing parts in the drawing. Place the identifying letter next to the name of the part.

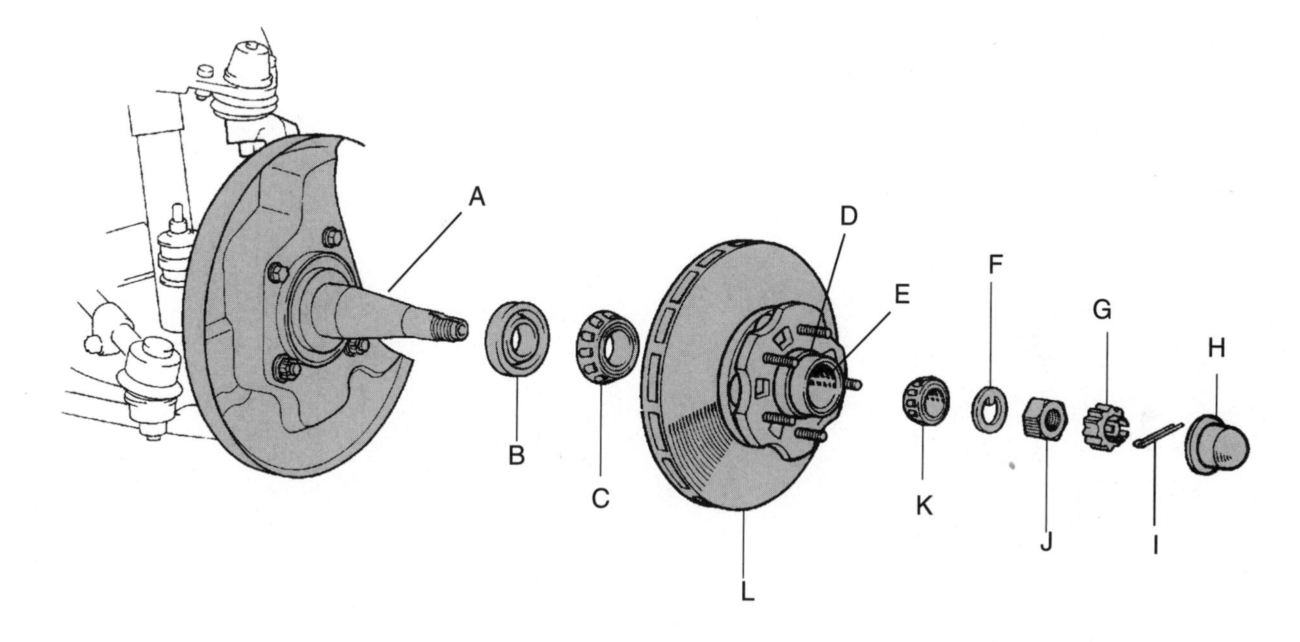

Outer Bearing	____	Nut Lock	____	Inner Bearing	____
Thrust Washer	____	Spindle	____	Oil Seal	____
Cap	____	Cotter Pin	____	Hub	____
Nut	____	Disc/Rotor	____	Bearing Cup	____

STOP

Instructor OK ______________ Score __________

Activity Sheet #73

IDENTIFY TAPERED AND DRIVE AXLE BEARING PARTS

Name______________________________ Class __________________

Directions: Identify the tapered and drive axle bearing parts in the drawing. Place the identifying letter next to the name of the part.

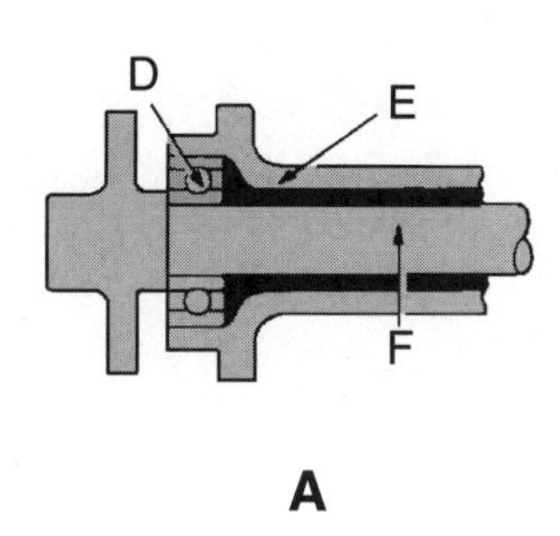

A

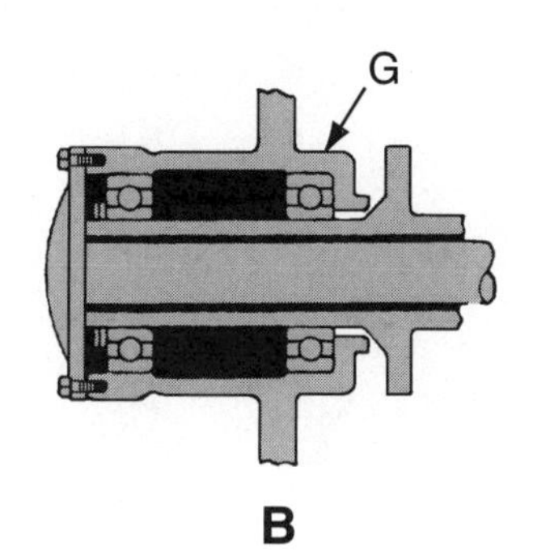

B

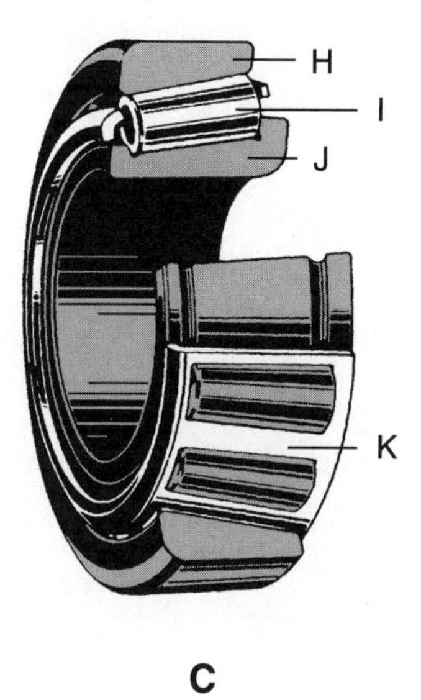

C

Enter the letter next to the correct bearing design.

____ Full-floating

____ Tapered Roller

____ Semi-floating

Enter the letter next to the correct part name.

Cage	____	Cup	____	Axle Shaft	____
Cone	____	Wheel Hub	____	Bearing	____
Roller	____	Axle Housing	____		

STOP

Instructor OK ______________ Score __________

Activity Sheet #74

BEARINGS AND GREASE

Name______________________________ Class ________________

Directions: Match the words on the left to the descriptions on the right. Write the letter for the correct word on the line provided. For the terms that you are not certain of, use the glossary in your textbook.

A. Bearing cage	___ The term for non-drive front- and rear-wheel bearings
B. Radial load	___ Indentations in the bearing or race from shock loads
C. End thrust	___ The term for bearings that are on live axles (those that drive wheels)
D. Thrust bearing	___ An axle design in which the bearings do not touch the axle but are located on the outside of the axle housing
E. Race	___ A stamped steel or plastic insert that keeps bearing balls or rollers properly spaced around the bearing assembly
F. Needle bearing	___ Side-to-side or front-to-rear force
G. Wheel bearings	___ Moving
H. Axle bearings	___ A grease of a consistency that allows it to be applied through a zerk fitting with a grease gun
I. Semi-floating axle	___ A bearing cup
J. Full-floating axles	___ When two parts have a pressed fit
K. Grease	___ A bearing surface at 90° to the load
L. NLGI	___ A very small roller bearing used to control thrust or radial loads
M. Chassis lubricant	___ Use one with the largest diameter that will fit into the hole
N. Dynamic	___ A bearing design that tends to be self-aligning
O. Static	___ An axle design in which the bearing rides on the axle

TURN

P.	Brinelling	___	A load in an up and down direction
Q.	Interference fit	___	A combination of oil and a thickening agent
R.	Tapered roller bearings	___	National Lubricating Grease Institute
S.	Cotter pin	___	At rest

Instructor OK ______________ Score ____________

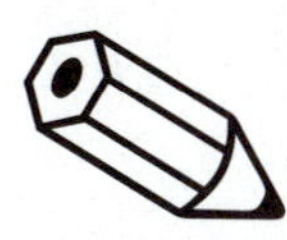

Activity Sheet #75

IDENTIFY SECTIONS OF THE TIRE LABEL

Name______________________________ Class ____________________

Directions: Identify the sections of the tire label in the drawing. Place the identifying letter next to the correct description.

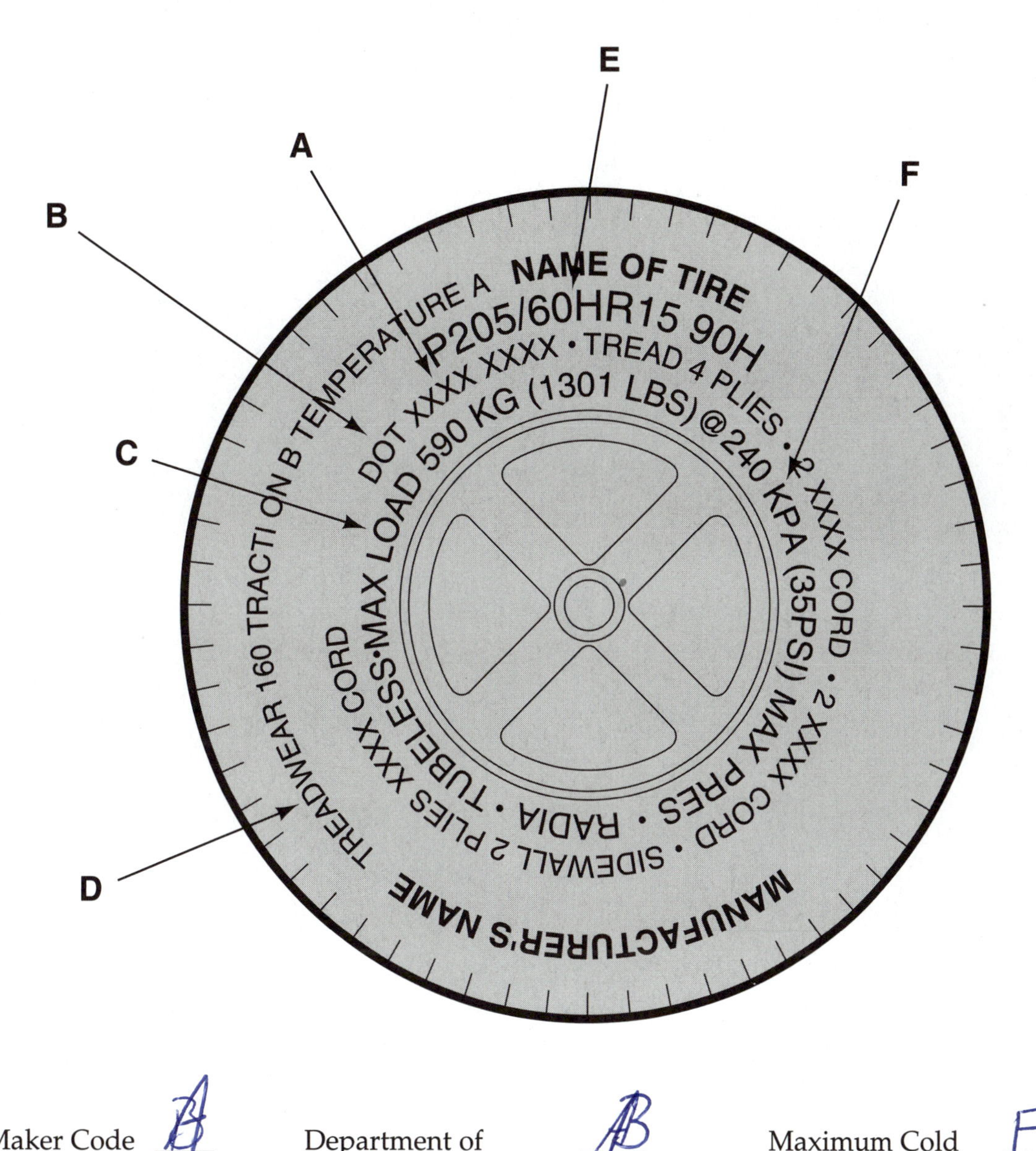

Tire Maker Code	A	Department of Transportation Code	AB	Maximum Cold Air Pressure	F
Tire Size	E	UTQG Rating	D	Load Maximum	C

STOP

Instructor OK ______________ Score ___________

Activity Sheet #76

IDENTIFY PARTS OF TIRE SIZE DESIGNATION

Name______________________ Class ______________

Directions: Identify the parts of this tire's size designation. List the correct part of the listed size designation on the line next to its matching description.

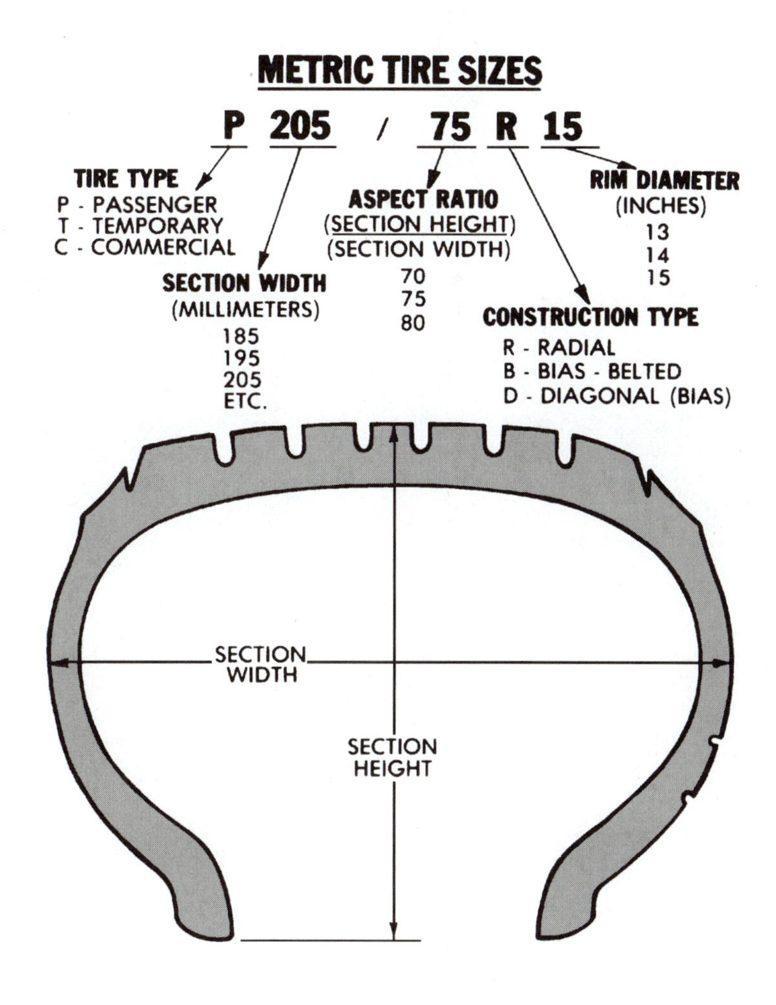

Tire Type P

Construction Type R

Section Width 205

Rim Diameter 15

Aspect Ratio 75

STOP

Instructor OK ______________ Score ____________

Activity Sheet #77

IDENTIFY PARTS OF THE TIRE

Name______________________________ Class ________________

Directions: Identify the parts of the tire in the drawing. Place the identifying letter next to the name of the part.

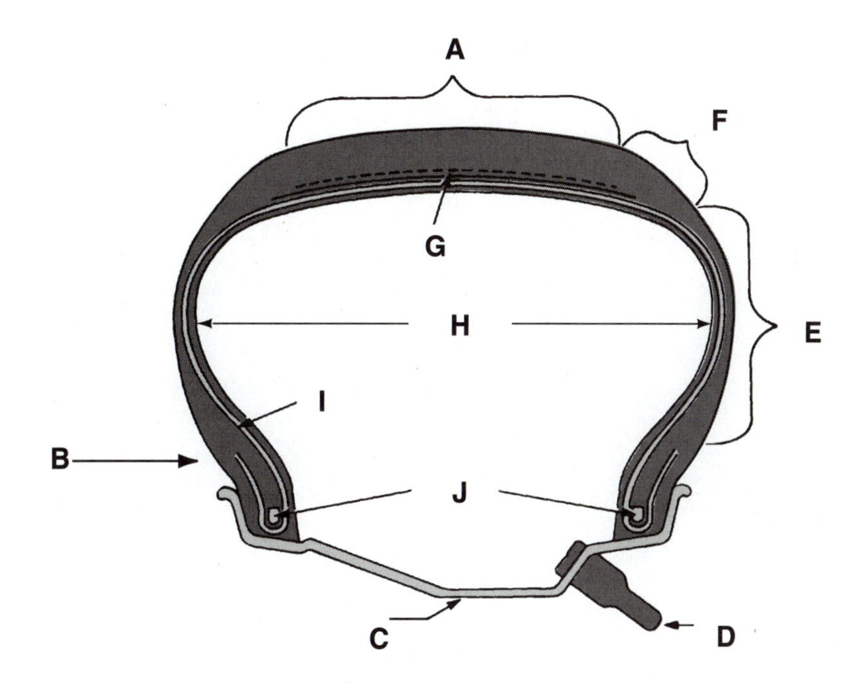

Sidewall	E	Shoulder	F	Valve	D
Tread	A	Belt	I	Rim	C
Casing	H	Bead	B	Plies	G
Chafing Strips	J				

STOP

Instructor OK ______________ Score __________

Activity Sheet #78

TIRES

Name______________________ Class ______________

Directions: Match the words on the left to the descriptions on the right. Write the letter for the correct word on the line provided. For the terms that you are not certain of, use the glossary in your textbook.

A. Radial runout	____ The speed at which the part of the tire that is contacting the ground is travelling
B. Lateral runout	____ The name for the up and down action of the tire that results in scalloped tire wear
C. Tire plug	____ A piece of rubber vulcanized to the inner liner of a tire to repair a leak
D. Patch	____ The combination of both static and couple imbalance
E. RMA	____ Material applied to the tire bead before installing a tire on a rim
F. Radial	____ Wobble of a part in a side-to-side direction
G. 4 to 8 psi	____ Type of wheel balance measured with the wheel stationary
H. 1/32″	____ A piece of rubber vulcanized into a hole in a tire
I. Wheel tramp	____ A tire design that has a bulging sidewall when properly inflated
J. Rubber lube	____ Rubber Manufacturers' Association
K. Static	____ Approximate pressure increase in a tire as it warms up
L. 0 mph	____ Wobble of a part in an up and down direction
M. Dynamic imbalance	____ The tread depth at which wear bars show up around the tire tread

STOP

Part I

Activity Sheets

Section Ten

Suspension, Steering, Alignment

Activity Sheet #79

IDENTIFY SHORT/LONG ARM SUSPENSION COMPONENTS

Name____________________________ Class ____________________

Directions: Identify the short/long arm suspension components in the drawing. Place the identifying letter next to the name of the component.

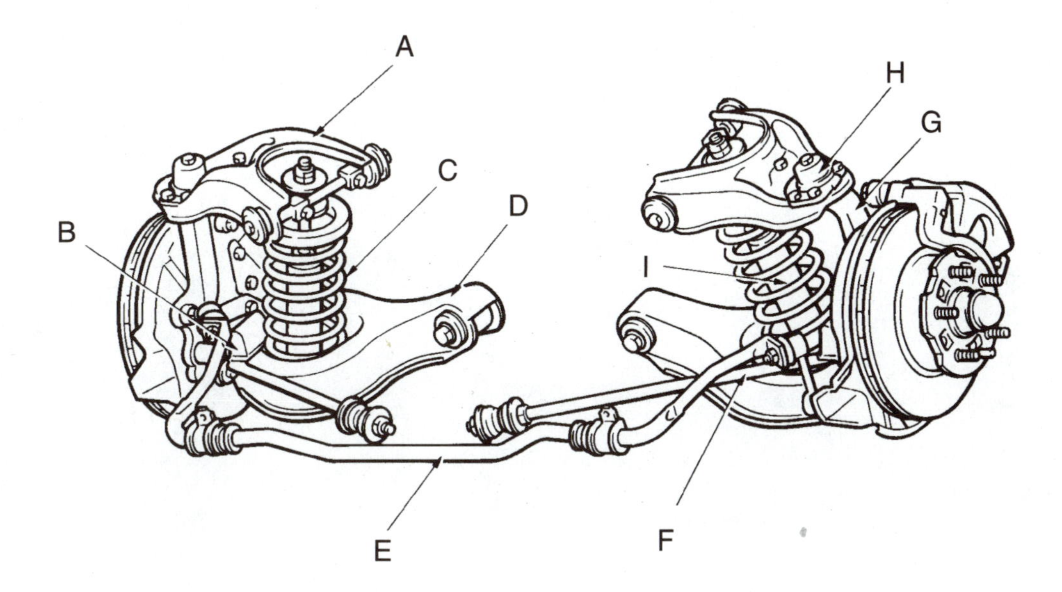

Steering Knuckle	G	Upper Control Arm	A	Upper Ball Joint	H
Lower Control Arm	D	Shock Absorber	I	Coil Spring	C
Bumper	B	Sway Bar	E	Strut Rod	F

STOP

Activity Sheet #80

IDENTIFY MACPHERSON STRUT SUSPENSION COMPONENTS

Name______________________ Class ______________

Directions: Identify the MacPherson strut suspension components in the drawing. Place the identifying letter next to the name of the component.

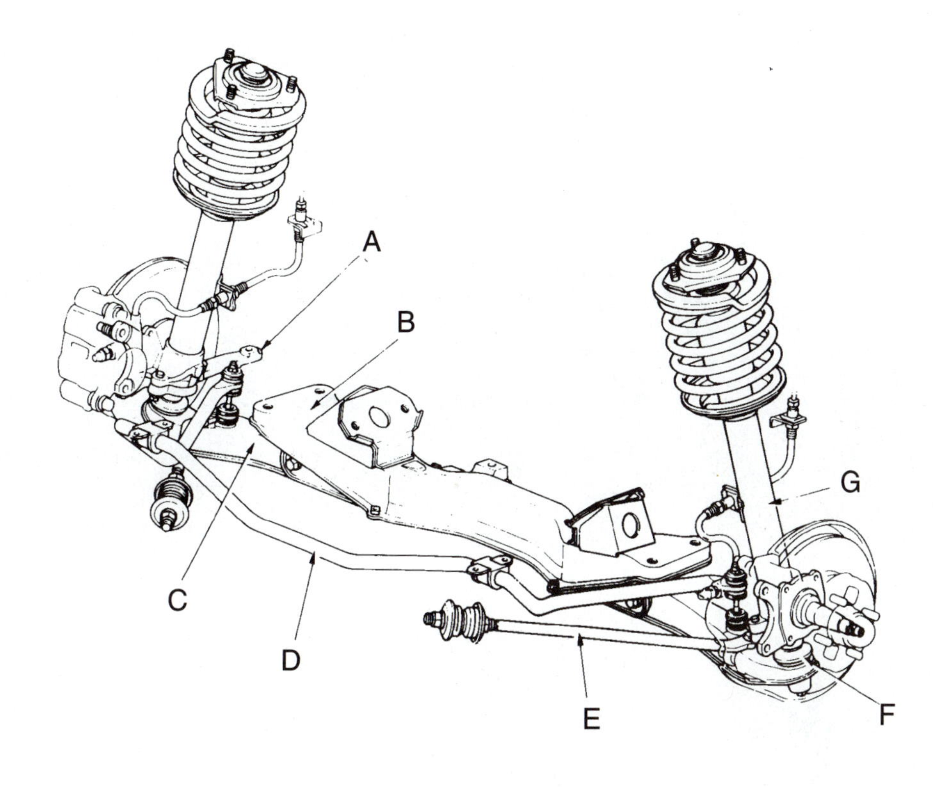

Steering Arm	A	Sway Bar	D	MacPherson Strut	G
Strut Rod	E	Control Arm	C	Ball Joint	F
Crossmember	B				

STOP

Instructor OK ______________ Score __________

Activity Sheet #81

SUSPENSION SYSTEMS

Name_________________________ Class _______________

Directions: Match the words on the left to the descriptions on the right. Write the letter for the correct word on the line provided. For the terms that you are not certain of, use the glossary in your textbook.

A. Chassis	E	Steel rod wound into a coil
B. Suspension	____	Spring with a smooth ride that allows for heavier carrying capacity too
C. Rigid axle	____	Automatic suspension that keeps the car body level during all driving conditions
D. Independent suspension	____	Type of suspension system in which only one wheel will deflect
E. Coil spring	____	When the wheel moves up as the spring compresses
F. Variable rate spring	____	Weight not supported by springs
G. Torsion bar	____	The group of parts that includes the frame, shocks and springs, steering parts, tires, brakes, and wheels
H. Overload spring	____	A group of parts that supports the vehicle and cushions the ride
I. Sprung weight	____	A suspension design that incorporates the shock absorber into the front suspension
J. Unsprung weight	____	When hydraulic fluid becomes mixed with air
K. Compression or jounce	____	Also called aeration
L. Rebound	____	Straight rod that works as a spring
M. Short/long arm (SLA)	____	Found on heavy trucks
N. Shock absorber	____	An additional spring that only works under a heavy load
O. Shock ratio	____	Weight supported by springs
P. Aeration	____	Suspension leveling system that keeps the vehicle at the same height when weight is added to parts of the car
Q. Cavitation	____	When the wheel moves back down after compression

TURN

R. Gas shock	____	Suspension design that uses two control arms of unequal length
S. MacPherson strut	____	Dampens spring oscillations
T. Adaptive suspension system	____	The difference between the amount of control on compression and extension
U. Active suspension	____	Pressurized to keep the bubbles from forming in the fluid

STOP

Instructor OK ______________ Score __________

Activity Sheet #82

IDENTIFY STEERING COMPONENTS

Name______________________________ Class ____________________

Directions: Identify the steering components in the drawing. Place the identifying letter next to the name of the component.

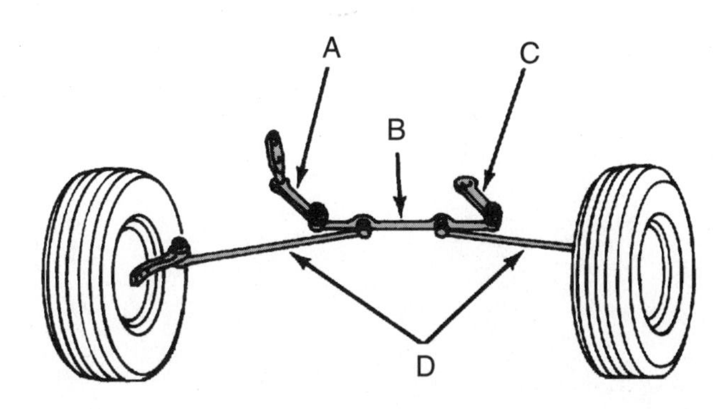

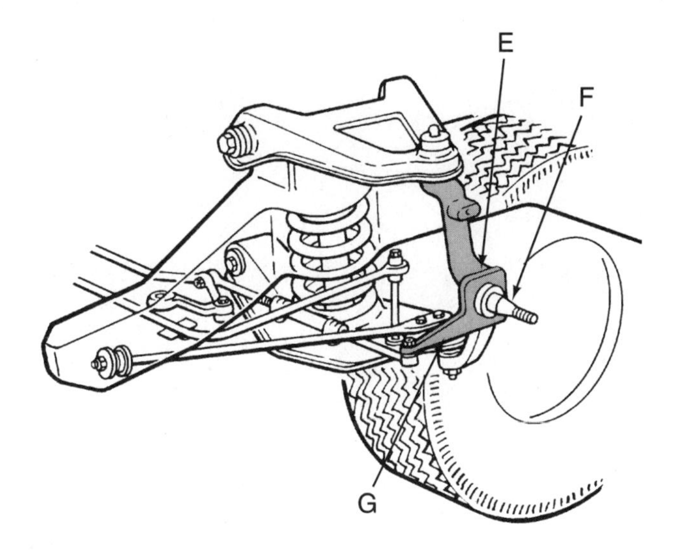

Tie Rods	____	Steering Arm	____	Center Link	____
Pitman Arm	____	Idler Arm	____	Steering Knuckle	____
Spindle	____				

Instructor OK ______________ Score ____________

Activity Sheet #83
IDENTIFY POWER STEERING COMPONENTS

Name______________________ Class ________________

Directions: Identify the power steering components in the drawing. Place the identifying letter next to the name of the component.

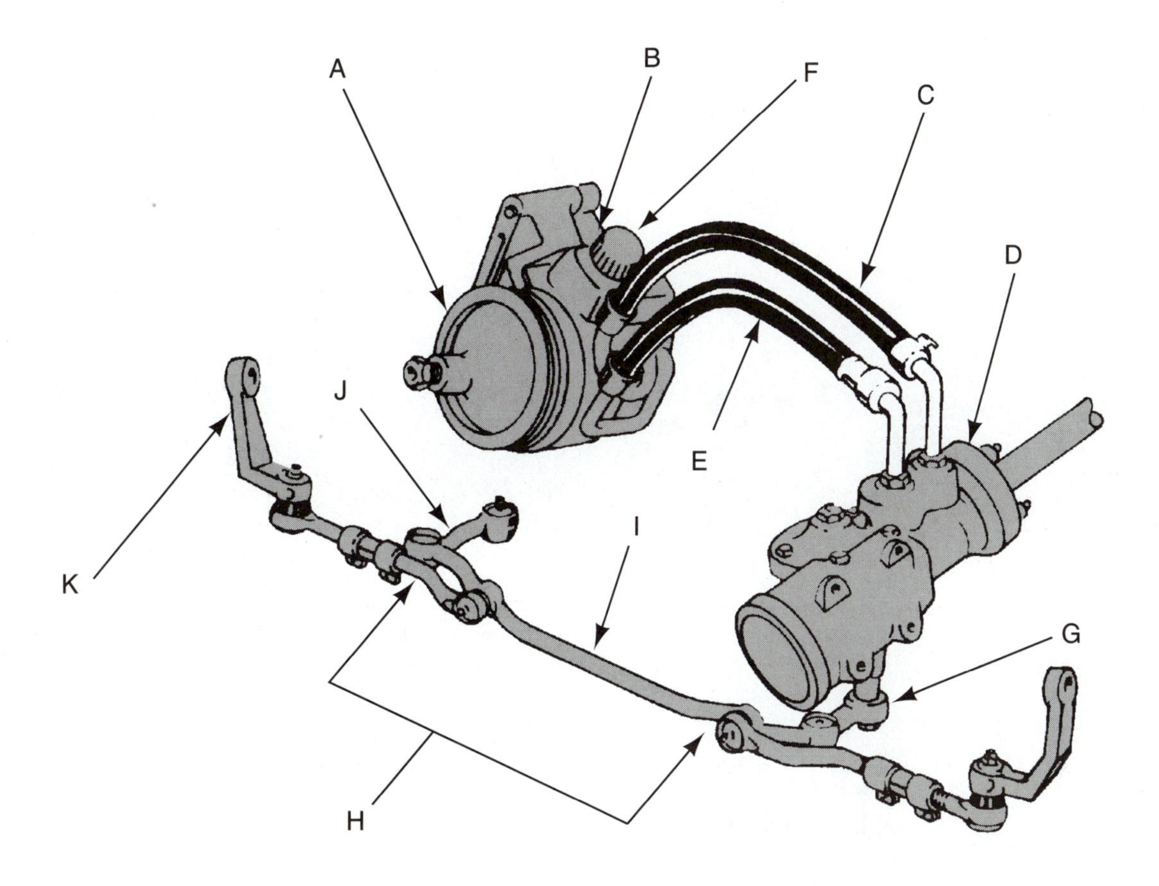

Pulley	A	Pressure Line	____	Pump	B
Return Line	____	Steering Gear	____	Reservoir	F
Pitman Arm	____	Steering Arm	____	Tie-Rods	____
Idler Arm	____	Center Link	____		

STOP

Instructor OK ______________ Score __________

Activity Sheet #84
IDENTIFY STEERING GEAR COMPONENTS

Name______________________________ Class ____________________

Directions: Identify the steering gear components in the drawing. Place the identifying letter next to the name of the component.

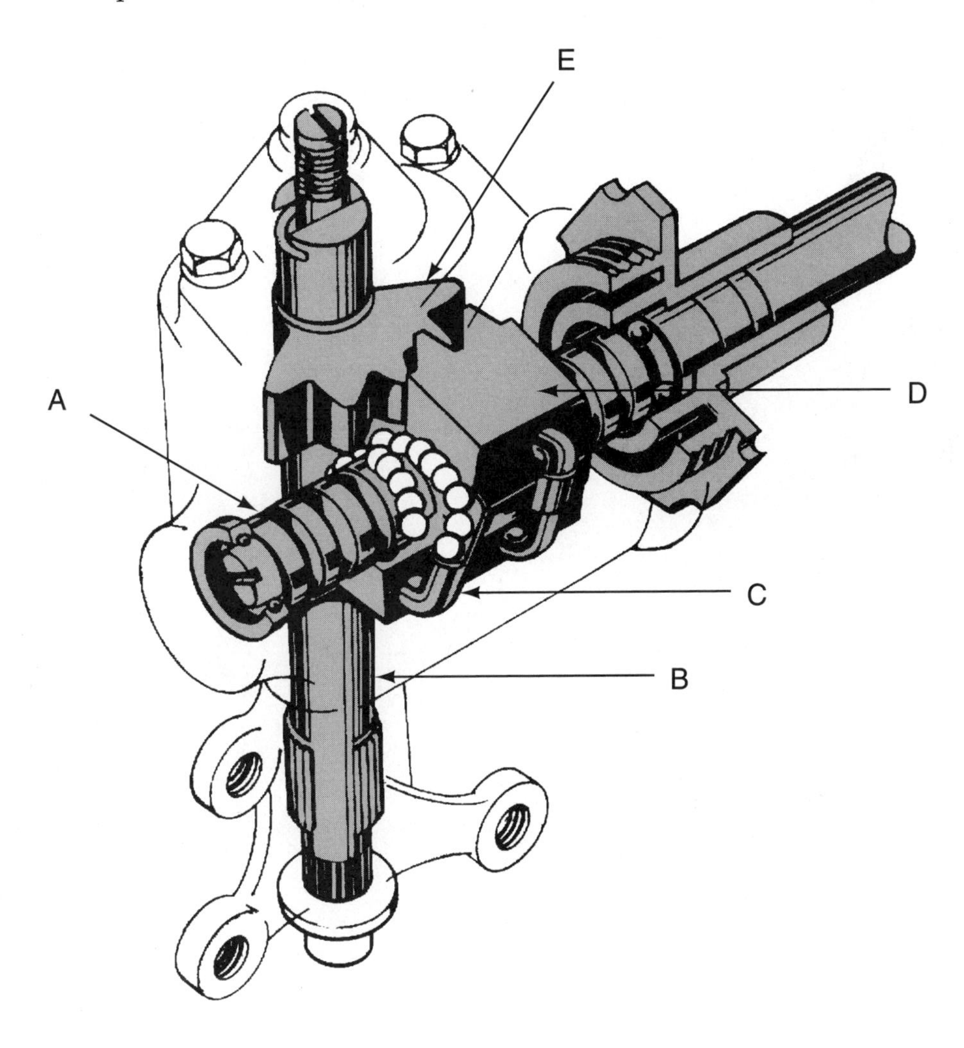

Ball Return Tubes	____	Sector Gear	____	Ball Nut	____
Sector Shaft	____	Worm Shaft	____		

STOP

Instructor OK ______________ Score __________

Activity Sheet #85

IDENTIFY RACK AND PINION STEERING GEAR COMPONENTS

Name______________________________ Class ____________________

Directions: Identify the rack and pinion steering gear components in the drawing. Place the identifying letter next to the name of the component.

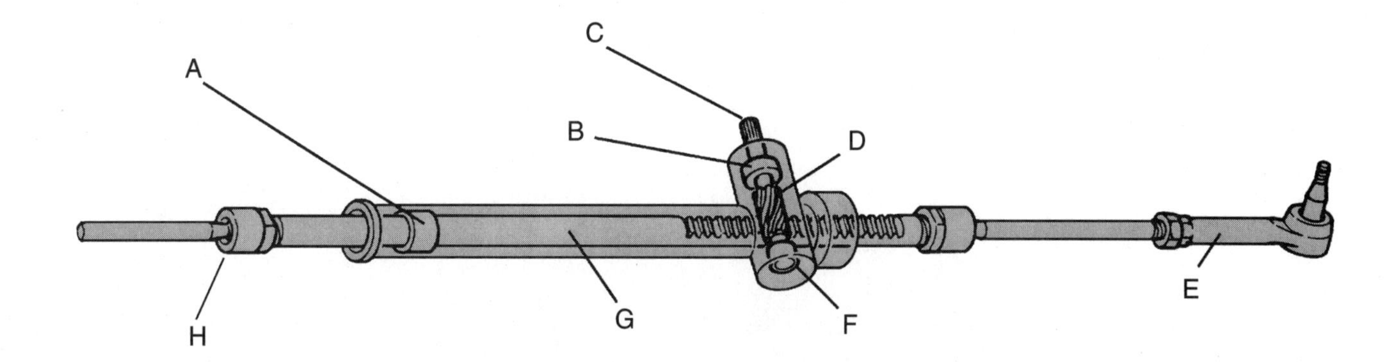

Rack Support Bushing	____	Lower Pinion Bearing	____	Upper Pinion Bearing	____
Inner Tie-Rod	____	Pinion Gear	____	Outer Tie-Rod	____
Rack Gear	____	Pinion Shaft	____		

STOP

Instructor OK ______________ Score __________

Activity Sheet #86

STEERING SYSTEMS

Name______________________ Class ______________

Directions: Match the words on the left to the descriptions on the right. Write the letter for the correct word on the line provided. For the terms that you are not certain of, use the glossary in your textbook.

A. Lock to lock	____ Most common steering on new vehicles
B. Steering ratio	____ Used to shorten and lengthen shafts or rods
C. Recirculating ball	____ Power steering with a piston attached to the steering linkage
D. Rack and pinion steering	____ Angled so that the front wheels toe out during a turn
E. Steering damper	____ Parts that connect the steering gear to the wheels
F. Steering linkage	____ Shock absorber on the steering linkage
G. Parallelogram steering	____ Number of teeth on the driving gear compared to the number of teeth on the driven gear
H. Ball sockets	____ Steering gear used with parallelogram steering
I. Turnbuckle	____ Most common type of power steering pump
J. Pressure relief valve	____ Type of steering shaft coupling
K. Steering arm	____ The faster the vehicle is driven, the ____ power assist is needed.
L. Roller, vane, and slipper	____ Hydraulic valve that bleeds off excess pressure
M. Flex coupling	____ Steering linkage that uses a steering box
N. Vane	____ Wheel that turns sharper during a turn
O. Linkage power steering	____ Allow steering linkage parts to pivot
P. Integral power steering	____ When the steering wheel is turned all the way from one direction to the other
Q. Inner	____ Types of steering pumps
R. Less	____ Power steering design that has the power steering components contained within the steering gear

STOP

Instructor OK ______________ Score ____________

Activity Sheet #87

IDENTIFY CAUSES OF TIRE WEAR

Name______________________________ Class ____________________

Directions: Choose the most probable cause of the following types of tire wear from the list. Place the letter next to the cause.

A

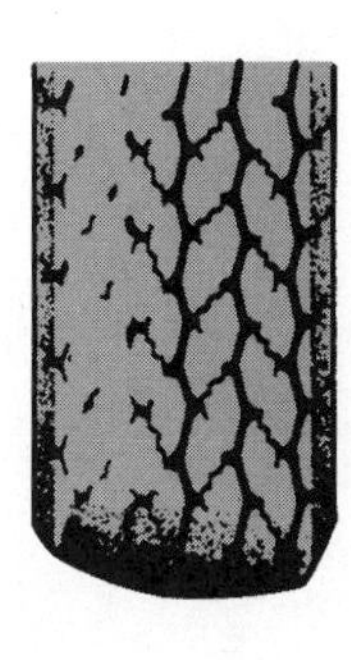

B

C

Toe Wear ____

Camber Wear ____

Loose Parts or Wheel Balance ____

STOP

Instructor OK _______________ Score __________

Activity Sheet #88

IDENTIFY CORRECT WHEEL ALIGNMENT TERM

Name______________________________ Class ____________________

Directions: Identify the correct wheel alignment term and place the letter that matches it on the line next to it on page 202.

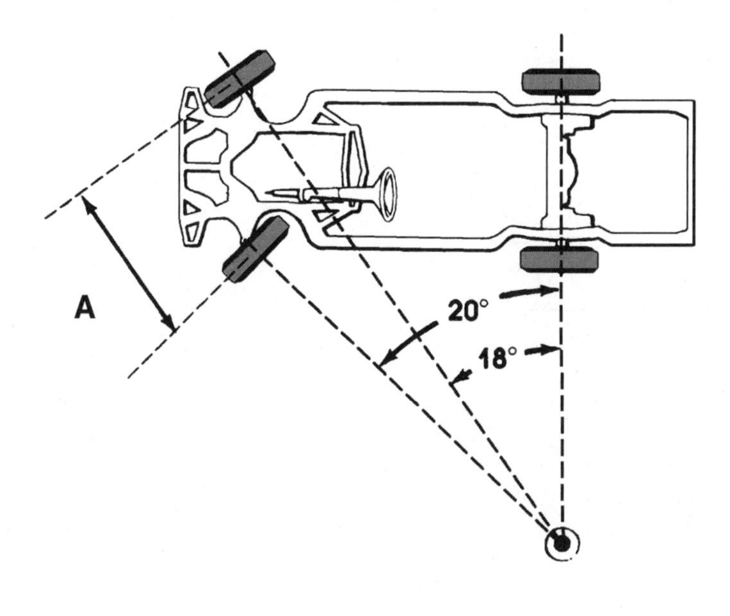

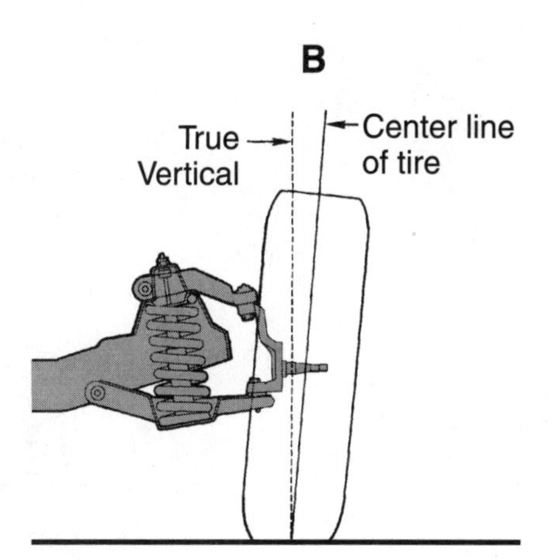

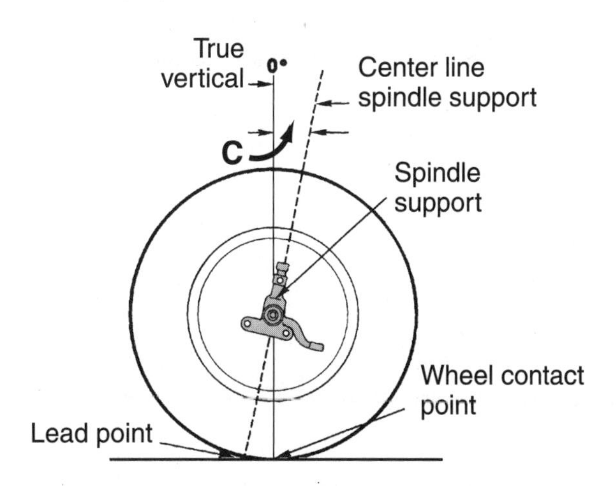

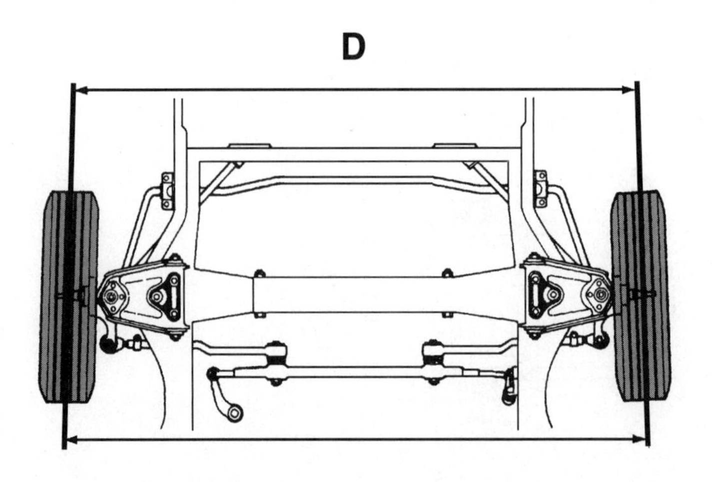

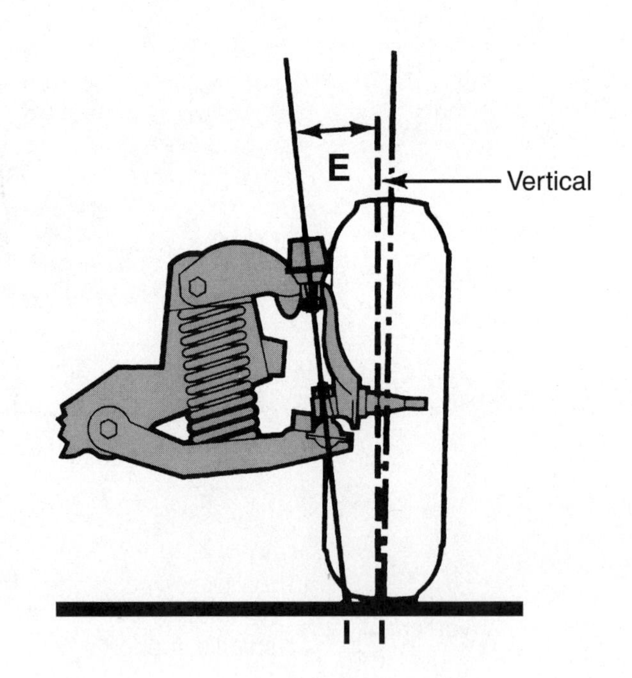

Caster ____

Camber ____

Steering Axis Inclination ____

Turning Radius ____

Toe ____

Instructor OK ______________ Score __________

Activity Sheet #89
ALIGNMENT

Name__________________________ Class ________________

Directions: Match the words on the left to the descriptions on the right. Write the letter for the correct word on the line provided. For the terms that you are not certain of, use the glossary in your textbook.

A. Toe	____	Inward or outward tilt of a tire at the top
B. Toe-in	____	The distance between the front and rear tires
C. Toe-out	____	The amount that the spindle support arm leans in at the top
D. Scuff	____	Also called turning radius or toe-out on turns
E. Camber	____	Forward or rearward tilt of the spindle
F. Positive camber	____	Also called the crossmember
G. Negative camber	____	When a car does not seem to respond to movement of the steering wheel during a hard turn
H. Camber roll	____	When a car turns too far in response to steering wheel movement
I. Caster	____	Comparison of the distances between the fronts and the rears of a pair of tires
J. Positive caster	____	The tendency during a turn for a tire to continue to go in the direction it was going before
K. Negative caster	____	When a tire is tilted out at the top
L. Included angle	____	When a tire is tilted in at the top
M. Scrub radius	____	The amount that one front wheel is behind the one on the other side of the car
N. Crossmember	____	A term that refers to the relationship between the average direction that the rear tires point and the average direction that the front tires point
O. Cradle	____	The forward tilt of the steering axis
P. Toe-out-on-turns	____	When the tires are closer together at the rear
Q. Ackerman angle	____	Tire wear resulting from incorrect toe adjustment
R. Wheel base	____	The steering axis pivot center line
S. Track	____	The large steel part of the frame beneath the engine and between the front wheels

TURN

T.	Tracking	____	When the tires are closer together at the front
U.	Set-back	____	The side-to-side distance between an axle's tires
V.	Slip angle	____	The rearward tilt of the steering axis
W.	Understeer	____	SAI and camber together
X.	Oversteer	____	A term describing how a tire rolls in a circle like a cone
Y.	Steering axis inclination (SAI)	____	A term describing how tires toe out during a turn because the steering arms are bent at an angle

Part I
Activity Sheets

Section Eleven

Drivetrain

Instructor OK ______________ Score __________

Activity Sheet #90

IDENTIFY CLUTCH COMPONENTS (ASSEMBLED)

Name______________________________ Class ____________________

Directions: Identify the clutch components in the drawing. Place the identifying letter next to the name of the component.

Related Question: Is the clutch shown here engaged or disengaged?

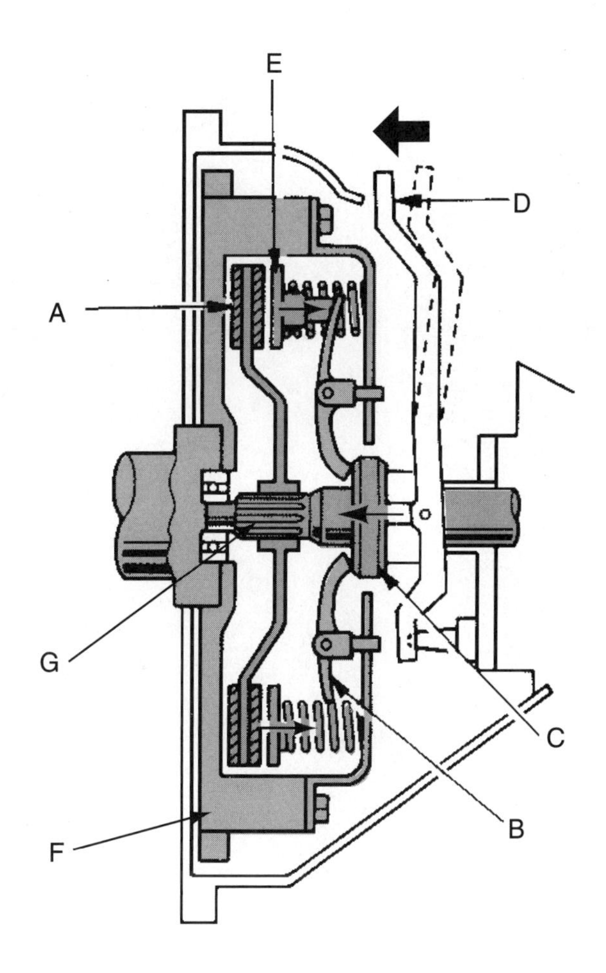

Clutch Disc	____	Input Shaft	____	Release Fork	____
Release Bearing	____	Flywheel	____	Pressure Plate	____
Release Lever	____				

STOP

Instructor OK ______________ Score __________

Activity Sheet #91

IDENTIFY CLUTCH COMPONENTS (EXPLODED VIEW)

Name______________________________ Class ____________________

Directions: Identify the clutch components in the drawing. Place the identifying letter next to the name of the component.

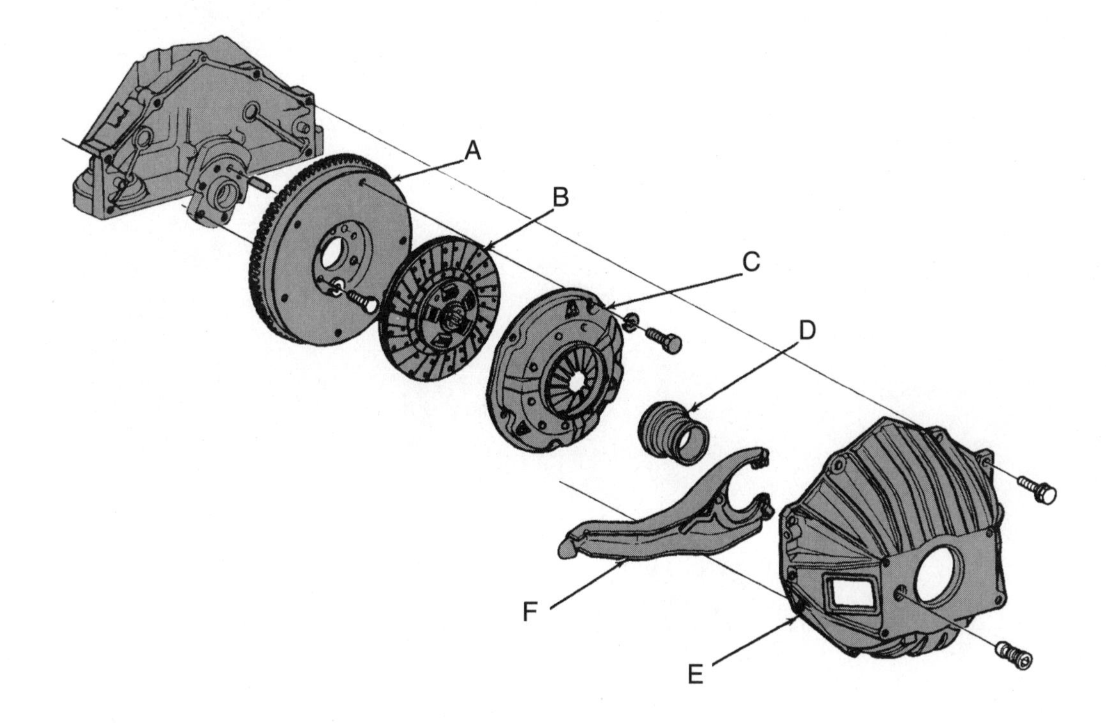

Release Bearing	____	Clutch Disc	____	Clutch Housing	____
Flywheel	____	Pressure Plate	____	Release Fork	____

STOP

Instructor OK ______________ Score __________

Activity Sheet #92

CLUTCH

Name______________________________ Class ________________

Directions: Match the words on the left to the descriptions on the right. Write the letter for the correct word on the line provided. For the terms that you are not certain of, use the glossary in your textbook.

A. Friction disc	____	It has splines and connects the clutch disc to the transmission
B. Clutch hub	____	It connects the release bearing to the clutch cable or linkage
C. Dampened hub	____	It contacts the rotating clutch to release the disc
D. Clutch facings	____	A type of spring that replaces the release levers and coil springs in a diaphragm clutch
E. Clutch cushion plate	____	Another name for the pressure plate assembly
F. Release levers	____	The clutch pedal return spring
G. Diaphragm spring	____	The output piston in a hydraulic clutch
H. Release bearing	____	A bearing or bushing in the crankshaft that supports the transmission input shaft
I. Throwout bearing	____	The part that presses the clutch disc against the flywheel
J. Clutch fork	____	What the clutch does when you apply the clutch pedal
K. Overcenter spring	____	The parts of a coil spring clutch that pull the pressure plate away from the flywheel
L. Slave cylinder	____	The inner part of a clutch disc
M. Clutch freeplay	____	Clutch part that absorbs shock during engagement
N. Input shaft	____	Another name for a throwout bearing
O. Clutch cover	____	A metal cushion that lets the clutch facings compress
P. Pilot	____	Driven member of a clutch
Q. Releases	____	The friction material part of the clutch disc
R. Pressure plate	____	A term that describes movement measured at the clutch pedal

STOP

Instructor OK ______________ Score __________

Activity Sheet #93

IDENTIFY MANUAL TRANSMISSION COMPONENTS

Name______________________________ Class ____________________

Directions: Identify the manual transmission components in the drawing. Place the identifying letter next to the name of the component.

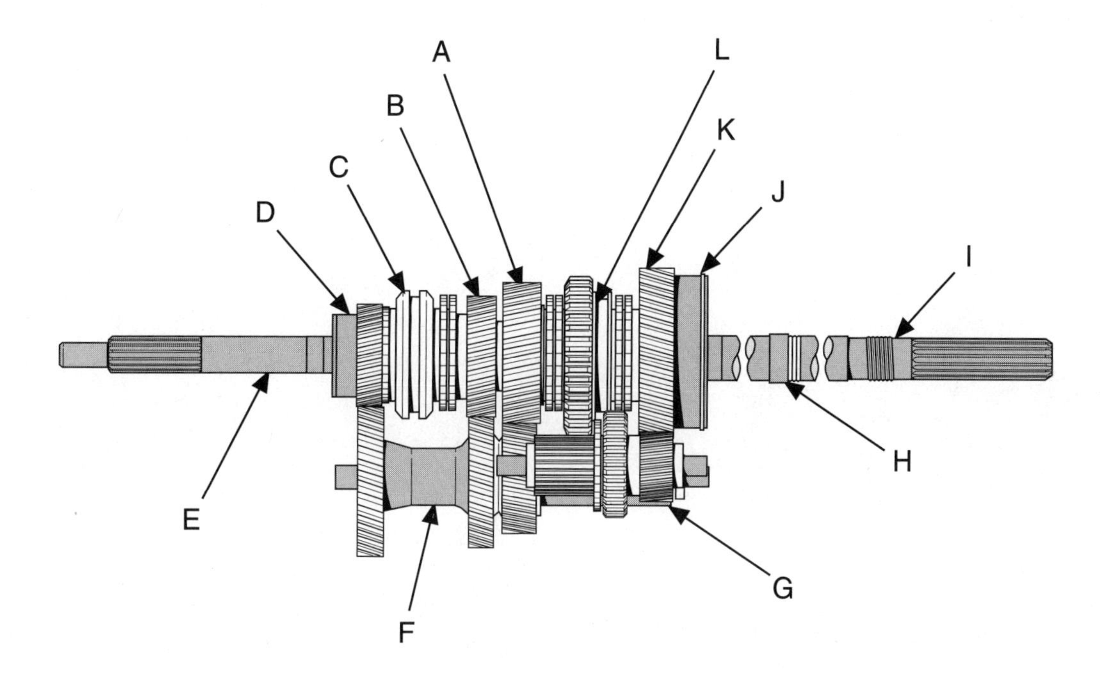

Second Gear	____	Input Shaft	____	Output Shaft Bearing	____
Output Shaft	____	Reverse Idler Gear	____	Input Shaft Bearing	____
Speedometer Gear	____	First Gear	____	1st/2nd Synchronizer	____
Third Gear	____	3rd/4th Synchronizer	____	Counter Gears	____

STOP

Instructor OK ______________ Score ____________

Activity Sheet #94

TRACE MANUAL TRANSMISSION POWER FLOW

Name______________________________ Class ____________________

Directions: Trace the power flow through each of the gear ratios of the transmission in the drawing. Place the identifying letter next to the gear range on page 216.

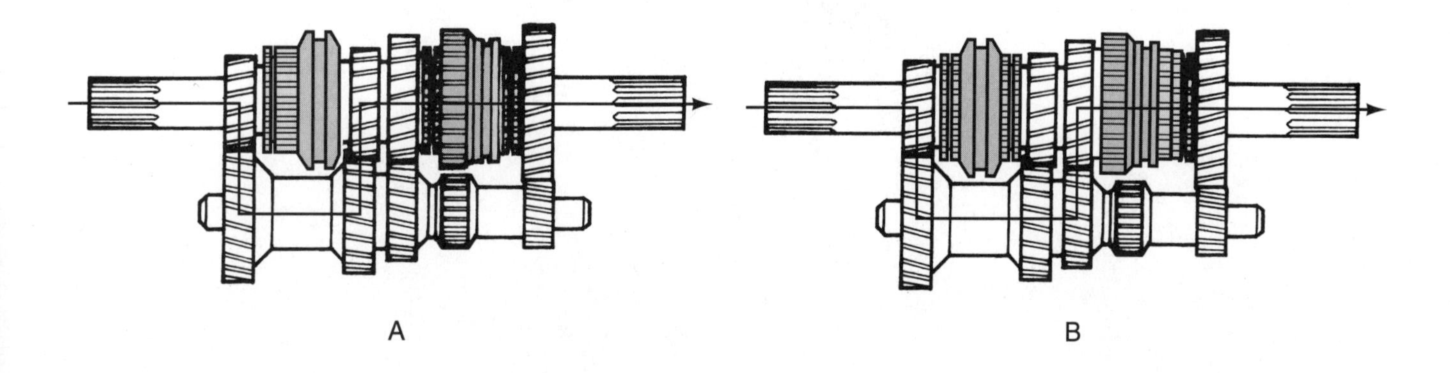

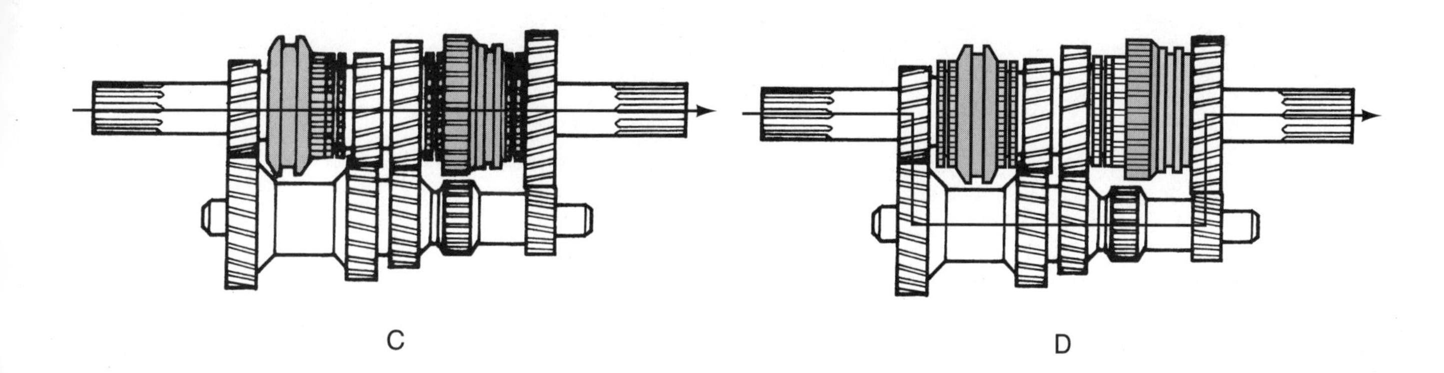

TURN

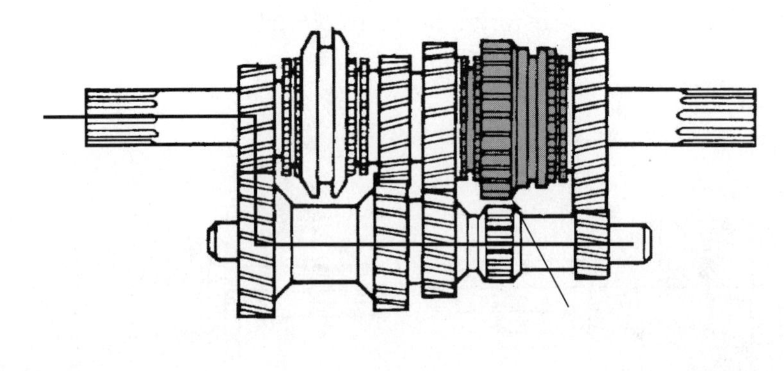

E

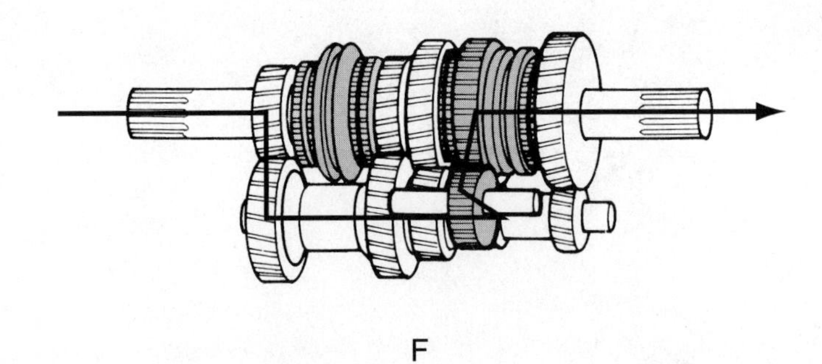

F

First Gear	____	Third Gear	____	Reverse	____
Second Gear	____	Fourth Gear	____	Neutral	____

STOP

Instructor OK ______________ Score __________

Activity Sheet #95

MANUAL TRANSMISSION

Name______________________________ Class ____________________

Directions: Match the words on the left to the descriptions on the right. Write the letter for the correct word on the line provided. For the terms that you are not certain of, use the glossary in your textbook.

A. Lower gear ratio	____	The ratio between the transmission output shaft and the differential ring gear
B. Gear ratio	____	The clearance between meshing gear teeth
C. Overdrive	____	A gear used to change direction of rotation
D. Granny gear	____	Another name for the countergear
E. Final drive ratio	____	The most popular style of synchronizer
F. Meshed	____	When there is a large difference between the ratios of a transmission's forward gears
G. Spur gears	____	Gear ratio for high gear in a manual transmission
H. Backlash	____	When there is a small difference between the ratios of a transmission's forward gears
I. Helical gears	____	When a transmission has a very low first gear
J. Idler gear	____	Approximate ratio for first gear
K. Synchronizer	____	Calculated by dividing the number of teeth on the driven gear by the number of teeth on the driving gear
L. Blocker ring	____	Gears designed with straight cut teeth
M. Countergear	____	When the output shaft turns faster than the input shaft
N. Dog teeth	____	A quieter operating gear design than spur gears
O. Final drive	____	A simple gear design with straight cut teeth
P. Spur gear	____	Lubricant used in many manual transmissions
Q. 2:1	____	Keeps two meshing gears from clashing during a shift
R. Clutch shaft	____	Output speed is slower
S. Cluster	____	Little teeth around the circumference of a gear

TURN

T.	1:1	____	Output from the differential ring gear
U.	Torque goes up	____	Another name for the input shaft
V.	3:1	____	One assembly made up of a series of gears (not cluster)
W.	SAE 90	____	Result when a small gear drives a larger gear
X.	Reverse	____	Requires an idler gear
Y.	Wide ratio transmission	____	When two gears are engaged
Z.	Close ratio transmission	____	Common gear ratio for second gear

STOP

Instructor OK ________ Score ________

Activity Sheet #96

IDENTIFY AUTOMATIC TRANSMISSION COMPONENTS

Name________________ Class ________________

Directions: Identify the automatic transmission components in the drawing. Place the identifying letter next to the name of the component.

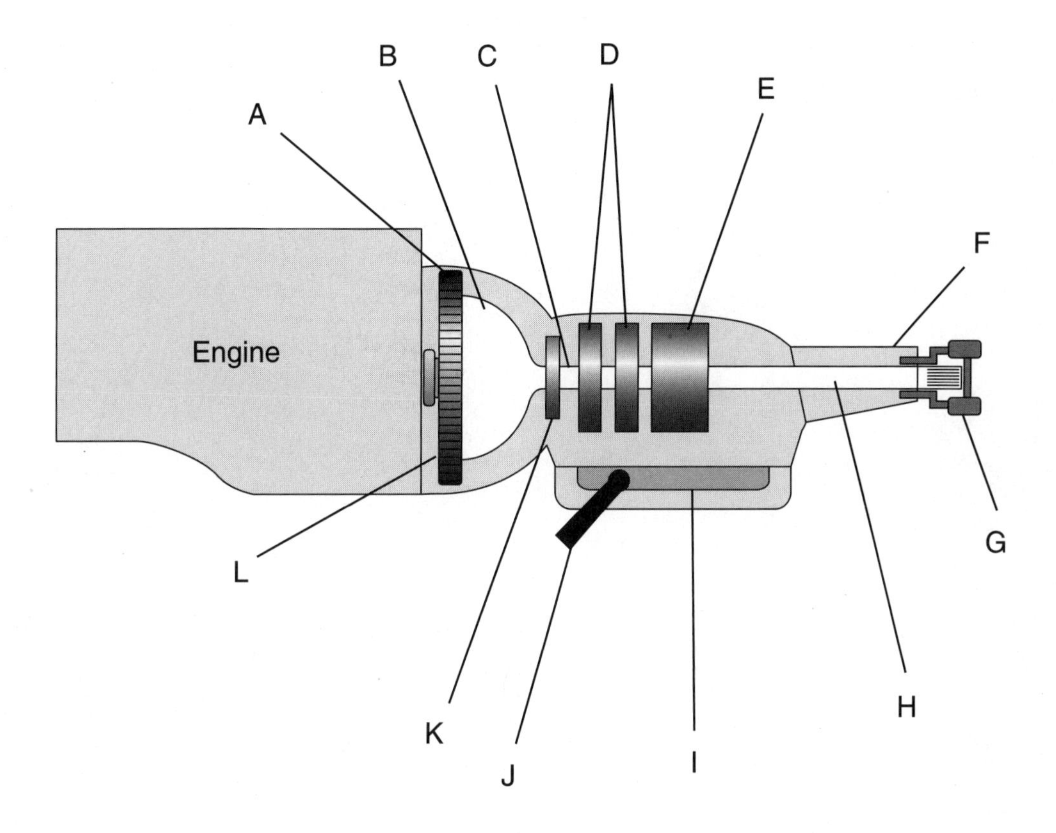

Extension Housing	____	Starter Ring Gear	____	Drive Shaft Yoke	____
Clutch Packs	____	Pump	____	Input Shaft	____
Valve Body	____	Torque Converter	____	Planetary Gears	____
Flexplate	____	Output Shaft	____	Shift Lever	____

STOP

Instructor OK ______________ Score ___________

Activity Sheet #97

IDENTIFY AUTOMATIC TRANSMISSION PARTS

Name______________________________ Class ____________________

Directions: Identify the automatic transmission parts in the drawing. Place the identifying letter next to the name of the part listed below.

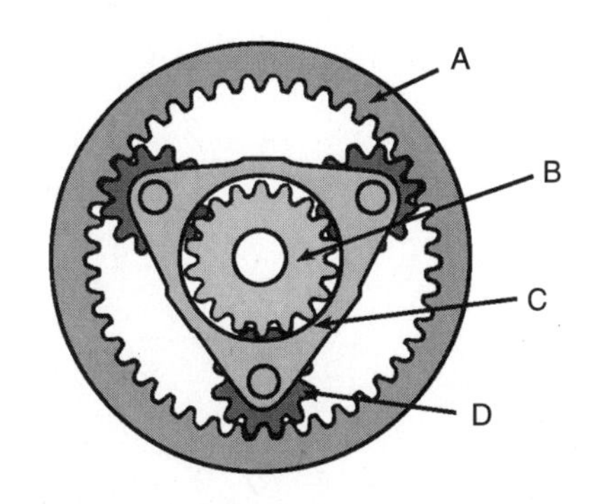

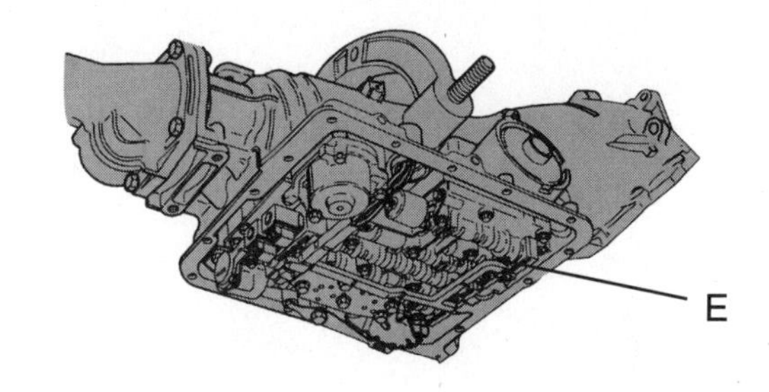

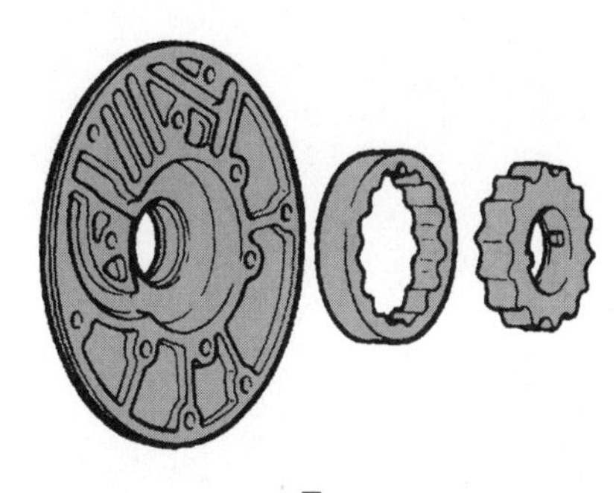

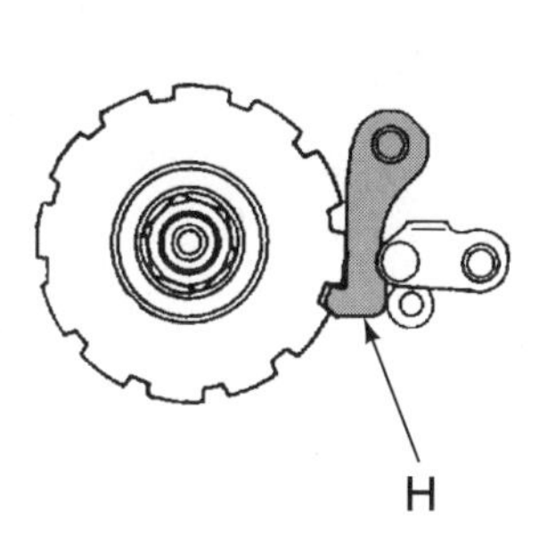

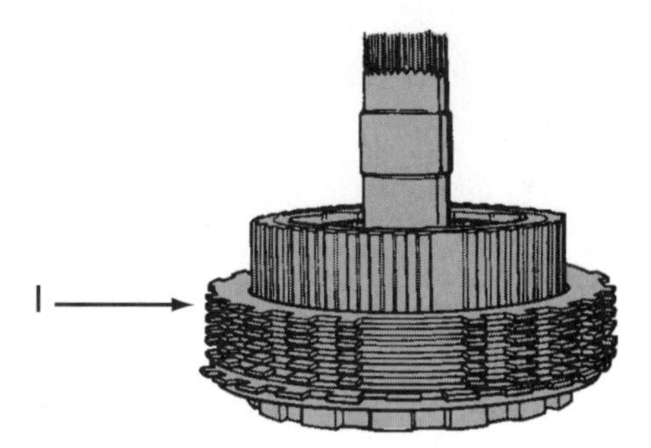

Clutch Pack	____	Carrier	____	Oil Pump	____
Pinion Gear	____	Torque Converter	____	Valve Body	____
Ring Gear	____	Park Pawl	____	Sun Gear	____

Instructor OK ______________ Score ___________

Activity Sheet #98

AUTOMATIC TRANSMISSION

Name______________________________ Class ___________________

Directions: Match the words on the left to the descriptions on the right. Write the letter for the correct word on the line provided. For the terms that you are not certain of, use the glossary in your textbook.

A. Fluid coupling	____	A restriction in a passage to slow down the flow of fluid
B. Torque converter	____	A valve that has lands, valleys, and faces
C. One-way clutch	____	The device that locks the output shaft of the transmission when the shift lever is placed in park
D. Overrunning clutch	____	Two planetary gearsets combined to provide more gear ratio possibilities
E. Coupling speed	____	Forced kickdown
F. Stall speed	____	Another name for a fluid clutch
G. Brake band	____	A torque converter with a friction disc that locks the impeller and turbine together
H. Accumulator	____	A continuously variable transmission
I. Spool valve	____	When the transmission shifts from a low gear to a higher gear; second to third, for instance
J. Orifice	____	When the transmission shifts to a lower gear
K. Valve body	____	The highest engine rpm that can be obtained when the vehicle is being prevented from moving while the engine is accelerated
L. Shift quadrant	____	A fluid coupling that multiplies torque
M. Upshift	____	A compound planetary gear design that shares the same sun gear between the gearsets
N. Downshift	____	A planetary gear design with long and short pinions
O. Throttle pressure	____	Another name for an overrunning clutch
P. Governor pressure	____	The hydraulic control assembly of the transmission

TURN

Q.	WOT	____	The readout on the gear selector that selects what gear the transmission is in
R.	Detent	____	The point at which all of the converter parts and ATF all turn as a unit
S.	Park pawl	____	Pressure that results in response to engine load
T.	CVT	____	A reservoir used in timing and cushioning gear shifts
U.	Impeller	____	A device that locks in one direction and freewheels in the other
V.	Modulator	____	Pressure that results from increases in vehicle speed
W.	Lock-up torque converter	____	An external brake planetary holding device
X.	Compound planetary gears	____	This part is actually part of the converter housing
Y.	Simpson geartrain	____	Wide-open throttle
Z.	Ravigneaux geartrain	____	A vacuum-operated diaphragm that controls shift points

STOP

Instructor OK ______________ Score __________

Activity Sheet #99
IDENTIFY DIFFERENTIAL COMPONENTS

Name______________________________ Class ____________________

Directions: Identify the differential components in the drawing. Place the identifying letter next to the name of the component.

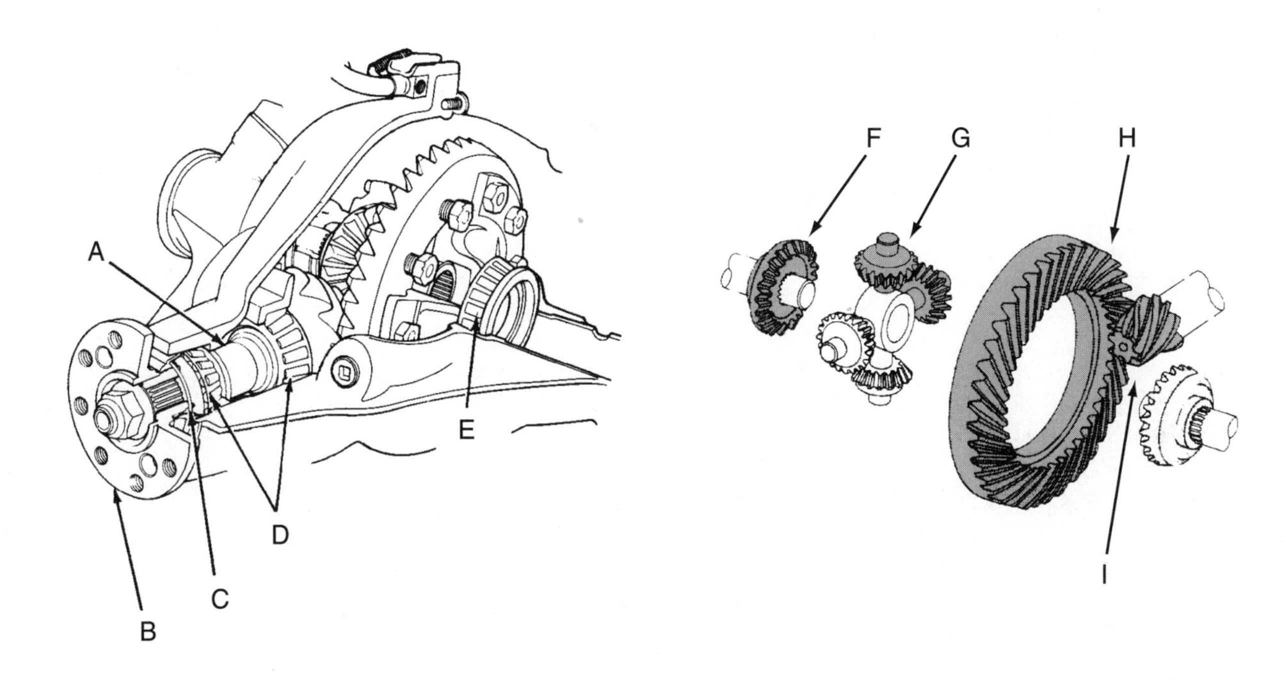

Drive Pinion Bearings	____	Companion Flange	____	Differential Bearing	____
Differential Pinion	____	Pinion Seal	____	Crush Sleeve	____
Drive Pinion	____	Ring Gear	____	Side Gear	____

STOP

Instructor OK ______________ Score __________

Activity Sheet #100

DRIVELINE AND DIFFERENTIAL

Name____________________________ Class ____________________

Directions: Match the words on the left to the descriptions on the right. Write the letter for the correct word on the line provided. For the terms that you are not certain of, use the glossary in your textbook.

A. Drive shaft	____ A universal joint whose output and input speed are constant
B. C-lock or C-clip axle	____ An axle with a groove on the inside that a clip fits into to keep it in place
C. Slip yoke	____ Another name for side gears and differential pinions
D. Constant velocity joint	____ A differential gearset where the pinion gear is lower than the centerline of the ring gear
E. Bearing retained axle	____ A pressed fit axle with a bearing retainer ring
F. Cardan joint	____ A type of rear axle in which the differential is not removable as an assembly
G. EP additives	____ The part of the drive shaft assembly that slides in and out of the transmission
H. Salisbury axle	____ A universal joint used with RWD
I. Spider gears	____ It locks up the spider gears when one wheel starts to lose traction
J. Hypoid gears	____ The assembly that transfers power from the transmission to the rear wheels
K. Limited slip	____ Part of the lubricant package that prevents welding between metal surfaces

STOP

Instructor OK ______________ Score __________

Activity Sheet #101

IDENTIFY FRONT-WHEEL DRIVE (FWD) AXLE SHAFT COMPONENTS

Name______________________________ Class ____________________

Directions: Identify the front-wheel drive (FWD) axle shaft components in the drawing. Place the identifying letter next to the name of the component.

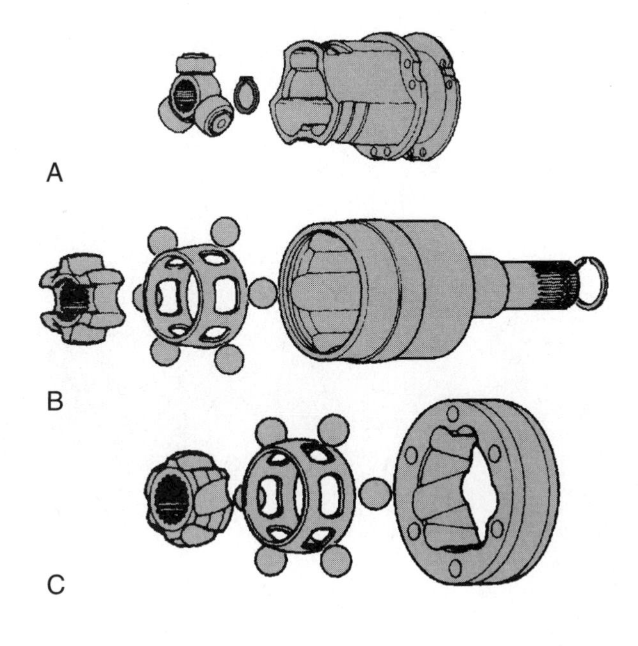

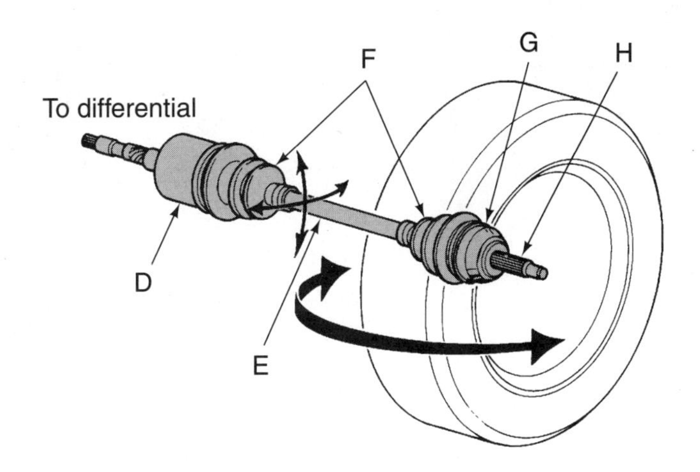

Cross Groove	____	Tripod Tulip	____	Double Offset	____
CV Joint Boots	____	Outboard Joint	____	Inboard Joint	____
Stub Axle	____	Drive Axle	____		

Instructor OK ______________ Score __________

Activity Sheet #102

IDENTIFY FOUR-WHEEL DRIVE COMPONENTS

Name______________________________ Class ____________________

Directions: Identify the four-wheel drive components in the drawing. Place the identifying letter next to the name of the component listed below.

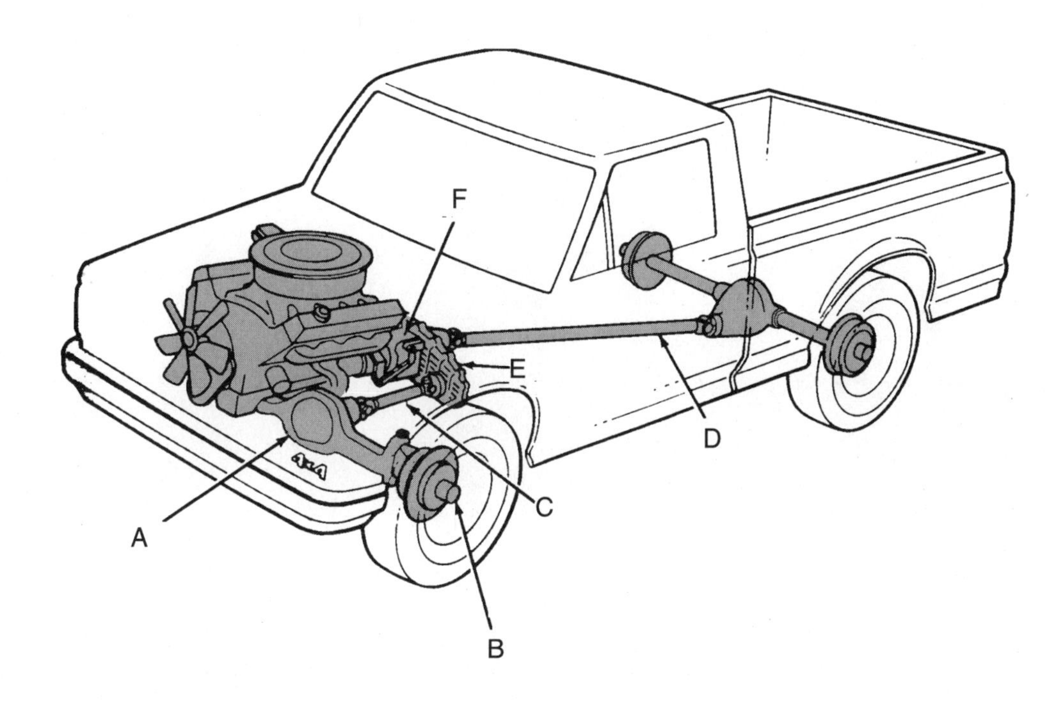

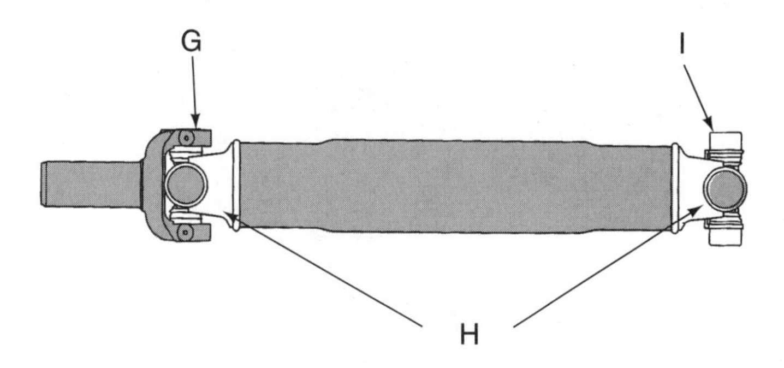

Front Drive Shaft	____	Front Differential	____	Rear Drive Shaft	____
Transfer Case	____	Locking Hub	____	Transmission	____
Drive Shaft Yokes	____	Universal Joint	____	Slip Yoke	____

Instructor OK ______________ Score __________

Activity Sheet #103
IDENTIFY TRANSAXLE COMPONENTS

Name______________________________ Class ____________________

Directions: Identify the transaxle components in the drawing. Place the identifying letter next to the name of the component.

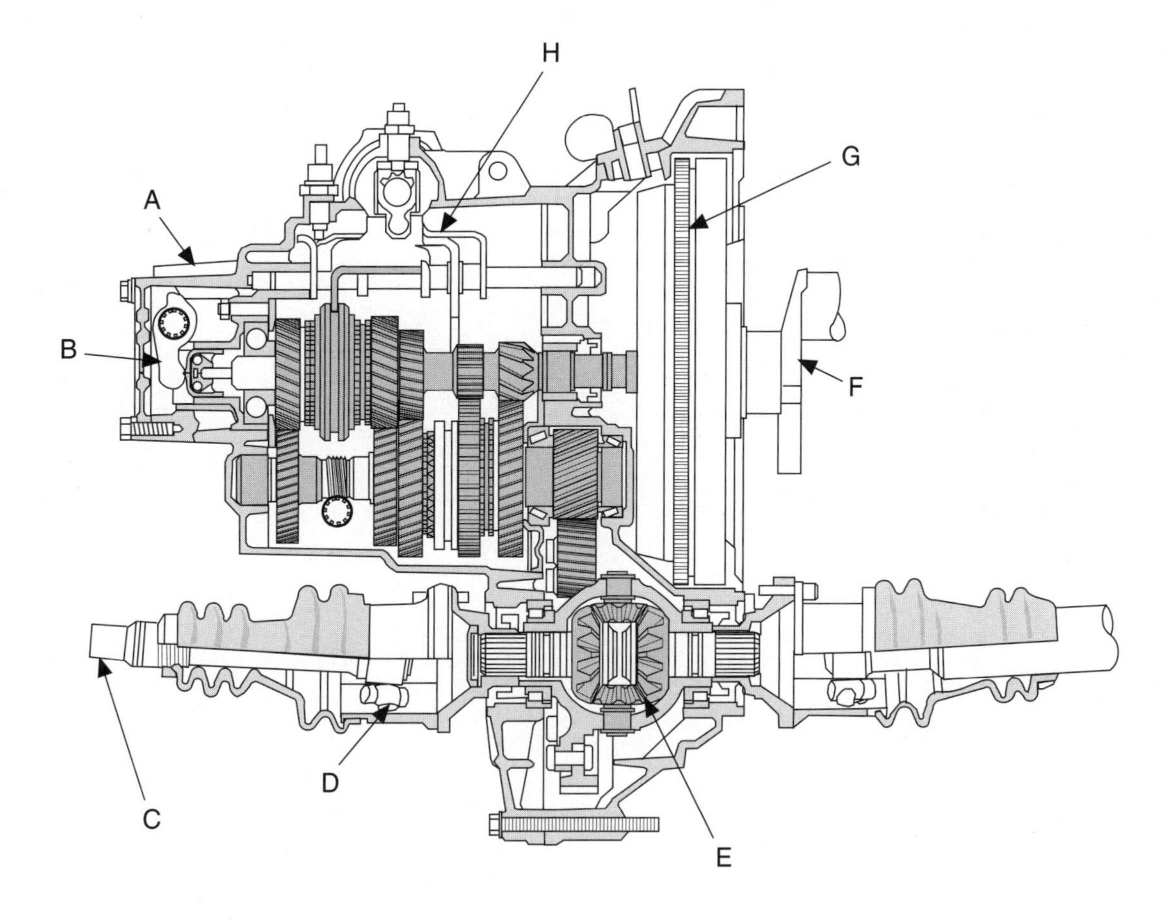

Axle Shaft	____	Transmission	____	CV Joint	____
Differential	____	Gear Shifter	____	Flywheel	____
Engine Crankshaft	____	Clutch Release Linkage	____		

STOP

Instructor OK ______________ Score __________

Activity Sheet #104
FRONT-WHEEL DRIVE

Name______________________________ Class ____________________

Directions: Match the words on the left to the descriptions on the right. Write the letter for the correct word on the line provided. For the terms that you are not certain of, use the glossary in your textbook.

A. Transaxle	____ The most commonly replaced part of a front axle assembly
B. Half shaft	____ The drive wheels on most cars currently produced
C. CV joint	____ A CV joint that allows for a change in the angle but not the length
D. Inboard joint	____ This is done to the stub shaft nut after it is torqued
E. Fixed joint	____ The sound a worn outboard CV joint makes
F. Rzeppa CV joint	____ These allow the front wheels to turn at different speeds during a turn
G. Front wheels	____ The name for a drive axle assembly on a front-wheel drive vehicle
H. Click	____ Constant velocity joint (abbreviation)
I. Clunk	____ The most common type of outboard, fixed CV joint
J. Boot	____ A combination transmission and differential
K. Staked	____ The inside CV joint on a FWD car
L. Torque steer	____ The sound a worn inboard CV joint makes
M. Differential gears	____ When the car pulls to one side during hard acceleration

STOP

Part I
Activity Sheets

Section Twelve

Comfort Systems and Vehicle Electronics

Instructor OK ______________ Score __________

Activity Sheet #105

IDENTIFY COOLING AND HEATING SYSTEM COMPONENTS

Name______________________________ Class ____________________

Directions: Identify the cooling and heating system components in the drawing. Place the identifying letter next to the name of the component.

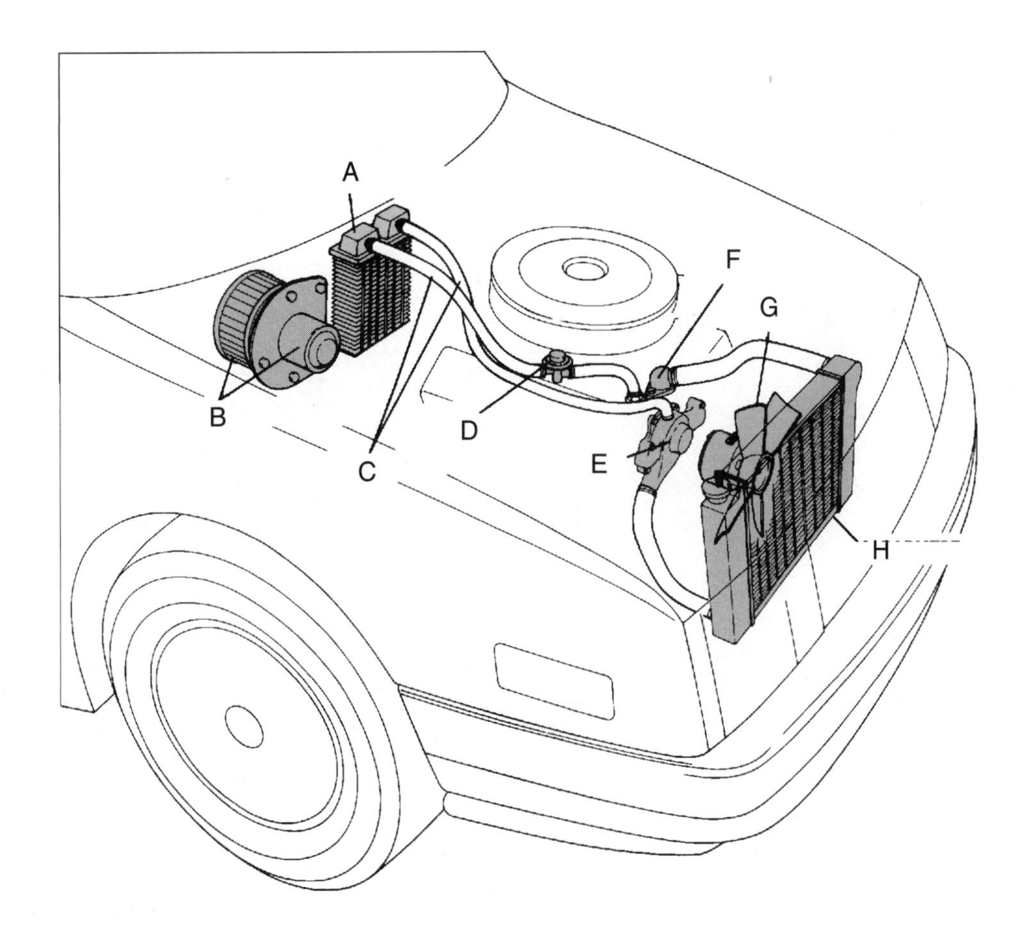

Fan and Blower Motor	____	Heater Core	____	Heater Valve	____
Heater Hoses	____	Thermostat	____	Radiator	____
Water Pump	____	Cooling Fan	____		

STOP

Instructor OK ______________ Score __________

Activity Sheet #106

HEATING AND AIR CONDITIONING

Name______________________________ Class ____________________

Directions: Match the words on the left to the descriptions on the right. Write the letter for the correct word on the line provided. For the terms that you are not certain of, use the glossary in your textbook.

A. Air conditioning	____ This is also called freon (earlier refrigerant)
B. Heat transfer	____ Protects the earth's surface from ultraviolet rays
C. Convection	____ Heat transfer where moisture is vaporized as it absorbs heat
D. Radiation	____ When the air is totally saturated with moisture
E. Evaporation	____ A term used to describe heat bouncing off a surface
F. Humidity	____ The process in which air inside of the passenger compartment is cooled, dried, cleaned, and circulated
G. Condensation	____ When air becomes warmer and moves upward
H. CFC	____ When a vapor changes to a liquid
I. Montreal Protocol	____ The movement of heat when there is a difference in temperature between two objects
J. R-12 refrigerant	____ The moisture content of the air
K. R-134A refrigerant	____ Chlorofluorocarbon (abbreviation)
L. 100% humidity	____ A refrigerant that is used in newer vehicles (hydrofluorocarbon)
M. Ozone	____ Sets limits on the production of ozone-depleting chemicals

STOP

Instructor OK ______________ Score __________

Activity Sheet #107

IDENTIFY AIR-CONDITIONING SYSTEM COMPONENTS

Name______________________________ Class ____________________

Directions: Identify the air-conditioning system components in the drawing. Place the identifying letter next to the name of the component.

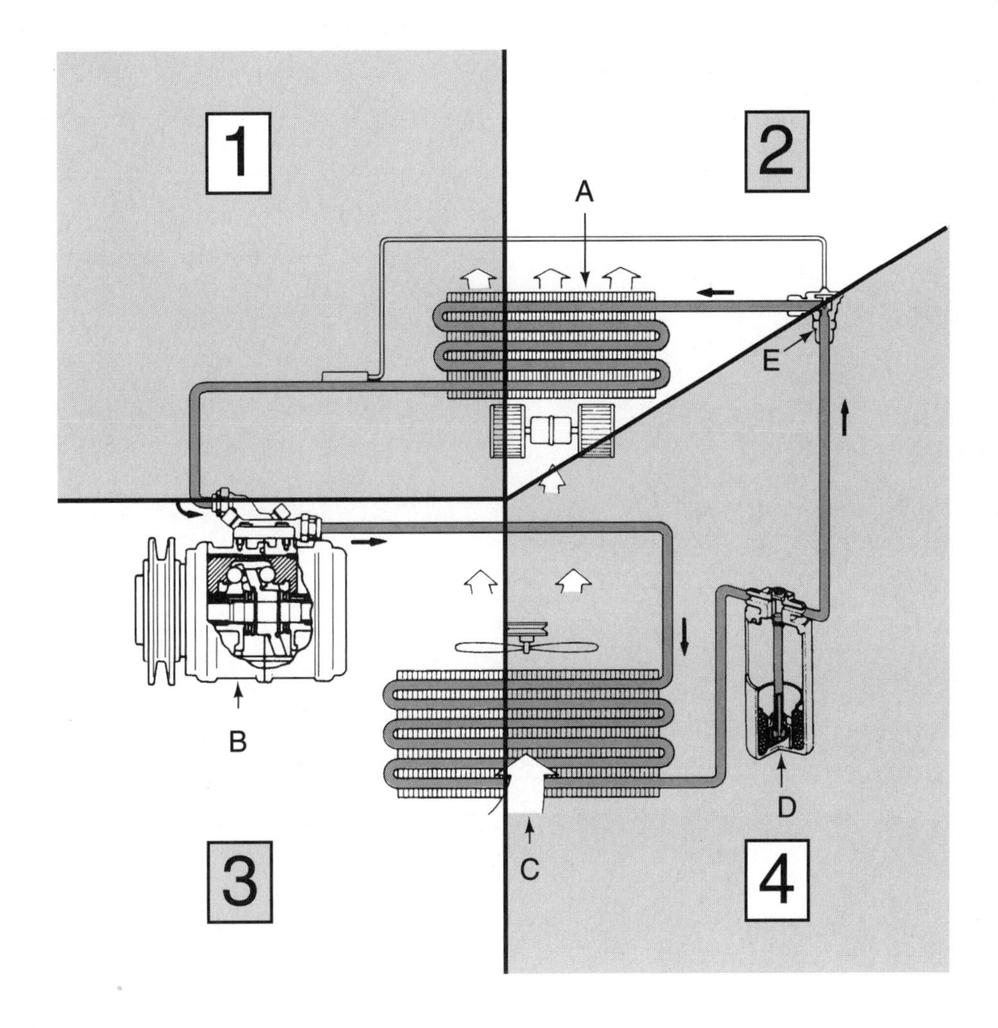

Receiver/Dryer ____ Expansion Valve ____ Evaporator ____

Compressor ____ Condenser ____

Refer to the drawing above and select the correct number for each of the following:

High-Pressure Liquid? ________________ Low-Pressure Gas? ________________

Low-Pressure Liquid? ________________ High-Pressure Gas? ________________

Instructor OK ______________ Score ____________

Activity Sheet #108

IDENTIFY AIR-CONDITIONING PARTS

Name______________________________ Class ____________________

Directions: Identify the air-conditioning parts in the drawing. Place the identifying letter next to the name of the parts.

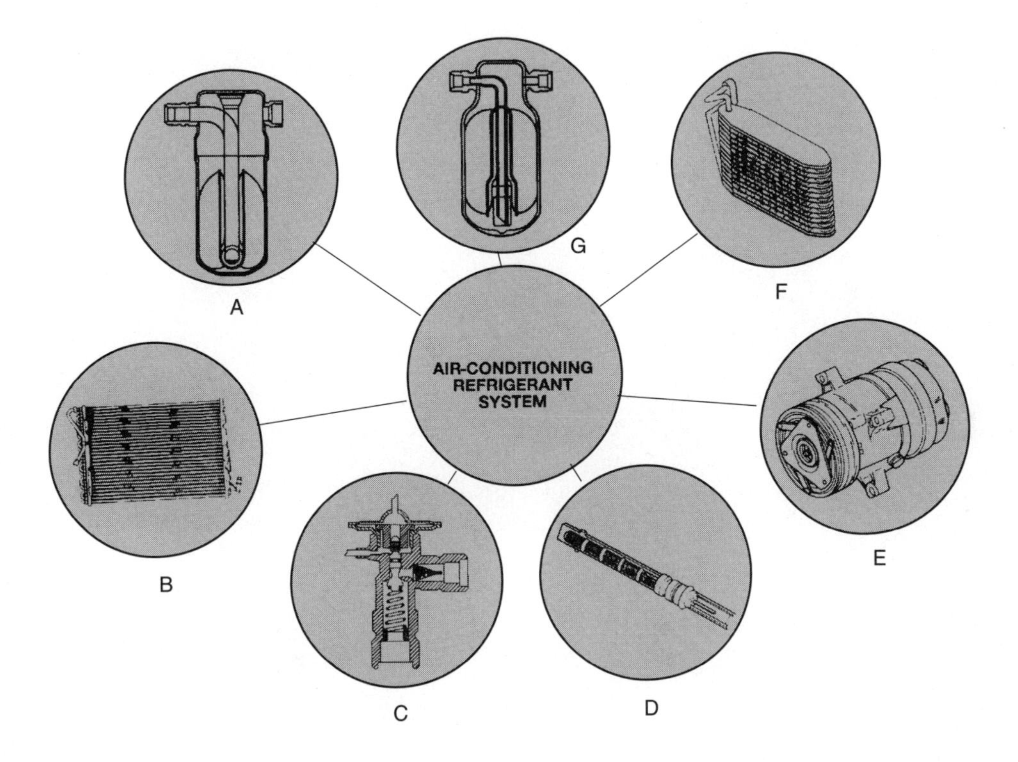

Condenser	____	Receiver/Dryer	____	Orifice Tube	____
Evaporator	____	Accumulator	____	Compressor	____
Expansion Valve	____				

STOP

Instructor OK ______________ Score __________

Activity Sheet #109

IDENTIFY PARTS OF A COMPUTER SYSTEM

Name______________________________ Class ____________________

Directions: Identify parts of a computer system in the drawing. Place the identifying letter next to the name of the part.

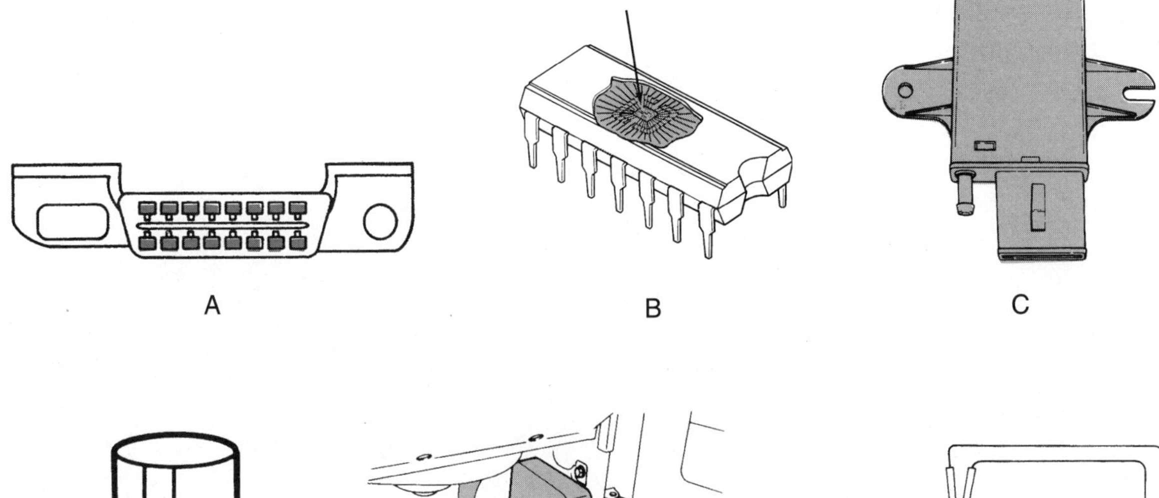

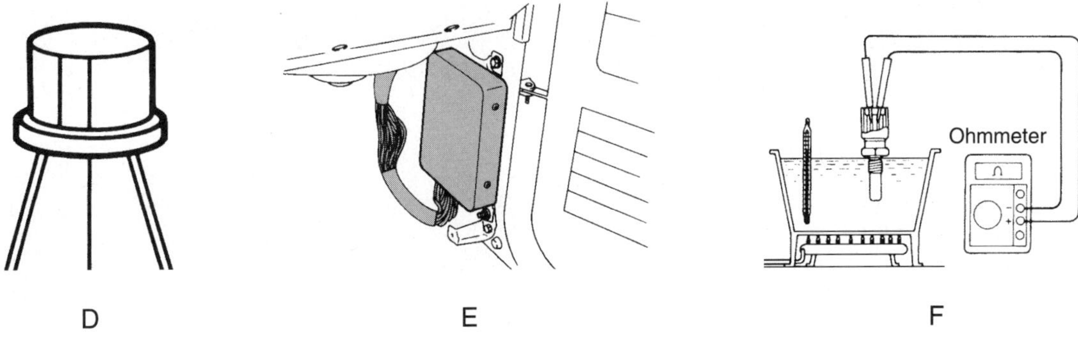

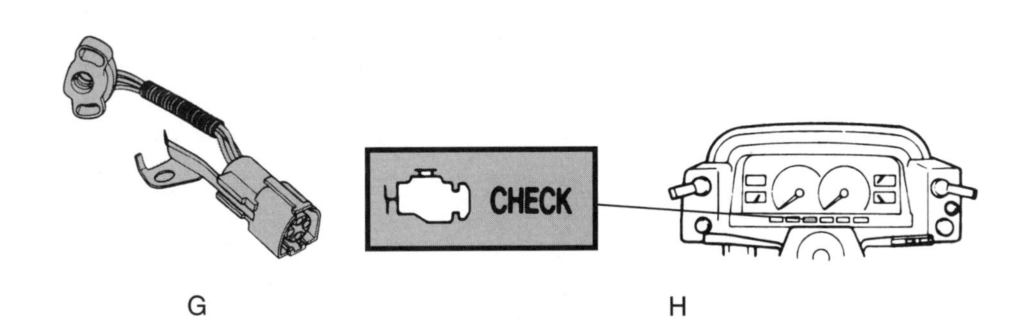

MIL	____	Thermistor	____	MAP Sensor	____
Chip (IC)	____	Computer	____	DLC	____
Transistor	____	TPS	____		

STOP

Instructor OK ______________ Score ___________

Activity Sheet #110

IDENTIFY COMPUTER CONTROLS

Name____________________________ Class ____________________

Directions: Identify computer *inputs* with an "S" on the line next to items that require a sensor. Identify computer *outputs* with an "A" next to the actuator or output.

____ Crankshaft position

____ Torque converter clutch

____ Manifold air temperature

____ Vehicle speed

____ Canister purge

____ Manifold pressure

____ Mass airflow

____ Spark timing

____ Idle air control

____ Exhaust oxygen

____ Air conditioning

____ Diagnostic data

____ Engine rpm

____ Engine cooling fan

____ Throttle position

____ Engine knock

____ Fuel injector

____ Coolant temperature

____ EGR

STOP

Instructor OK ______________ Score __________

Activity Sheet #111

COMPUTER AND ELECTRONICS

Name______________________________ Class ____________________

Directions: Match the words on the left to the descriptions on the right. Write the letter for the correct word on the line provided. For the terms that you are not certain of, use the glossary in your textbook.

A. Semiconductor	____ Information stored as electronic signals
B. Electron theory	____ The calculating and decision-making chip in the computer
C. Diode	____ A variable resistor used to measure linear or rotary motion
D. Zener diode	____ Cycles per second
E. LED	____ Can be either a conductor or an insulator
F. Integrated circuit	____ Malfunction indicator lamp
G. Microprocessor	____ Electronically erasable programmable read-only memory
H. Hardware	____ Electrons flow from – to +
I. Software	____ Its resistance changes as its temperature changes
J. RAM	____ Diodes that give off light
K. ROM	____ Used as an electronic voltage regulator
L. EEPROM	____ Guidelines that provide standardization of terms
M. Actuator	____ The mechanical parts of an electronic system
N. Thermistor	____ A device that uses a piezoelectric element to sense vibration
O. Potentiometer	____ P-type and N-type crystals back to back
P. Piezoelectric	____ This is like a notepad that you can read from and write to
Q. Hertz	____ Permanently programmed information
R. Knock sensor	____ Sharing of electrical circuit conductors
S. OBD II	____ A device that is controlled by the computer
T. MIL	____ A complete miniaturized electric circuit
U. Multiplexing	____ Crystals that develop a voltage on their surfaces when pressure is applied

STOP

Instructor OK ______________ Score __________

Activity Sheet #112

ADVANCED EMISSIONS AND ON BOARD DIAGNOSTICS (OBD)

Name______________________________ Class ____________________

Directions: Match the words on the left to the descriptions on the right. Write the letter for the correct word on the line provided. For the terms that you are not certain of, use the glossary in your textbook.

A. OBD	____	Society of Automotive Engineers
B. OBD II	____	Data link connector
C. Scan Tool	____	Measures emissions in grams per mile
D. Monitors	____	Standardization of terms
E. FTP standard	____	All five of the trip monitors must operate to complete
F. FTP	____	Powertrain control module
G. DLC	____	Malfunction indicator lamp
H. Post CAT O_2 Sensor	____	Used to access stored codes
I. SAE	____	Processed data used by the engine
J. SAE J1930	____	Used primarily to improve air quality
K. Standard Communication Protocol	____	Self-detects exhaust emission increase of over 50%
L. PCM	____	Senses catalytic converter efficiency
M. DLC (OBD II)	____	Used to look for malfunctions
N. VIN (OBD II)	____	Requires various emission monitors to operate to complete
O. PID	____	Used to indicate monitors are clear
P. Freeze frame data	____	Requires manufacturers to use the same computer language
Q. MIL	____	DTC that is always emissions related
R. Warm-up cycle	____	Found under the left side of the dash
S. Trip	____	Checks for leaks no larger than 0.040"
T. Drive cycle	____	Automatically transmitted to the scan tool
U. Pending Code	____	Used to look for malfunctions
V. Type "A" code	____	The speed of oxygen sensor oscillations

TURN

W. Monitor	____ Digital storage oscilloscope
X. Readiness indicators	____ Stored PIDs
Y. Comprehensive component monitor	____ Federal Test Procedure
Z. Evaporative monitor	____ Occurs every time the engine cools off and temperature rises to at least 40°F
AA. Switch ratio	____ Set after first time a fault is identified
AB. DSO	____ Checks devices not tested by other OBD II monitors

STOP

Instructor OK ______________ Score ______________

Activity Sheet #113

IDENTIFY ON BOARD DIAGNOSTIC II COMPONENTS

Name______________________________ Class ____________________

Directions: Identify the OBD II components in the drawings. Place the identifying letter next to the name of the component listed below.

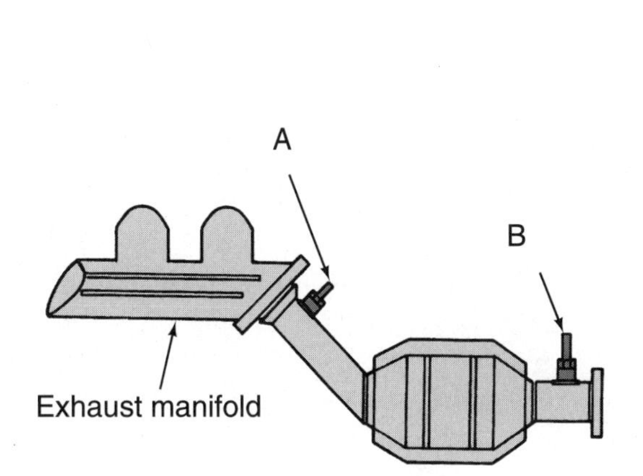

C

DTC	PO101
Engine SPD	2567 RPM
ECT (°)	108°F
VEHICLE SPD	54 MPH
ENGINE LOAD	18.8%
MAP	14.8 in Hg
FUEL STAT 1	OL
FUEL STAT 2	UNUSED
ST FT 1	3.1%
LT FT 1	-1.5%

D

Example: P0137 low voltage bank 1 sensor 2

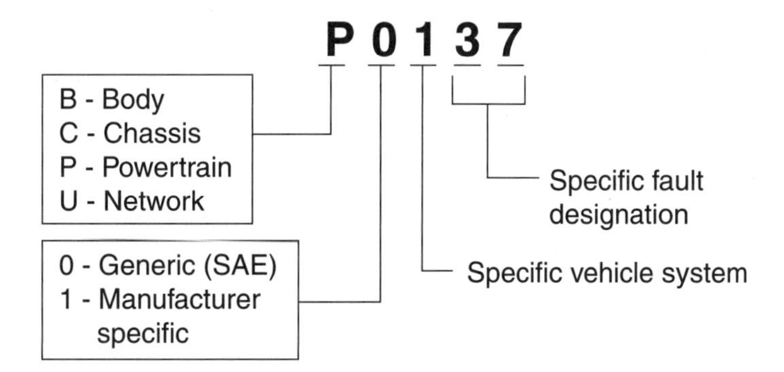

E

- Misfire
- Comp component
- Heated catalyst
- AIR
- O_2 sensor
- EGR system
- Fuel system
- Catalyst
- EVAP
- A/C refrigerant
- O_2 sensor heater

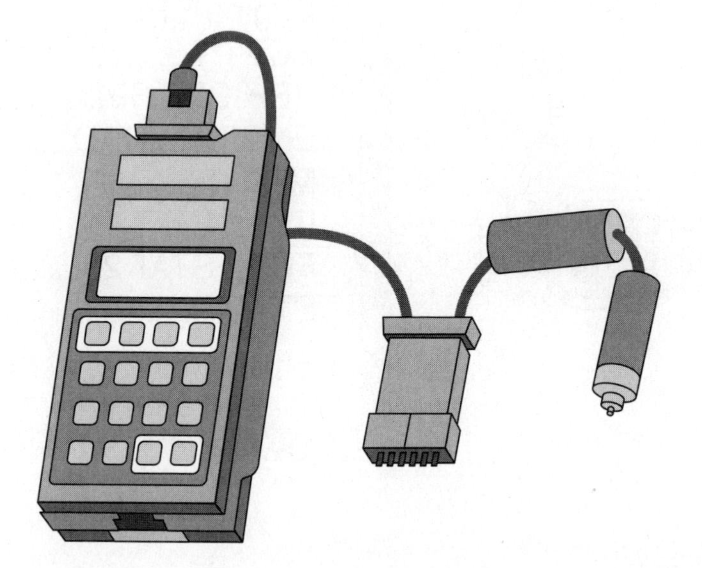

F

Readiness Status Categories ____

Upstream HO_2S Sensor ____

Freeze Frame Data ____

Scan Tool ____

Downstream HO_2S Sensor ____

OBD II Trouble Codes ____

STOP

Instructor OK ______________ Score __________

Activity Sheet #114
ANTILOCK BRAKES

Name______________________________ Class ____________________

Directions: Match the words on the left to the descriptions on the right. Write the letter for the correct word on the line provided. For the terms that you are not certain of, use the glossary in your textbook.

A. Teves	____ Produces a digital signal
B. ABS	____ Part of
C. EBCM	____ What twisted pairs prevent
D. CAB	____ Called remote or add-on ABS
E. EBTCM	____ Synthetic brake fluid
F. PMV	____ ABS problems only
G. Radio interference	____ Separate
H. Hall effect	____ Rear-wheel antilock
I. Lateral acceleration sensor	____ Test for moisture
J. EHCU	____ Domestic antilock brake system
K. Integral	____ Typical hydraulic system warning light
L. Non-integral	____ Traction control system
M. Accumulator	____ Electronic brake control module
N. Non-integral ABS	____ Accumulates water
O. RWAL	____ Necessary to bleed some ABS
P. RABS	____ Controller antilock brake
Q. Four channel	____ Electronic brake and traction control module
R. BPMV	____ Measures force encountered while turning
S. TCS	____ Creates AC voltage
T. ASR	____ Electrohydraulic control unit
U. Amber light	____ Pressure modulator valves
V. Red light	____ Rear antilock brake system
W. Brake fluid test strips	____ Antilock brake system
X. DOT 5	____ Most effective ABS

TURN

Y.	Hygroscopic	____	Brake pressure modulator valve
Z.	Scan tool	____	Stores brake fluid under very high pressure
AA.	Wheel speed sensor	____	Acceleration slip regulation

STOP

Instructor OK ______________ Score __________

Activity Sheet #115

IDENTIFY ANTILOCK BRAKE COMPONENTS

Name______________________________ Class __________________

Directions: Identify the antilock brake components in the drawing. Place the identifying letter next to the name of the component listed below.

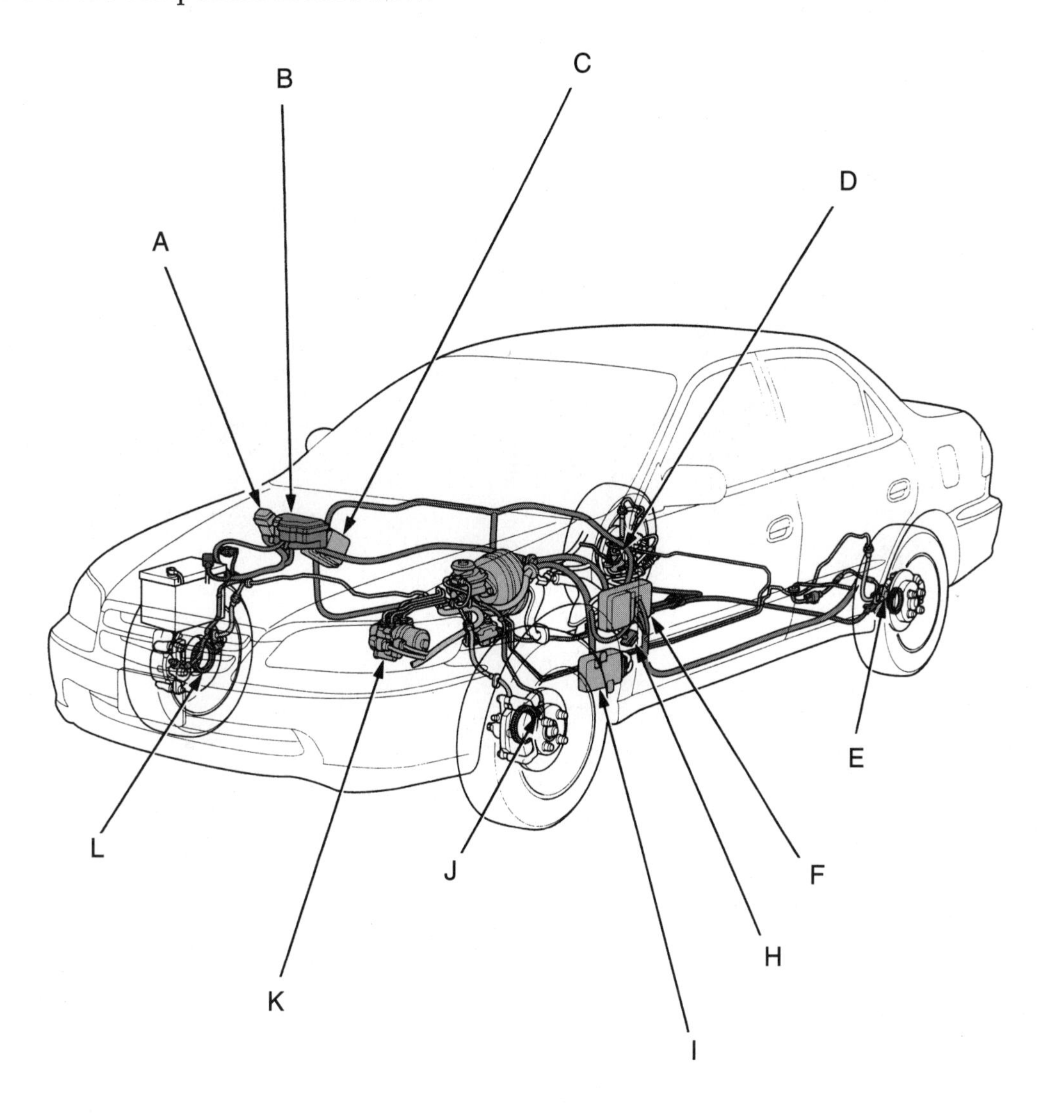

Left Rear-Wheel Sensor	____	Driver's Fuse Box	____	ABS Relay Box	____
ABS Modulator	____	Right Rear-Wheel Sensor	____	Passenger's Fuse Box	____
Right Front-Wheel Sensor	____	ABS Control Unit	____	Left Front-Wheel Sensor	____
Underhood Fuse Box	____	Data Link Connector	____		

Instructor OK ______________ Score __________

Activity Sheet #116

SAFETY, SECURITY, COMFORT SYSTEMS, AND ELECTRICAL ACCESSORIES

Name______________________________ Class ________________

Directions: Match the words on the left to the descriptions on the right. Write the letter for the correct word on the line provided. For the terms that you are not certain of, use the glossary in your textbook.

A.	Passive restraints	____	Supplemental inflatable restraints
B.	Active restraints	____	Pendulum-like device that locks during sudden deceleration
C.	Seat belts and air bags	____	Initiator or igniter
D.	Pretensioners	____	Supplemental restraint system
E.	Mechanical pretensioners	____	Located in the top of the dash
F.	Seat belt warning system	____	Manually buckled seat belt
G.	SIR	____	Global positioning system
H.	SAR	____	Automatic seat belts
I.	SRS	____	Produces high-frequency sounds
J.	Air bag	____	Heated gas inflators
K.	Passenger side air bag	____	Flexible nylon bag
L.	Driver side air bag	____	Used to inflate the air bag
M.	Discriminating sensors	____	Key fob transmitter
N.	Safing sensors	____	Dash light, with a bell or buzzer
O.	Squib	____	Resistance pellet is embedded in it
P.	Deployment time	____	Required on all cars and light trucks
Q.	Nitrogen gas	____	Used for power windows
R.	HGI	____	Automatically darken in response to sunlight
S.	Resistance key	____	Produces nondirectional sound
T.	Transponder key	____	Found in the front of the vehicle
U.	Keyless entry	____	Supplemental air restraints
V.	GPS	____	Located in the steering wheel
W.	A tweeter	____	Controls the slack in seat belts

TURN

X.	A woofer	____	100 milliseconds
Y.	Photochromatic mirrors	____	Found in the center console
Z.	Permanent magnet DC motors	____	Receives a radio signal

Instructor OK ______________ Score __________

Activity Sheet #117

IDENTIFY SUPPLEMENTAL RESTRAINT SYSTEM COMPONENTS

Name______________________________ Class __________________

Directions: Identify the air bag components in the drawing. Place the identifying letter next to the name of the component listed below.

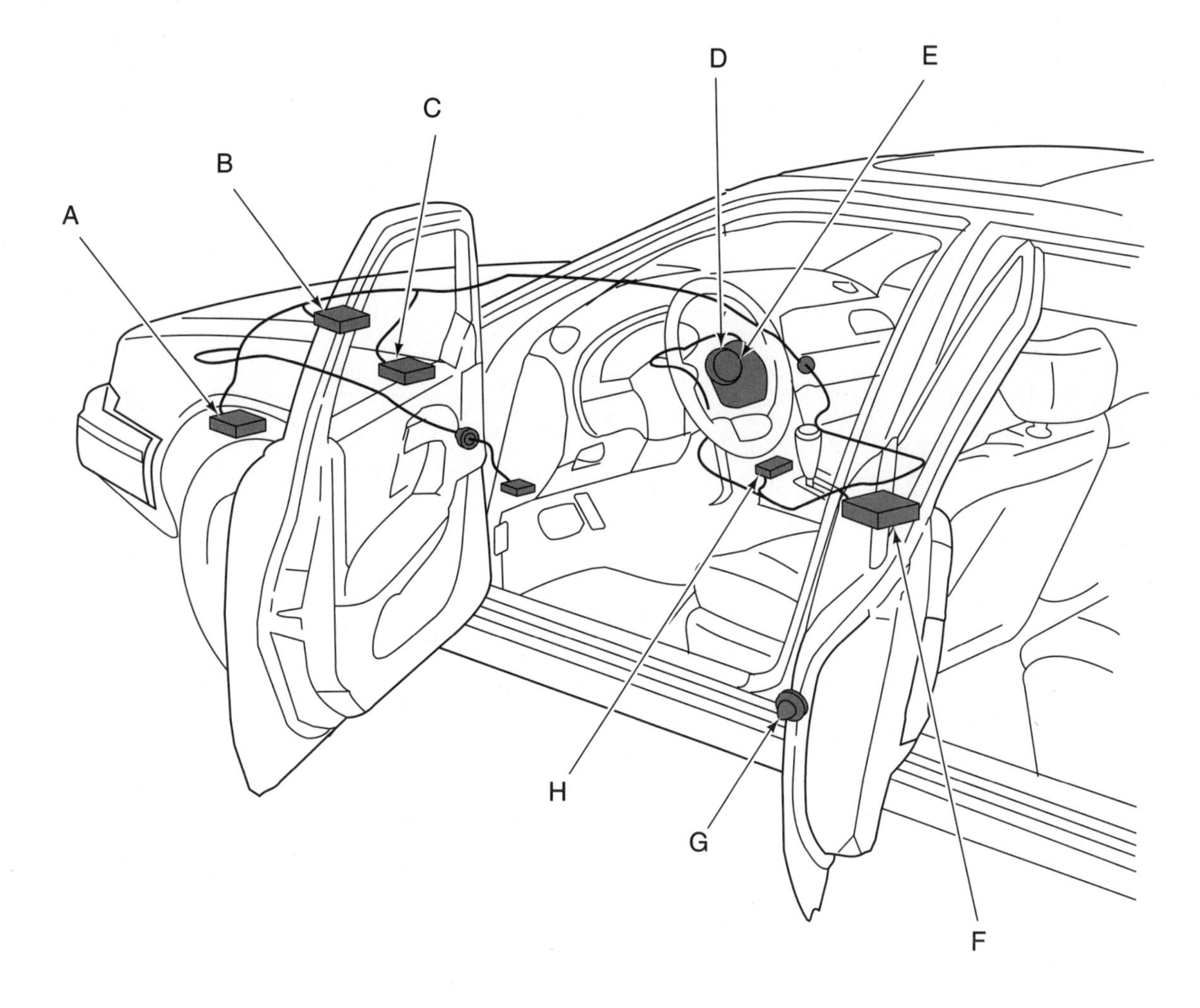

Driver Door Switch	____	Control Module	____	Right Impact Sensor	____
Center Impact Sensor	____	Clock Spring	____	Left Impact Sensor	____
Air Bag Module	____	Safing Sensor	____		

STOP

Part II

ASE Lab Preparation Worksheets

Introduction

Instructor OK ______________ Score __________

ASE Lab Preparation Worksheet #I-1

VEHICLE OWNER'S MANUAL

Name____________________________ Class __________________

OBJECTIVE:

Upon completion of this assignment, you should be able to use a vehicle owner's manual to locate maintenance service specifications. This task will help prepare you to pass the ASE certification examination in engine performance.

DIRECTIONS:

Review this worksheet completely before starting. Use your vehicle owner's manual or one provided by your instructor. Record the requested information in the spaces provided. If you are completing this worksheet on your personal vehicle, you may want to save it for future reference.

TOOLS AND EQUIPMENT REQUIRED:

Vehicle owner's manual

PROCEDURE:

Vehicle year ________ Make _______________ Model _________________

Engine: 4-cylinder ___ 6-cylinder ___ 8-cylinder ___

Transmission: Standard ___ Automatic ___

Is there an underhood label? Yes ___ No ___

Is there an underhood vacuum diagram? Yes ___ No ___

What is the name of the manual being used to complete this worksheet?

____

Capacities:

Battery ______ Cold Cranking Amps

Crankcase (with oil filter) ______ qt.

Oil filter capacity ______ qt.

Cooling system (without AC) ______ qt.

Cooling system (with AC) ______ qt.

Differential capacity ______ pt./qt.

Transmission capacity ______ pt./qt.

Fuel tank capacity ______ gallons

OIL SPECIFICATIONS:

SAE engine oil viscosity ____________________

API service rating ____________________

Transmission fluid type ____________________

Differential lubricant type ____________________

TIRES:

Pressures Front ______ psi

Rear ______ psi

Lug nut torque specification ______ ft.-lb

BELT TENSIONS:

Alternator ____________________

Power steering pump ____________________

Air-conditioning compressor ____________________

Smog pump ____________________

NOTES:

____

____

____

____

____

STOP

Instructor OK ______________ Score __________

ASE Lab Preparation Worksheet #I-2

VEHICLE IDENTIFICATION NUMBER (VIN)

Name______________________________ Class ____________________

OBJECTIVE:

Upon completion of this assignment, you should be able to identify the model year and country of manufacture of a vehicle. This task will help prepare you to pass the ASE certification examination in engine performance.

DIRECTIONS:

Before beginning this lab task, review the worksheet completely. Fill in the information in the spaces provided as you complete this assignment.

RELATED INFORMATION:

Each manufacturer used its own sequence and meaning for the numbers and letters of the vehicle identification number (VIN) through the 1980 model year. After 1980 the VIN codes were standardized. VINs were required to include seventeen digits. Each digit identifies a characteristic of the vehicle. Information identified by the VIN includes the country of origin, model year, body style, engine, manufacturer, vehicle serial number, and more. Each manufacturer can choose its own code for each digit of the VIN except the first and tenth digits. Those positions use a universal code that all manufacturers must use. The first digit of the VIN identifies the country where the vehicle was manufactured and the tenth identifies the model year of the vehicle. The following charts identify the meaning of the codes for each of these positions.

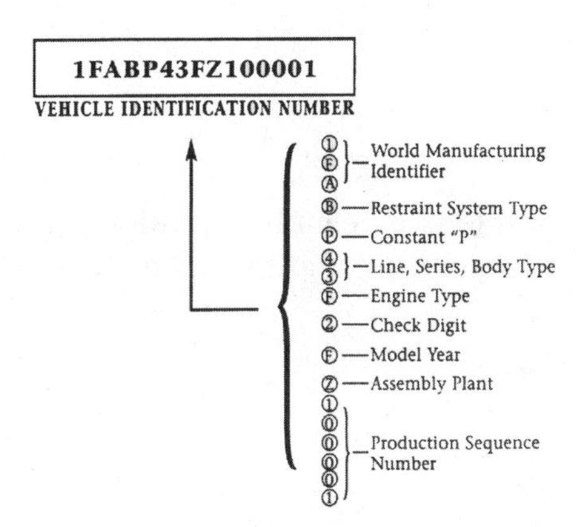

Code Charts:

	Country of Origin (1st digit)		Model Year (10th digit)		
1	United States	A	1980	S	1995
2	Canada	B	1981	T	1996
3	Mexico	C	1982	V	1997
4	United States	D	1983	W	1998
6	Australia	E	1984	X	1999
9	Brazil	F	1985	Y	2000
J	Japan	G	1986	1	2001
K	Korea	H	1987	2	2002
L	Taiwan	J	1988	3	2003
A	England	K	1989	4	2004
F	France	L	1990	5	2005
V	Europe	M	1991	6	2006
W	Germany	N	1992	7	2007
Y	Sweden	P	1993	8	2008
Z	Italy	R	1994	9	2009

PROCEDURE:

1. Use the code charts to identify the model year and country of manufacture for the following vehicle identification numbers.

VIN	Model Year	Country of Origin
a. 1G1CZ19H6HW135780	_______	____________________
b. J74VN13G9L5021104	_______	____________________
c. 1Y1SK5262TS005258	_______	____________________
d. 2MEPM6046KH637745	_______	____________________
e. 3T25V21E7WX055441	_______	____________________

2. Where is the VIN located on vehicles that are sold in the United States?

____

3. Locate the vehicle identification numbers on five vehicles manufactured after 1982. List the VIN, the model year, and the country of origin below.

VIN	Model Year	Country of Origin
a.____________________	_______	____________________
b.____________________	_______	____________________
c.____________________	_______	____________________
d. ____________________	_______	____________________
e. ____________________	_______	____________________

STOP

Instructor OK ______________ Score __________

ASE Lab Preparation Worksheet #I-3

IDENTIFYING SHOP EQUIPMENT

Name______________________________ Class ____________________

DIRECTIONS:

Write your answers in the spaces provided. In column A, list ten pieces of automotive equipment that are available in your school shop. In column B, briefly describe the purpose of each piece of equipment.

Column A	Column B
1. ____________	____________________
2. ____________	____________________
3. ____________	____________________
4. ____________	____________________
5. ____________	____________________
6. ____________	____________________
7. ____________	____________________
8. ____________	____________________
9. ____________	____________________
10. ____________	____________________

Use the list to identify the equipment pictured. Write the answer number in the space provided.

1. Drill press	2. Bench vise	3. Valve grinder	4. Solvent tank
5. Tire changer	6. Battery charger	7. Tire balancer	8. Arbor press
9. Parts cleaner	10. Grinder	11. Oscilloscope	12. Grease gun

A _____

B _____

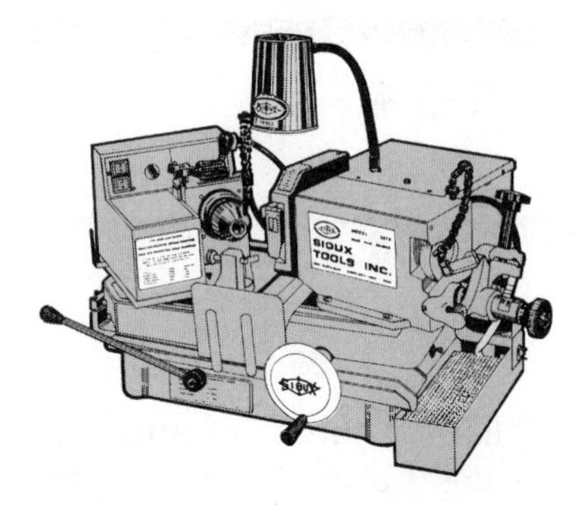
SIOUX
TOOLS INC.

C _____

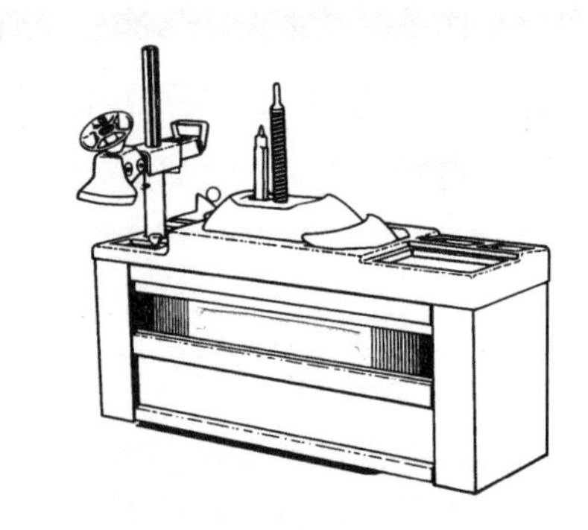

D _____

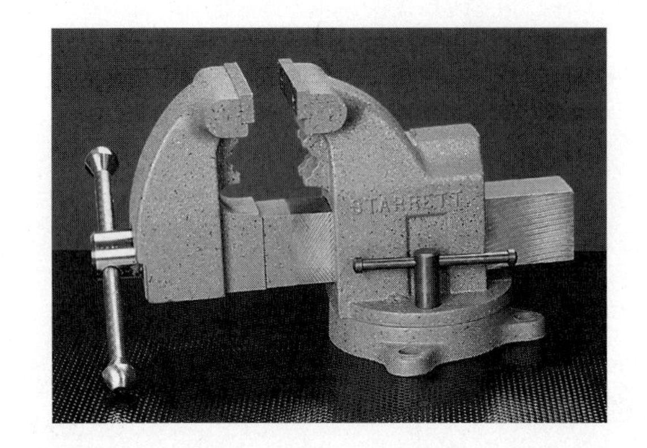
STARRETT

E _____

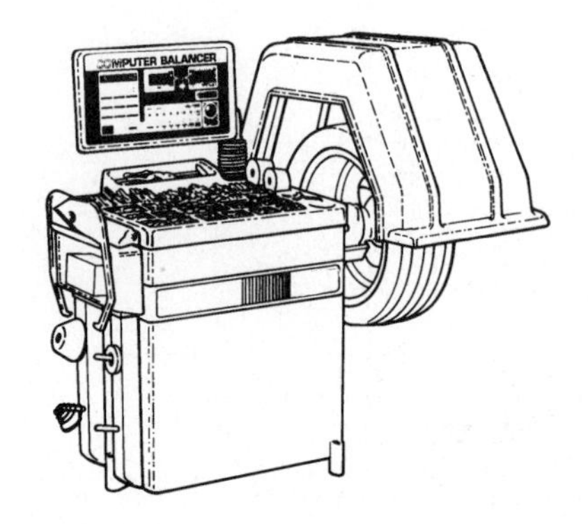
COMPUTER BALANCER

F _____

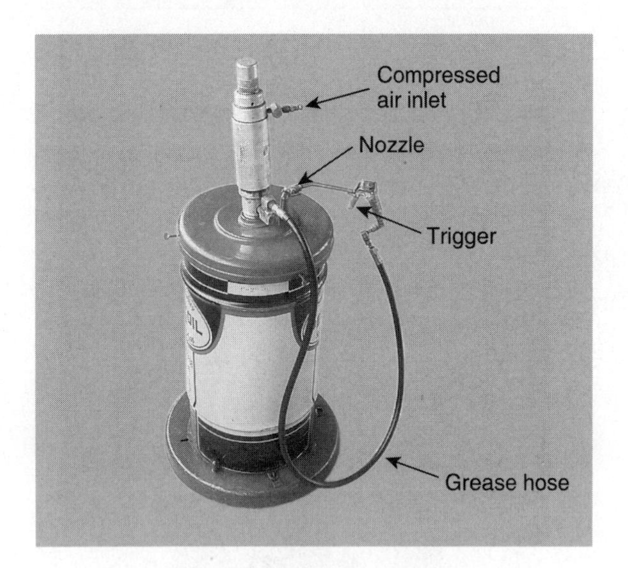
Compressed
air inlet
Nozzle
Trigger
Grease hose

G _____

H _____

STOP

Instructor OK ______________ Score __________

ASE Lab Preparation Worksheet #I-4

SAFETY TEST

Name______________________ Class ______________

OBJECTIVE:

Upon completion of this assignment, you should have an understanding of the hazards that are present in the automotive shop environment.

Most accidents are caused by impatience, carelessness, or poor judgment. The most common accidents in an automotive repair shop involve eye injuries and fires.

DIRECTIONS:

Choose the word from the list that best completes each statement. The words or phrases may be used more than once and not all words or phrases are used. Write your choice in the space provided.

vapor	radiator hose	rags	acid
clean sweep	clothes	black	blue
extinguisher	explosions	CO_2	eye protection
electrical	baking soda	water	one foot
CO	positive	hydraulic jack	jack
creeper	air	skin	tires
liquid	dust	fan belt	dressed
tool rest	side	ground	green
negative	battery	instructor	friend
gasoline	vehicle		

1. Gasoline in its ______________________ form is the most dangerous.
2. Place dirty __________________ in an approved receptacle.
3. Never use ______________________ to clean parts.
4. Wipe up or use ________________________ on all oil, brake fluid, or grease spills.

5. A fire ______________________________ should be used on fuel fires.
6. Two types of common fire extinguishers are dry powder and _______________________ .
7. Before an ______________ fire can be extinguished, the electrical system must be disconnected.
8. Before opening a radiator, test for pressure in the system by squeezing the ________________ .
9. Battery______________________________ is a chemical combination of sulfuric acid and water.
10. Battery acid will cause holes in __________________________ .
11. The most common cause of battery ______________________________ is the battery charger.
12. ______________________________ must be worn around batteries, air-conditioning machinery, compressed air, and other hazardous situations.
13. Battery acid may be neutralized with ______________________ .
14. If acid gets on skin or eyes, immediately flush with __________________ for at least 15 minutes.
15. Before raising a vehicle all the way on a hoist, raise it about __________________ and shake the vehicle to be certain it is properly placed.
16. Use a ______________________________ to raise and lower a vehicle only.
17. A jacked-up vehicle should be placed firmly on _____________ stands.
18. When a ________________ is not in use, it should be stored against a wall in a vertical position.
19. Compressed __________________ is useful but dangerous.
20. Compressed air or grease can penetrate _______________ .
21. Exercise caution when inflating _______________________ that have been remounted on rims.
22. The keys should be out of the ignition any time a _____________ is being inspected or adjusted.
23. Mushroomed tools or chisels should be _________________________________ before use.
24. The ________________________ must be positioned as close to the grinding wheel as possible.
25. Stand to the ______________________ when starting a grinder.
26. The third terminal on electrical equipment is for ____________________ .
27. The color of the electrical ground wire on an extension cord is ______________________ .
28. When disconnecting a car battery, disconnect the __________________________ cable first.
29. Before removing the starter or alternator, disconnect the ______________________________ .
30. If you should become injured while working in the shop, inform your _________________immediately.

Student Signature ______________________________ Date ______________________

Instructor OK _______________ Score __________

ASE Lab Preparation Worksheet #I-5

SHOP SAFETY LAYOUT

Name_____________________________ Class ____________________

OBJECTIVE:

Upon completion of this assignment, you should know the location of the shop emergency equipment and be aware of emergency procedures.

PROCEDURE:

1. On the next page, sketch the layout of your school's automotive shop area. Use the letters that precede each of the items listed to note their location in the shop.

   A. All doors marked with an exit sign
   B. Fire extinguishers
   C. First-aid kits
   D. Emergency telephone
   E. Floor mops
   F. Emergency eyewash
   G. Hand brooms
   H. Dust pans
   I. Push brooms
   J. Water hose
   K. Hazardous materials poster
   L. Air hoses
   M. Sink
   N. Exhaust ventilation hoses

2. Describe two major types of fires that may ignite in an automotive shop environment.

   1. ____
   2. ____

3. Check off the types of fire extinguishers in your school's shop.

   CO_2 ____ Dry powder ____ Foam ____ Water ____

4. Which of the following hazardous wastes may be encountered in your school's shop?

   Coolant ____ Gasoline ____ Solvent ____ Motor oil ____
   Brake dust ____ Freon (R-12) ____ Dirty water ____

TURN

5. In the event that the building had to be evacuated, where outside the building would you meet with your instructor?

____

Sketch the layout of your school's automotive shop.

Instructor OK ______________ Score __________

ASE Lab Preparation Worksheet #I-6

LOCATE VEHICLE LIFT POINTS

Name________________________ Class ________________

OBJECTIVE:

Upon completion of this assignment, you should be able to locate the correct lift points to safely raise a vehicle.

DIRECTIONS:

Before beginning this lab assignment, review the worksheet completely. Fill in the information in the spaces provided as you complete each task.

PROCEDURE:

Vehicle year __________ Make ________________ Model ________________

1. Before you start working, check the service manual for proper lift points.

   Manual ________________________________ Page # ______

   On the drawings put an X to indicate the proper lifting points.

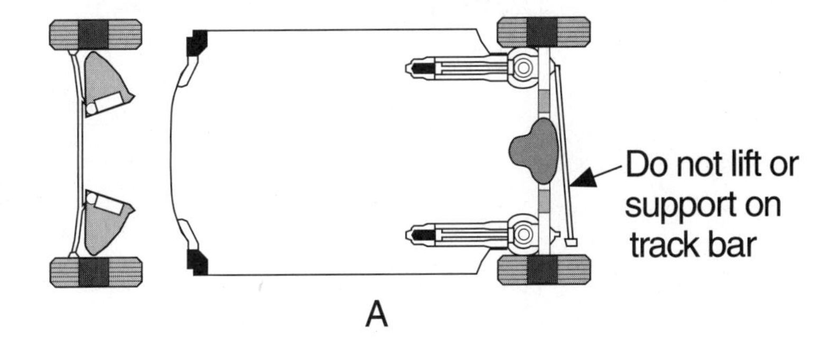

A

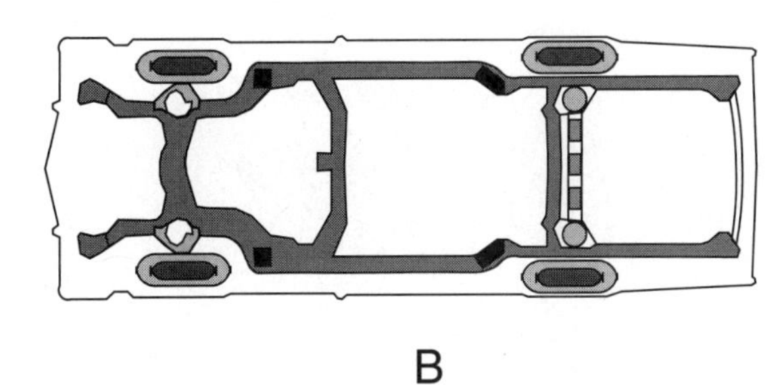

B

TURN

2. Place the letter in the space provided that best identifies the equipment in the drawings.

______ Single-post frame-contact lift

______ Two-post frame-contact lift

______ Surface mount frame-contact lift

______ Surface mount wheel-contact lift

A.

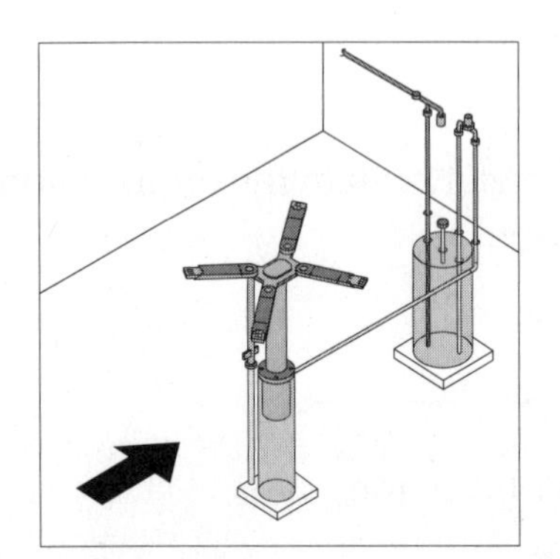

B.

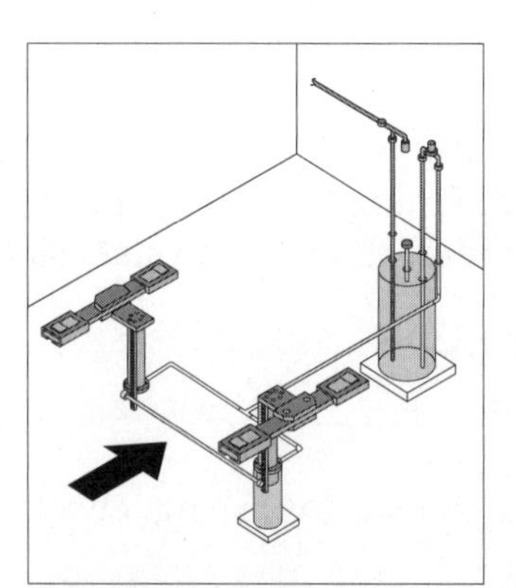

C.

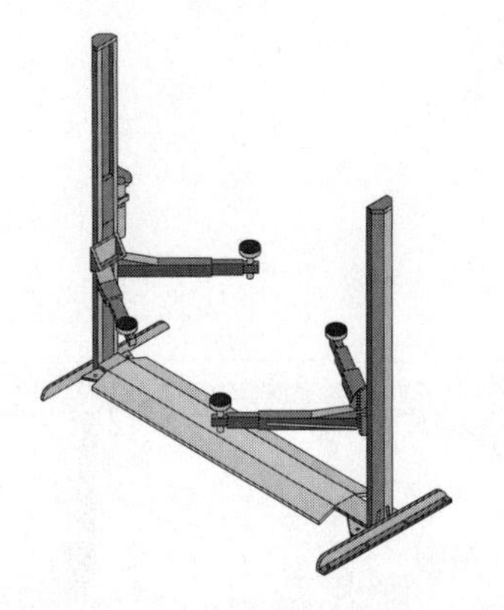

D.

STOP

Instructor OK ______________ Score ______________

ASE Lab Preparation Worksheet #I-7

COMPLETING A REPAIR ORDER (R.O.)

Name______________________________ Class ____________________

OBJECTIVE:

Upon completion of this assignment, you should be able to complete a repair order for required repairs. This task will help prepare you for employment in the automotive industry.

DIRECTIONS:

Before beginning this assignment, review the worksheet completely. Fill in the information in the spaces provided as you complete each task.

RELATED INFORMATION:

Repair orders are used to keep an accurate record of work completed and parts used during a vehicle repair. They serve as an agreement between the customer and the repair shop as to what repairs are to be completed and the cost of the parts and repairs. Agreeing to the specific repairs and their cost before work is started prevents many surprises and disagreements when the vehicle is picked up by the customer.

The repair order is filled out by the repair shop when the customer brings the vehicle in for repair. The customer's name, address, description of the vehicle, and the needed service or repairs are recorded. An estimate of the time the job will be finished and of the cost of repairs is included.

Note: An estimate of the cost of repairs is required by law in several states.

PROCEDURE:

Complete a repair order for the following customer:

Customer #1

Mrs. Jane Pollano	1989 Ford, Taurus
456 Willard Drive	License # GHY 385
Susanwash, CA 93004	145,467 miles
Phone # 123-2145	Vin # 1G1CZ19H6KW135675
Service requested:	Sixty thousand mile service $225.00
Parts needed:	5 quarts oil $2.50 each
	1 oil filter 8.50

1. Complete a repair order using the above information.
   a. Enter the customer's name, address, and telephone number in the spaces provided.
   b. Fill in the date and your name (the person writing the repair order).
   c. Insert the related information about the vehicle in the correct spaces.
   d. List the services to be performed in the section labeled *labor instructions.*
   e. Estimate the total price and write it in the space marked *original estimate.*
2. The repair order is now ready to be signed by the customer. Ask your instructor to review and sign the repair order in the space for customer authorization.

3. The repairs have been completed.
   a. Record any parts used. The charges for parts and labor are then priced and totaled.
   b. Calculate the sales tax on the parts only. (Some states charge tax on the labor.) Multiply the parts total by the percentage of the tax (6%).

      Example: $22.00 (Parts)

      × .06 (% of tax)

      $1.320 Tax

   c. Total the parts, labor, and tax. The vehicle is now ready for delivery to the customer.

Complete a repair order for the following customer:
Customer #2

Mr. William Black comes in to your shop with his Honda Accord. He is concerned because the engine speed increases but the vehicle does not maintain speed whenever he drives up the hill leading to his house at 1435 Hill Street, Oakhill, CA 93005. This problem has become more pronounced lately. Yesterday he had to call his wife to pick him up at the bottom of the hill. After hearing his concerns, it is explained that it would be necessary to fill out a repair order before diagnosing his problem. The service writer told Mr. Black that there would be a half-hour labor charge at $50.00 per hour for the diagnosis. Mr. Black agreed to the charge and the 7% tax on the parts and labor. He said that he could be reached by phone at home at 876-4876 or at work at 456-9834. The vehicle identification number (1Y1SK5262TR0076546), license number (YTR 746), and mileage (75,736) were obtained from the vehicle.

Prepare the repair order (estimate) before proceeding any further. Have your instructor review it and sign in the space for the customer authorization.

The technician checked the Honda and found that the clutch was slipping and needed to be replaced. The technician checked the parts and labor guide and found that the job would require 6.5 hours to complete. He also made a list of the required parts and got prices from a local parts store.

Pressure plate	$124.98
Clutch disc	65.68
Throwout bearing	24.58

A call was made to Mr. Black at 2:45 PM and he approved the repairs. The repairs were completed. It is time for you to complete the repair order.

STOP

THOMSON
DELMAR LEARNING™

DATE	TIME REQUESTED	WRITTEN BY
NAME		
ADDRESS		
CITY	ZIP	
HOME PHONE	BUSINESS PHONE	

YEAR	MAKE	MODEL	LICENSE NO.	MILEAGE
VEHICLE ID #				

Part or Lubricant Description:

TOTAL PARTS		

Labor Description **Labor Time:** **HRS.** **MINS.**

Labor Description	Labor Time	HRS.	MINS.

I authorize the listed labor and materials required for this repair in an amount not to exceed
You are authorized to operate this vehicle on streets for testing purposes

$ ______ Original estimate

Customer Signature ______

ALL PARTS WILL BE DISCARDED UNLESS INSTRUCTED OTHERWISE SAVE ☐ DISCARD ☐

Revised Estimate ______ Date ______ Time ______ Person Contacted ______ Contacting Person ______ BY PERSON ☐ BY PHONE ☐

I acknowledge being notified and approving an increase in the original estimate.

Customer Signature ______

RECOMMENDED SERVICE & COMMENTS	EST. COST

TOTAL LABOR TIME		
x SHOP RATE		
LABOR TOTAL		
TOTAL LABOR AND PARTS		
LABOR		
PARTS		
SALES TAX		
TOTAL		

STUDENT TECHNICIANS

DISCLAIMER STATEMENT

The School District does not assume responsibility or accept liability for vehicles and vehicle parts or work performed by students. The proper performance of tasks by students is a slow, methodical process which cannot be accelerated. Student work is done as an educational experience and is observed by the instructor to see that it is done properly. However, instances do occur where work has to be re done because the student failed to follow instructions.

Parts installed are not warranted beyond that given by respective manufacturers. No other warranties are made.

I understand that all jobs done in this laboratory are for student learning experience, with no expressed warranty or specific completion time.

I hereby authorize the repair work listed hereon, including sublet work to be done along with the purchase of necessary materials. The automotive staff and/or students may operate the described vehicle for testing and inspection at owner's risk. An express lien is acknowledged on said vehicle to secure the amount of repairs thereto. I hereby agree to hold the School District, the Board of Trustees, and all District offices, agents, and employees free from any loss, damages, liability, cost or expense caused in any way that may arise during or as a result of this vehicle being repaired by or used in the school Auto Shop.

NOTES (specs, procedures, additional service or repair information):

ADDITIONAL RECOMMENDATIONS FOR SERVICE OR REPAIRS:

THOMSON
DELMAR LEARNING™

DATE	TIME REQUESTED	WRITTEN BY
NAME		
ADDRESS		
CITY	ZIP	
HOME PHONE	BUSINESS PHONE	
YEAR / MAKE	MODEL	LICENSE NO. / MILEAGE
VEHICLE ID #		

Part or Lubricant Description:

TOTAL PARTS		

Labor Description

Labor Description	Labor Time:	HRS.	MINS.

I authorize the listed labor and materials required for this repair in an amount not to exceed
You are authorized to operate this vehicle on streets for testing purposes

$ ________ Original estimate

Customer Signature________________________

ALL PARTS WILL BE DISCARDED UNLESS INSTRUCTED OTHERWISE SAVE ☐ DISCARD ☐

Revised Estimate Date Time Person Contacted Contacting Person BY PERSON ☐ BY PHONE ☐

I acknowledge being notified and approving an increase in the original estimate.

Customer Signature ________________________

RECOMMENDED SERVICE & COMMENTS	EST. COST

TOTAL LABOR TIME		
x SHOP RATE		
LABOR TOTAL		
TOTAL LABOR AND PARTS		
LABOR		
PARTS		
SALES TAX		
TOTAL		

STUDENT TECHNICIANS

DISCLAIMER STATEMENT

The School District does not assume responsibility or accept liability for vehicles and vehicle parts or work performed by students. The proper performance of tasks by students is a slow, methodical process which cannot be accelerated. Student work is done as an educational experience and is observed by the instructor to see that it is done properly. However, instances do occur where work has to be re done because the student failed to follow instructions.

Parts installed are not warranted beyond that given by respective manufacturers. No other warranties are made.

I understand that all jobs done in this laboratory are for student learning experience, with no expressed warranty or specific completion time.

I hereby authorize the repair work listed hereon, including sublet work to be done along with the purchase of necessary materials. The automotive staff and/or students may operate the described vehicle for testing and inspection at owner's risk. An express lien is acknowledged on said vehicle to secure the amount of repairs thereto. I hereby agree to hold the School District, the Board of Trustees, and all District offices, agents, and employees free from any loss, damages, liability, cost or expense caused in any way that may arise during or as a result of this vehicle being repaired by or used in the school Auto Shop.

NOTES (specs, procedures, additional service or repair information):

ADDITIONAL RECOMMENDATIONS FOR SERVICE OR REPAIRS:

STOP

Part II

ASE Lab Preparation Worksheets

Service Area 1

Oil Change Service

ASE Lab Preparation Worksheet #1-1

MAINTENANCE SPECIFICATIONS

Name____________________________ Class ________________

OBJECTIVE:

Upon completion of this assignment, you should be able to use a vehicle maintenance guide to locate maintenance service specifications. This task will help prepare you to pass the ASE certification examination in engine performance.

DIRECTIONS:

Review this worksheet completely before starting. Use your own vehicle or one provided by your instructor. Use a Car Care Guide, service manual, or computer program to locate the requested information. Record the information in the spaces provided. If the information is not available or does not apply, write N/A in the answer space.

Note: If you are completing this worksheet on your personal vehicle, you may want to save it for future reference.

TOOLS AND EQUIPMENT REQUIRED:

Car Care Guide or vehicle service manual

PROCEDURE:

Vehicle year ________ Make ______________ Model ________________

Engine: 4-cylinder ___ 6-cylinder ___ 8-cylinder ___

Transmission: Standard ___ Automatic ___

Is there an underhood label? Yes ___ No ___

Is there an underhood vacuum diagram? Yes ___ No ___

What is the name of the manual or program being used to complete this worksheet?

____

Capacities:

Battery _____ Cold Cranking Amps

Crankcase (without oil filter) _____ qt.

Oil filter capacity _____ qt.

Cooling system (without AC) _____ qt.

Cooling system (with AC) _____ qt.

Differential capacity _____ pt./qt.

Transmission capacity _____ pt./qt.

Fuel tank capacity _____ gallons

Oil Specifications:

SAE engine oil viscosity ____________________

API service rating ____________________

Transmission fluid type ____________________

Differential lubricant type ____________________

Tires:

Pressures Front ______ psi

Rear ______ psi

Lug nut torque specification ______ ft.-lb

Belt Tension Specifications:

Alternator ____________________

Power steering pump ____________________

Air-conditioning compressor ____________________

Smog pump ____________________

NOTES:

____

____

____

____

____

STOP

Instructor OK ______________ Score ________

ASE Lab Preparation Worksheet #1-2

RAISE AND SUPPORT A VEHICLE (JACK STANDS)

Name_________________________ Class __________________

OBJECTIVE:

Upon completion of this assignment, you should be able to safely raise a vehicle with a jack and support it on jack stands.

DIRECTIONS:

Before beginning this lab assignment, review the worksheet completely. Fill in the information in the spaces provided as you complete each task.

RELATED INFORMATION:

Before starting this worksheet, complete Worksheet 6 (Locate Vehicle Lift Points) on page 277.

PROCEDURE:

Vehicle year ___________ Make ________________ Model _________________

1. Raise a Car at the Front:
   a. Center the jack under the front crossmember or frame. *Do not jack on the radiator, oil pan, or front steering linkage!*

      Jack centered? Yes ____ No ____

   b. Raise the vehicle until both front wheels are about 6″ off the ground. Are both wheels leaving the ground equally?

      Yes ____ No ____

   c. Place the jack stand in the recommended position.
   d. Lower the vehicle onto jack stands.
   e. The front wheels should still be off the ground after lowering the vehicle onto stands.

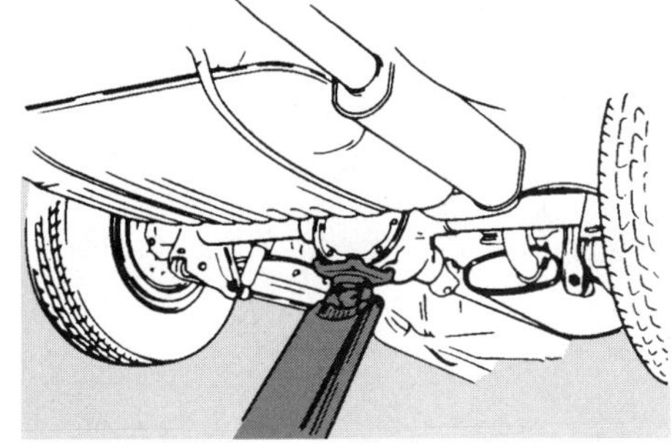

2. Raise All Four Wheels:

   Note: When all four wheels are to be raised off the ground, raise the rear and support it first. Then the front can be raised and supported.

   Note: When raising the rear wheels, be careful not to damage the fuel tank. Positioning the jack handle so it is off to the side and behind the rear wheel is sometimes a good option.

Rear Wheels:

a. Center the jack on the rear axle or crossmember so that both sides of the vehicle are raised equally.
b. Raise the vehicle until the tires are about a foot off the ground.
c. In front of the rear wheels, there is a bend in the frame. Place the jack stands there. If the car has leaf springs, place them in front of the spring eye.

d. Ask your instructor to check your work.

Instructor OK ________________________

e. Lower and remove the jack.

Note: When rear wheels are to be removed from a vehicle, the vehicle should be supported by the frame with the suspension system hanging free. Otherwise, there may not be enough clearance for the wheel to be removed.

Front Wheels:

a. Center the jack under the front crossmember or frame. *Do not jack on the radiator, oil pan, or front steering linkage!*

b. Raise the vehicle until both front wheels are about 6″ off the ground. Be sure both wheels are leaving the ground equally.

c. Place the jack stands on the frame, just behind the front wheels.

d. Ask your instructor to check your work.

Instructor OK ________________________

e. Lower the jack and return it so other students can use it.

To lower the vehicle, reverse the procedure that was used to raise the vehicle.

When you are finished, clean your work area and put the tools in their proper places.

STOP

Instructor OK ______________ Score __________

ASE Lab Preparation Worksheet #1-3

RAISE A VEHICLE USING A FRAME-CONTACT LIFT

Name____________________________ Class ____________________

OBJECTIVE:

Upon completion of this assignment, you should be able to raise a vehicle using a frame-contact lift.

DIRECTIONS:

Before beginning this lab assignment, review the worksheet completely. Fill in the information in the spaces provided as you complete each task.

PROCEDURE:

Vehicle year ___________ Make _________________ Model ___________________

Note: Get assistance and approval from your instructor before lifting a vehicle more than 6" off the ground.

1. Before you start working, check the service manual for proper lift points.

   Manual ____ Page # _______

2. Prepare to lift the vehicle:

   a. Center the vehicle over the lift.

   b. Turn off the engine.

   c. Put the shift lever in neutral position.

   d. Did you apply the parking brake? Yes ____ No ____

   e. Adjust the lift pads to contact the appropriate lift points on the vehicle. Be careful that the vehicle's center of gravity is over the posts of the lift.

   Vehicle centered? Yes ____ No ____

Note: The vehicle center of gravity is not always the center of the vehicle. The center of gravity is dependent on the vehicle weight distribution.

3. Lifting the vehicle:

   a. Raise the lift slowly until the pads contact the lift points. Double-check to see that they are centered under the lift points. Also check that the lift is not going to contact the exhaust system or any lines or cables.

   Are the lift points centered? Yes ____ No ____

   Is the lift clear of the exhaust system? Yes ____ No ____

   Is the lift clear of any cables or lines? Yes ____ No ____

   b. Raise the vehicle until all tires leave the ground.

TURN

Be certain that the lift arms or contact pads do not contact the vehicle's tires when the vehicle is raised.

c. Shake the vehicle to be sure it will not fall off the lift when raised further.

d. Before proceeding further have the instructor check your work.

Instructor OK ____

e. Raise the vehicle to the desired height and set any safety devices that apply.

4. Lowering the vehicle:

   a. Be certain all toolboxes, air or electrical hoses, or lubrication equipment are removed from under the vehicle before lowering it.

   b. Release any safety devices (if the lift is so equipped).

   c. Lower the lift *all the way* to the floor.

   d. Move the lift arms to clear the vehicle.

   e. Back the vehicle off the lift.

5. When you are finished, clean the work area and put the tools in their proper places.

Instructor OK ________________ Score ____________

ASE Lab Preparation Worksheet #1-4

CHECK ENGINE OIL LEVEL

Name________________________________ Class ____________________

OBJECTIVE:

Upon completion of this assignment, you should be able to check a vehicle's engine oil level.

DIRECTIONS:

Before beginning this lab assignment, review the worksheet completely. Fill in the information in the spaces provided as you complete each task.

TOOLS AND EQUIPMENT REQUIRED:

Safety glasses, shop towel

PROCEDURE:

Vehicle year ___________ Make __________________ Model ___________________

Engine size ________________ # of Cylinders _____

1. Locate the following oil specifications for the vehicle:

   Recommended viscosity _________

   Recommended API rating _________

2. Engine oil level is checked with the engine off and at normal operating temperature.

   Is the engine at normal operating temperature? Yes ____ No ____

   Is the engine off? Yes ____ No ____

3. When possible, the oil should be checked after the engine has been off for about 5 minutes. Oil remaining in other parts of the engine will have a chance to return to the oil pan.

   Has the engine been off for 5 minutes? Yes ____ No ____

4. Pull the dipstick out and wipe it clean with a shop towel.

5. Push the dipstick *all the way* into the dipstick tube and then pull it back out. Hold a towel under it so that oil does not accidentally drip onto the vehicle's fender.

6. Read the oil level on the dipstick. If the reading is unclear, flip the dipstick over and repeat the test (read the back side of the dipstick).

   Dipstick reading: Clear ____ Unclear _____

7. The correct oil level is between the "add" and the "full" lines. When the level is below the "add" line, one quart of oil is added.

   Note: It is not necessary to add oil when the level is between the "add" and "full" marks.

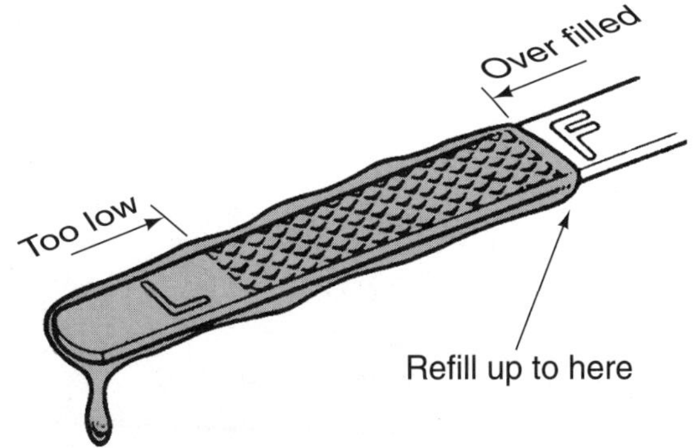

What is the oil level?

Oil level: Full _____ Oil needed _____

8. If the oil level is low, check the service sticker to see if the car is due for servicing.

   Mileage at last oil change __________________

   Service required? Yes ____ No ____

   Note: Be sure that the dipstick is correctly seated on the dipstick tube. If it is not, the crankcase ventilation system can draw in dirty air. This can lead to premature engine failure. Also, some fuel-injected cars will not run properly unless the crankcase is sealed.

9. Close the hood and dispose of the shop towel in the proper manner.

WINDOW SERVICE STICKER

OIL CHANGE	
DATE	MILES
FILTER CHANGE	
☐ OIL ☐ AIR ☐ FUEL	
DATE	MILES
LUBRICATION	
DATE	MILES
OTHER SERVICE	
DATE	MILES
DESCRIPTION	
OIL CHANGE RECORD (USE UNDER HOOD)	
CUSTOMER'S NAME	BRAND-WEIGHT
DATE	MILES

DOOR SERVICE STICKER

STOP

Instructor OK ______________ Score __________

ASE Lab Preparation Worksheet #1-5

OIL CHANGE

Name______________________ Class ________________

OBJECTIVE:

Upon completion of this assignment, you should be able to do an engine oil change.

DIRECTIONS:

Before beginning this lab assignment, review the worksheet completely. Fill in the information in the spaces provided as you complete each task.

RELATED INFORMATION:

The oil and oil filter are customarily changed at the same time. Sometimes the oil is changed without changing the filter, but rarely is the filter changed without changing the oil. If the intent is to change just the oil, use this worksheet. If you plan to change the oil and the filter, use both this worksheet and the Change the Filter Worksheet 1-6.

TOOLS AND EQUIPMENT REQUIRED:

Safety glasses, fender covers, jack and jack stands or vehicle lift, drain pan, filter wrench, shop rag, combination wrench

PARTS AND SUPPLIES REQUIRED:

Motor oil

PROCEDURE:

Vehicle year __________ Make ________________ Model ________________

Repair Order # _______ Engine size ______________ # of Cylinders __________

1. Start a repair order for the vehicle. Place the shop copy under the windshield wiper.
2. For best results drain the crankcase when the engine is warm. The oil will flow faster and more will drain out, carrying with it deposits that have collected in the pan.

   Engine warm? Yes ____ No ____
3. Open the hood and place fender covers on the fenders and front body parts.
4. Before starting to work on the car, remove the *oil filler cap* and put it on top of the air cleaner or in a conspicuous place. This will remind you (or one of the people working with you) to refill the crankcase before the engine is started after the oil change.

   Oil filler cap removed? Yes ____ No ____
5. Raise the vehicle on a hoist or with a jack. (Be sure to support it with jack stands.)

   Note: When draining the crankcase, it is important that the vehicle be level. This will prevent oil from being trapped in the oil pan.
6. Position an oil drain pan under the oil pan drain plug.

TURN

Be environmentally aware! Properly dispose of used oil. Drain oil into a container and save it for recycling. More than 220 million gallons of oil are disposed of improperly each year. Used oil can damage the water supply and kill plants and wildlife. One gallon of improperly disposed oil can pollute 1 million gallons of drinking water.

7. Loosen the drain plug with the correct size wrench.

   Which way did you turn the wrench to loosen the plug?

   Clockwise ______

   Counterclockwise ______

8. What size wrench was used?

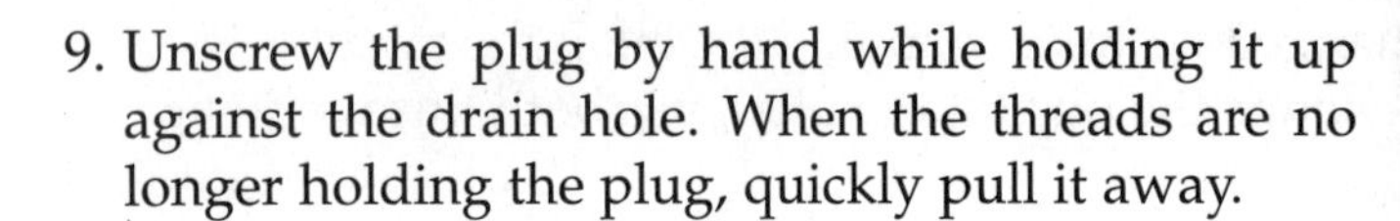

   ____

9. Unscrew the plug by hand while holding it up against the drain hole. When the threads are no longer holding the plug, quickly pull it away.

   Note: A commercial oil drain tank includes a catch screen for the plug.

10. Allow the oil to drain until it no longer drips from the oil pan. This usually takes about 5 minutes.

11. Check the condition of the plastic or copper drain plug gasket. Replace it if necessary.

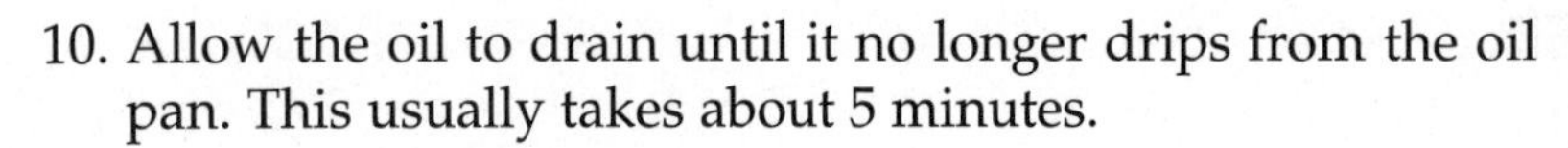

    Gasket condition: OK ____ Replace ____

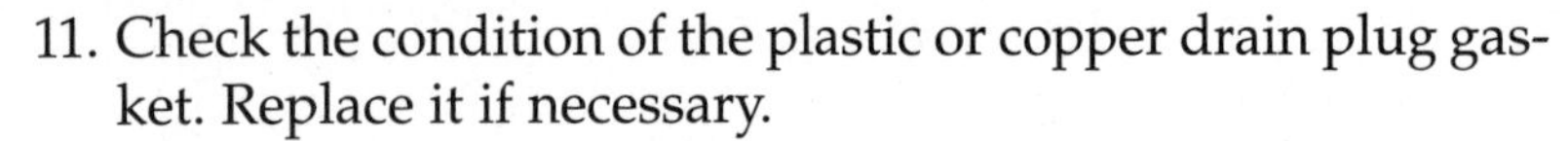

    Note: Those annoying little leaks on the garage floor are often caused by leaking oil drain plug gaskets.

12. Thread the drain plug into the pan *all the way* by hand.

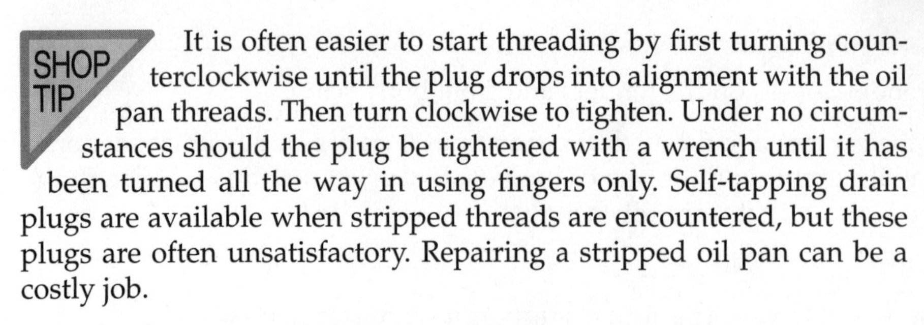

It is often easier to start threading by first turning counterclockwise until the plug drops into alignment with the oil pan threads. Then turn clockwise to tighten. Under no circumstances should the plug be tightened with a wrench until it has been turned all the way in using fingers only. Self-tapping drain plugs are available when stripped threads are encountered, but these plugs are often unsatisfactory. Repairing a stripped oil pan can be a costly job.

Thread condition: OK ____ Damaged ____

13. Tighten the oil pan drain plug with the correct size wrench.

14. Lower the vehicle.

15. Refill the crankcase and replace the fill cap.

    How much oil was added to refill the crankcase? ________ qt.

16. Wipe up any oil that may have spilled.
17. Start the engine. Let the engine idle. Stay in the car until the oil pressure light goes out or until pressure is indicated on the gauge.

    Does the vehicle have a: Light _____ or a Gauge _____

    Did the light go out after the engine was started? Yes ____ No ____ N/A ____
18. Check for leaks. Leaks ____ No leaks ____
19. Double-check the crankcase oil level and top off if needed.
20. Before completing the paperwork, clean your work area, put the tools in their proper places, and wash your hands.
21. Fill out a service sticker and install it.

    Service sticker installed? Yes ____ No ____
22. Mileage recommendation for next service ________________________
23. Write recommendation for the next oil change on the repair order.
24. Were there any other recommendations for needed service or unusual conditions that you noticed while changing the oil?

    Yes ____ No ____
25. Record your recommendations for additional service or repairs on the repair order.
26. Complete the repair order.

SANTA BARBARA CITY COLLEGE
Automotive Technology
721 Cliff Drive
SANTA BARBARA, CA. 93109
NEXT SERVICE DUE
OR
DATE MILES

WINDOW SERVICE STICKER

OIL CHANGE	
DATE	MILES
FILTER CHANGE	
☐ OIL ☐ AIR ☐ FUEL	
DATE	MILES
LUBRICATION	
DATE	MILES
OTHER SERVICE	
DATE	MILES
DESCRIPTION	
OIL CHANGE RECORD (USE UNDER HOOD)	
CUSTOMER'S NAME	BRAND-WEIGHT
DATE	MILES

DOOR SERVICE STICKER

Instructor OK ______________ Score __________

ASE Lab Preparation Worksheet #1-6

OIL FILTER CHANGE

Name______________________________ Class ____________________

OBJECTIVE:

Upon completion of this assignment, you should be able to change a spin-on-type oil filter.

DIRECTIONS:

Before beginning this lab assignment, review the worksheet completely. Fill in the information in the spaces provided as you complete each task.

RELATED INFORMATION:

The oil and oil filter are customarily changed at the same time. Sometimes the oil is changed without changing the filter, but rarely is the filter changed without changing the oil. If the intent is to change just the oil, use the Oil Change Worksheet 1-5. If the intent is to change the oil and the filter, use both Worksheets 1-5 and 1-6.

TOOLS AND EQUIPMENT REQUIRED:

Safety glasses, fender covers, jack, and jack stands or vehicle lift, drain pan, filter wrench, shop rag

PROCEDURE:

Vehicle year ___________ Make _________________ Model ___________________

Repair Order # ________ Engine _________________ # of Cylinders ___________

Start a repair order for the vehicle. Place the shop copy under the windshield wiper.

1. Obtain the oil filter before starting to work.
2. What is the brand of the filter? ____________________

   What is the part number of the filter? __________________
3. Locate the following specifications:

   Crankcase oil capacity _______________

   Approximate oil filter capacity _______________

   SAE oil viscosity _______________

   API service rating _______________

4. Raise the vehicle with a lift or place it on jack stands to gain access to the filter.
5. Position a drain pan under the oil filter.
6. Use a filter wrench. Install it all the way against the base of the oil filter.

Failure to position the filter wrench at the base of the filter can result in a torn or distorted filter, which can be difficult to remove.

7. Remove the filter. Be careful! Oil may spill as the filter is removed.
8. Clean the filter mounting surface. Be especially sure that the O-ring from the old filter is not stuck to the engine block.
9. Compare the sealing surfaces of the old and new filters to be sure they are identical.

   Same _______ Different _______

   Note: The new filter may not be exactly the same size as the old filter, but the sealing surfaces must be identical.

10. Put a few drops of oil on the filter's rubber O-ring.
11. Install the filter by hand until it contacts the filter mounting surface.
12. Refer to the filter instructions on the filter or its box for tightening specifications. The usual recommendation is to turn the filter an additional 1/2 to 1 revolution.

    What is the tightening recommendation for the filter that is being installed?

    ____

    Note: Do not use a filter wrench to tighten the filter.

13. Add the correct type and amount of oil to the engine.

    Note: Remember to include the filter capacity when figuring the amount of oil needed to refill the crankcase.

14. How much oil did you add to the engine? ___________ qt(s).
15. Run the engine until the oil pressure gauge or indicator light registers oil pressure.
16. Shut off the engine and check the oil level on the dipstick. Full _____ Low _____ High _____

    Recheck the oil level after 5 minutes to be sure it is correct. Full _____ Low _____ High _____
17. Check for oil leaks. If any leaks were noticed, where did the oil appear to be coming from?

    ____

18. Before completing the paperwork, remove the fender covers and put the tools in their proper places. Clean your work area and wash your hands.

ENVIRONMENTAL NOTE

Be environmentally aware and properly dispose of the used oil filter. Ask your instructor for the proper disposal method in your area.

19. Fill out a service sticker using a pencil. (A pen can run or fade during humid weather.) Install it on the driver's door jamb for future reference.

    Is the door record installed? Yes ____ No ____

CONTINUED

20. Be sure to note the mileage recommendation for the next filter service on the repair order (RO).
21. Note the mileage recommendation for the next service. ______________________________
22. Was the recommendation for the next oil change indicated on the repair order?

    Yes _____ No _____
23. Were there any other recommendations for needed service or unusual conditions that you noticed while changing the oil?

    Yes _____ No _____
24. Record your recommendations for additional service or repairs on the repair order.
25. Complete the repair order.

STOP

Part II

ASE Lab Preparation Worksheets

Service Area 2

Underhood Inspection

Instructor OK ______________ Score __________

ASE Lab Preparation Worksheet #2-1

CHECK THE BRAKE MASTER CYLINDER FLUID LEVEL

Name_________________________ Class _______________

OBJECTIVE:

Upon completion of this assignment, you should be able to check and refill a brake master cylinder to the proper level. This task will help prepare you to pass the ASE certification examination in brake service.

DIRECTIONS:

Before beginning this lab assignment, review the worksheet completely. Fill in the information in the spaces provided as you complete each task.

TOOLS AND EQUIPMENT REQUIRED:

Safety glasses, fender covers, shop towel

> **Note:** Brake fluid rapidly absorbs moisture when exposed to air. Do not leave the lid off the brake fluid container, or the cover off the master cylinder. Also, remember that brake fluid can damage the vehicle's paint.

PROCEDURE:

Vehicle year __________ Make _______________ Model _______________

1. Install fender covers on the vehicle.
2. Remove the master cylinder cover.

   Note: Before removing the cover from the master cylinder, clean around it to prevent dirt from entering the system.

   Type of cover: Screw-type cap ____

   Pry-off bale-type ____

Caps

6 mm (1/4 inch) from top of opening

Master cylinder

3. Inspect the fluid level.

   (Within 1/4″ of top) Ok _____ Low _____

4. Inspect the fluid condition.

   Clear _____ Cloudy _____ Dark _____

   Note: Vehicles have had two chambers (tandem) since 1967 for safety reasons. Many vehicles incorporate a dual reservoir as well.

5. Does the vehicle have a single or dual chamber master cylinder? Single _____ Dual _____

   How many reservoirs are used on the master cylinder? One _____ Two _____

TURN

Note: If there are two reservoirs and one is larger, the larger one is for disc brakes. As disc brakes wear, the level drops as more fluid is required. If the large chamber is low, check the disc brake linings for wear. Refer to Worksheet 9-4.

The larger reservoir is: Full _____ Low _____ N/A _____

6. Add the proper brake fluid as necessary to fill the brake master cylinder reservoir to the proper level.

   Note: Clean any brake fluid spills immediately. Remember, brake fluid will damage the vehicle's paint. Work carefully and dilute accidental spills with water.

7. Reinstall the reservoir cover.
8. Look for any leakage or dampness around the master cylinder. Dampness can indicate that the cylinder might need to be rebuilt or replaced.

   Are there any signs of leakage or dampness? Yes ____ No ____

   If so, where on or near the master cylinder is the leakage or dampness located?

   At the cover _____

   Front of the cylinder _____

   Rear of the cylinder _____

   Other _____
9. Was it necessary to add any brake fluid to the system? Yes ____ No ____
10. What type of brake fluid does this system require?

    DOT 3 _____

    DOT 4 _____

    DOT 5 _____
11. Before completing the paperwork, clean your work area, put the tools in their proper places, and wash your hands.

STOP

Instructor OK ______________ Score __________

ASE Lab Preparation Worksheet #2-2

CHECK CLUTCH MASTER CYLINDER FLUID LEVEL

Name______________________________ Class ____________________

OBJECTIVE:

Upon completion of this assignment, you should be able to check and refill a clutch master cylinder to the proper level. This task will help prepare you to pass the ASE certification examination in manual transmissions.

DIRECTIONS:

Before beginning this lab assignment, review the worksheet completely. Fill in the information in the spaces provided as you complete each task.

TOOLS AND EQUIPMENT REQUIRED:

Safety glasses, fender covers, shop towel

PROCEDURE:

Vehicle year ___________ Make _________________ Model ________________

Note: The brake fluid used in clutch master cylinders rapidly absorbs moisture when exposed to air. Do not leave the lid off the fluid container or the cover off the master cylinder. Also, remember that brake fluid can damage the vehicle's paint. Clean spills by diluting them with water.

1. Install fender covers on the vehicle.
2. Remove the clutch cylinder reservoir cover.

Note: Before removing the cover, clean around the clutch master cylinder cover to prevent dirt from entering the system.

Type of cover:

Screw-type cap ______ Plug-type ______

3. Inspect the fluid level.

   (Within 3/8″ of top) OK _____ Low ______

4. Inspect the fluid condition.

   Clear ______ Cloudy ______ Dark ______

5. Add brake fluid if necessary.

   Yes _____No _____

6. What type of brake fluid does this system require?

   DOT 3 ______ DOT 4 ______ DOT 5 ______

7. Reinstall the reservoir cover.

Clutch master cylinder
Clutch slave cylinder
Fluid level
Freeplay

8. Look for any leakage or dampness around the master cylinder. Dampness can indicate that the cylinder needs to be rebuilt or replaced.

   Is there any leakage or dampness? Yes ____ No ____

   If so, where on or near the clutch master cylinder is the leakage or dampness located?

   ____

9. Before completing the paperwork, clean your work area, put the tools in their proper places, and wash your hands.

STOP

Instructor OK ________________ Score ____________

ASE Lab Preparation Worksheet #2-3

CHECK POWER STEERING FLUID LEVEL

Name________________________________ Class ________________

OBJECTIVE:

Upon completion of this assignment, you should be able to check the power steering fluid and refill it to the proper level. This task will help prepare you to pass the ASE certification examination in steering and suspension.

DIRECTIONS:

Before beginning this lab assignment, review the worksheet completely. Fill in the information in the spaces provided as you complete each task.

TOOLS AND EQUIPMENT REQUIRED:

Safety glasses, fender covers, shop towel

PROCEDURE:

Vehicle year ____________ Make __________________ Model __________________

1. Perform this procedure when the fluid is warm. Start the engine and run it until it is at normal operating temperature.

   Engine warm? Yes ____ No ____

2. Cycle the system by turning the steering wheel through its complete range of travel from left to right. This will increase the temperature of the power steering fluid.

   Note: Do not hold the steering wheel at full right or left for more than 10 seconds. Damage to the system could result.

3. Shut off the engine and check the fluid level.

   OK _____ Low _____

   Fluid temperature:

   Hot _____ Cold _____

4. Add specified fluid as needed.

   Type of fluid specified: ______________________

5. Was additional fluid required?

   Yes _____ No _____

6. Is there evidence of a leak? Yes ____ No ____

   Note: When looking for a power steering fluid leak, it will be necessary to inspect the entire power steering system.

If a leak was noticed, where is it located?

Pressure hose _____

Return hose _____

Reservoir _____

Steering gear _____

7. Before completing the paperwork, clean your work area, put the tools in their proper places, and wash your hands.
8. Were there any other recommendations for needed service or unusual conditions that you noticed while you were checking the power steering fluid level?

   Yes _____ No _____

Recommendations: ____

____

____

____

Instructor OK ______________ Score __________

ASE Lab Preparation Worksheet #2-4

CHECK AND CORRECT COOLANT LEVEL

Name__________________________ Class ________________

OBJECTIVE:

Upon completion of this assignment, you should be able to check the radiator coolant level and add coolant to the proper level. This task will help prepare you to pass the ASE certification examinations in engine repair and engine performance.

DIRECTIONS:

Before beginning this lab assignment, review the worksheet completely. Fill in the information in the spaces provided as you complete each task.

TOOLS AND EQUIPMENT REQUIRED:

Safety glasses, fender covers, shop rag

PROCEDURE:

Vehicle year ___________ Make _________________ Model _________________

1. Open the hood and place fender covers over the fenders.
2. Inspect coolant level in the recovery tank. It should be filled to the "cold" line if the coolant is cold.

   Coolant temperature:

   Cold ____ Warm ____
3. What is the coolant condition/color?

   Clear ____ Green ____ Orange ____

   Yellow ____ Rusty ____ Red ____
4. Fill the recovery tank as needed.

   Water added ____ Coolant added ____

Atmospheric type pressure cap

Overflow tube

Overflow tank

Note: If the recovery tank was empty it will be necessary to check the coolant level in the radiator. Normally it is not necessary to remove the radiator cap to check the coolant level if the recovery tank has coolant in it.

6. Check the top of the radiator cap. What is the pressure rating of the cap? ____ psi
7. What radiator cap pressure is specified for the vehicle? ____ psi

   Is this the correct radiator cap for the vehicle? Yes ____ No ____
8. What is the temperature of the cooling system? Hot ____ Cold ____

The radiator should not be opened when there is pressure in the system. Before opening the radiator cap, squeeze the upper radiator hose to be sure the system is not under pressure.

9. Check the cooling system pressure. Is the upper radiator hose hard or soft when you squeeze it?

   Hard _____ Soft _____

10. If the top hose is soft, fold a shop rag and place it over the radiator cap.

11. Hold down firmly on the cap and turn it counterclockwise 1/4 turn until the cap is opened to the safety catch.

Wire type clamp

Squeeze hose

SAFETY NOTE Let up slowly on the pressure you are exerting on the cap. If coolant escapes, press the cap back down. (On most cars, cap pressure will be no more than 17 psi.) If pressure is allowed to escape from a hot system, the coolant boiling point will be lowered and it may boil.

Cover cap with rag

12. If coolant escapes, retighten the cap.

    Coolant escapes _____ No coolant escapes _____

    If coolant escaped while attempting to open the radiator cap it will be necessary to wait for the vehicle to cool, or consult with your instructor before proceeding.

13. If no coolant escapes, press down on the cap while turning it counterclockwise to remove it.

CAUTION After removing the radiator cap DO NOT look into the radiator for at least 30 seconds. Sometimes it takes the coolant several seconds before it begins to boil.

14. Observe the coolant level. Full _____ Low _____

15. Inspect the coolant condition and color.

    Clear _____ Green _____ Orange _____

    Yellow _____ Rusty _____ Red _____

16. Fill the radiator as needed.

17. Replace the radiator cap making sure that it is fully locked in place.

18. Before completing the paperwork, clean your work area, put the tools in their proper places, and wash your hands.

19. Were there any other recommendations for needed service or unusual conditions that you noticed while checking the coolant level?

    Yes _____ No _____

    Recommendations:____

    ____

STOP

Instructor OK ________________ Score ____________

ASE Lab Preparation Worksheet #2-5

IDENTIFY AND INSPECT ACCESSORY DRIVE BELTS (V-BELTS)

Name______________________________ Class ____________________

OBJECTIVE:

Upon completion of this assignment, you should be able to identify and inspect accessory drive belts for adjustment and condition. This task will help prepare you to pass the ASE certification examinations in suspension and steering and engine performance.

DIRECTIONS:

Before beginning this lab assignment, review the worksheet completely. Fill in the information in the spaces provided as you complete each task.

TOOLS AND EQUIPMENT REQUIRED:

Safety glasses, fender covers, shop towel

Note: Many late-model vehicles have a single V-ribbed accessory belt. **Select a vehicle with multiple V-belts for this exercise.**

PROCEDURE:

Vehicle year ____________ Make ________________ Model ______________

1. Open the hood and place fender covers over the fenders.
2. How many accessory belts are on the vehicle?

   One _____

   Two _____

   Three _____

   Four _____

3. In the spaces provided write the names of the accessories that are driven by belts on this engine. Belt #1 is the belt that is closest to the engine.

   Belt #1 ______________________________

   Belt #2 ______________________________

   Belt #3 ______________________________

   Belt #4 ______________________________

4. Inspect each of the belts and record its condition in the space provided.

	Good	Worn	Damaged	Glazed	Loose
Belt #1	_____	_____	_____	_____	_____
Belt #2	_____	_____	_____	_____	_____
Belt #3	_____	_____	_____	_____	_____
Belt #4	_____	_____	_____	_____	_____

TURN

5. Remove the fender covers from the vehicle and close the hood.
6. Put any tools that you used in their proper places and clean your work area.
7. What recommendations should be made to the customer concerning the condition of the drive belts?

____

____

____

____

____

____

STOP

Instructor OK _______________ Score __________

ASE Lab Preparation Worksheet #2-6

BATTERY VISUAL INSPECTION

Name_____________________________ Class ____________________

OBJECTIVE:

Upon completion of this assignment, you should be able to inspect a battery for condition and electrolyte level. This task will help prepare you to pass the ASE certification examinations in engine repair, electrical, and engine performance.

DIRECTIONS:

Before beginning this lab assignment, review the worksheet completely. Fill in the information in the spaces provided as you complete each task.

TOOLS AND EQUIPMENT REQUIRED:

Safety glasses, fender covers, shop towel

PROCEDURE:

Vehicle year ____________ Make __________________ Model ___________________

1. Open the hood and install fender covers on the vehicle.

- In addition to being dangerous to skin and eyes, battery acid can damage clothing and the car's paint. Work carefully and dilute accidental spills immediately with water.
- Batteries give off hydrogen gas when charging. Be careful to avoid an accidental spark.

2. Inspect the alternator belt. OK _____ Loose _____
3. Check the condition of the battery terminal clamps/posts.

   Tight _____ Loose _____ Clean _____ Corroded _____
4. What type of terminals does the battery have?

   Top _____ Side _____ "L" _____
5. Battery cable condition:

   Clean _____ Frayed _____ Corroded _____

   Well-insulated _____ Worn insulation _____
6. Check the battery's external condition.

   Clean _____ Dirty _____

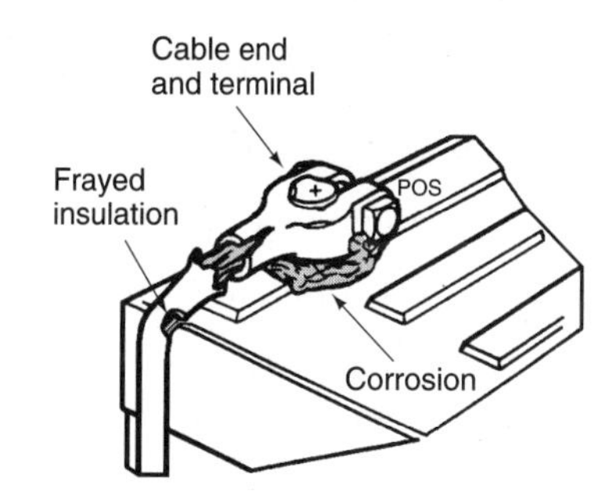

A

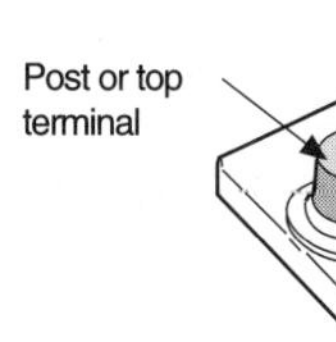

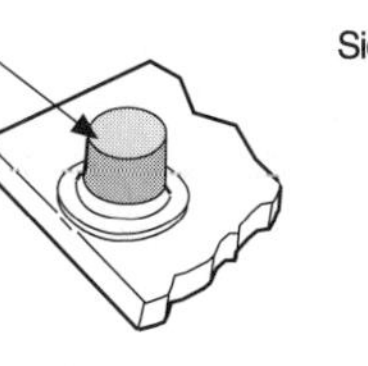

B

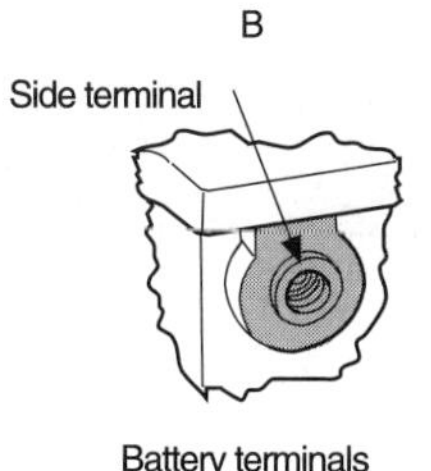

C

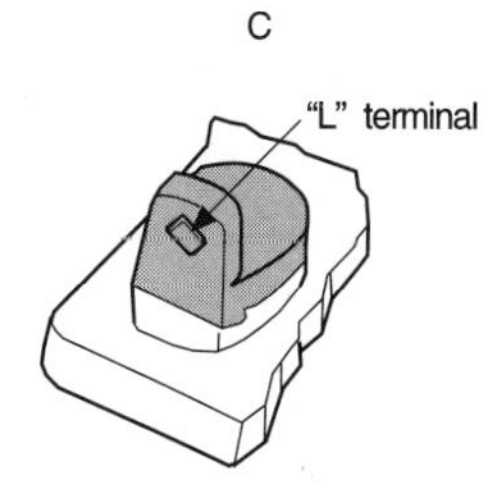

Battery terminals

TURN

Note: A mixture of baking soda and water may be used to clean the top of a battery. Battery acid may be neutralized with baking soda. Do not allow the baking soda mixture to enter the battery.

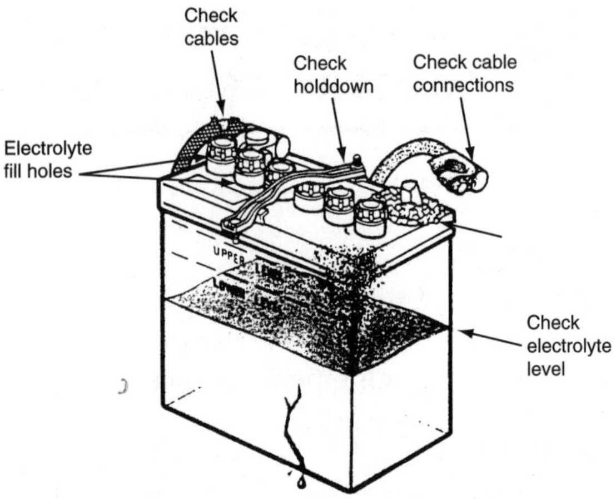

Check cables
Check holddown
Check cable connections
Electrolyte fill holes
Check electrolyte level

7. Check the condition of the battery holddown.

   OK _____ Needs service _____

8. Does the battery have a built-in hydrometer?

   Yes _____ No _____

   If it has a built-in hydrometer:

   a. What color is it? ____________________

   b. What does this color indicate? Charged _____ Low charge _____ Electrolyte low _____

9. Does the battery have removable cell caps? Yes _____ No _____

10. Check the level of the battery electrolyte by looking through the translucent case or by looking into the cells. The electrolyte should be at least 1/2″ above the separator plates.

    OK _____ Low _____

11. If the battery electrolyte level is low, fill the battery with water to just below the "split ring" full indicator or to the maximum level mark on the side of the battery.

    Note: If the electrolyte is low there might be a problem with the charging system.

12. Use a paper towel to clean up any water or electrolyte that has spilled. Battery acid will ruin shop towels and damage the vehicle's paint.

13. Before completing the paperwork, clean your work area, put the tools in their proper places, and wash your hands.

14. List any required service or repairs to the battery.

	Needs Service	
	Yes	No
Replace battery clamps	_____	_____
Replace battery cables	_____	_____
Replace battery holddown	_____	_____
Adjust AC generator belt	_____	_____
Check charging system	_____	_____
Other	_____	_____

15. Record any recommendations for needed service or unusual conditions that you noticed while checking the battery.

Recommendations: ____

____

____

____

STOP

Instructor OK ______________ Score __________

ASE Lab Preparation Worksheet #2-7

INSPECT OPERATION OF THE LIGHTING SYSTEM

Name____________________________ Class ____________________

OBJECTIVE:

Upon completion of this assignment, you should be able to inspect the operation of a vehicle's lighting systems. This task will help prepare you to pass the ASE certification examination in electrical systems.

DIRECTIONS:

Before beginning this lab assignment, review the worksheet completely. Fill in the information in the spaces provided as you complete each task.

PROCEDURE:

Vehicle year ___________ Make _________________ Model __________________

High beam
Low beam

Low beam
High beam

Lighting System			Okay	Problem	N/A
1. Headlights	Low beam	Left	_____	_____	_____
		Right	_____	_____	_____
	High beam	Left	_____	_____	_____
		Right	_____	_____	_____
	Aim		_____	_____	_____

Note: The following systems are usually checked with the *key on* and *engine off* (KOEO).

2. License plate lights		_____	_____	_____
3. Turn signals	Front	_____	_____	_____
	Rear	_____	_____	_____
4. Emergency flashers	Front	_____	_____	_____
	Rear	_____	_____	_____

TURN

		Okay	Problem	N/A
5. Back-up lights		_____	_____	_____
6. Brake lights	Right	_____	_____	_____
	Left	_____	_____	_____
	Center	_____	_____	_____
7. Running lights (side marker lights)	Left	_____	_____	_____
	Right	_____	_____	_____
8. Interior courtesy lights		_____	_____	_____
9. Interior dome light		_____	_____	_____
10. Dash indicator lights		_____	_____	_____
a. Turn signal dash indicators		_____	_____	_____
b. Oil pressure indicator		_____	_____	_____
c. Water temperature indicator		_____	_____	_____
d. Brake warning light		_____	_____	_____
(emergency brake applied to check bulb)		_____	_____	_____
e. Malfunction indicator light (MIL)		_____	_____	_____
f. Maintenance reminder light		_____	_____	_____
g. Air bag (SLR)		_____	_____	_____
h. Antilock brakes system (ABS)		_____	_____	_____
i. Other		_____	_____	_____
11. Other lighting systems (describe)		_____	_____	_____
a. ____________________		_____	_____	_____
b. ____________________		_____	_____	_____
c. ____________________		_____	_____	_____
d. ____________________		_____	_____	_____

12. Record any recommendations for service or repair that you noticed while checking the lighting system.

Recommendations: ____

____

____

____

Instructor OK ______________ Score __________

ASE Lab Preparation Worksheet #2-8

VISIBILITY CHECKLIST

Name____________________________ Class ____________________

OBJECTIVE:

Upon completion of this assignment, you should be able to inspect the windows, windshield wipers, and mirrors to ensure safe vehicle operation.

DIRECTIONS:

Before beginning this lab assignment, review the worksheet completely. Fill in the information in the spaces provided as you complete each task.

PROCEDURE:

Vehicle year ___________ Make __________________ Model ___________________

1. Inspect the condition of the following:
   a. Windshield glass: Fogged _____ Cracked _____ Chipped _____ Pitted _____ Good _____
   b. Windshield rubber molding: Good _____ Cracked _____ Evidence of leaks _____
   c. Rear window glass: Fogged _____ Cracked _____ Chipped _____ Pitted _____
   d. Rear window rubber molding: Good _____ Cracked _____ Evidence of leaks _____
   e. How are the side windows operated? Hand crank _____ Power _____
2. Check the operation and condition of the wiper blades.
   a. Do they operate properly? Yes _____ No _____
   b. Do they operate in the proper range (without going off the window or hitting the trim)?
      Yes _____ No _____
   c. Is the blade rubber soft or torn? Yes _____ No _____
   d. Is the tension spring good? Yes _____ No _____
3. Check the condition and operation of the windshield washer.
   a. Check the reservoir liquid level. OK _____ Low _____
   b. Are the washer nozzles aimed correctly? Yes _____ No _____
   c. Are the washer nozzles plugged? Yes _____ No _____
   d. Is there a sufficient volume of fluid? Yes _____ No _____
4. Check the condition of the side window glass.
   a. Right front window:
      Is the glass broken? Yes _____ No _____
      Will the window roll up and down? Yes _____ No _____
      Condition of the window molding: Good _____ Cracked _____ Evidence of leaks _____
   b. Left front window:
      Is the glass broken? Yes _____ No _____

TURN

Will the window roll up and down? Yes _____ No _____

Condition of the window molding: Good _____ Cracked _____ Evidence of leaks _____

c. Right rear window:

Is the glass broken? Yes _____ No _____ N/A _____

Will the window roll up and down? Yes _____ No _____

Condition of the window molding: Good _____ Cracked _____ Evidence of leaks _____

d. Left rear window:

Is the glass broken? Yes _____ No _____ N/A _____

Will the window roll up and down? Yes _____ No _____

Condition of the window molding: Good _____ Cracked _____ Evidence of leaks _____

5. Check the condition of the mirrors.

a. Interior mirror: Tight _____ Glass clear _____

b. Driver side exterior mirror: Loose _____ Missing _____ Cracked _____ Good _____

c. Passenger side mirror:

Loose _____ Missing _____ Cracked _____ N/A _____ Good _____

d. Are the outside mirrors power operated? Yes _____ No _____

If so, do they both move up and down, as well as right to left? Yes _____ No _____

6. Check the front window defroster.

a. Does the blower motor work?

At all speeds _____ Only middle speed _____ Not working _____

Only high speed _____ Only slow speed _____

b. Do the heater controls operate smoothly without binding? Yes _____ No _____

c. Does air blow from the ducts? Yes _____ No _____

d. Do the windows fog when the defroster is turned on? Yes _____ No _____

Note: This could be due to a leaking heater core.

7. Check the rear window defroster.

a. Does the vehicle have a rear window defroster? Yes _____ No _____

b. If it has a defroster, check the electrical strips in the window. Are any of them scratched, torn, or obviously damaged?

Yes _____ No _____ N/A _____

8. Record any recommendations for additional repair or service to the windows, wipers, washers, or defrosters.

Recommendations: ____

____

____

____

STOP

Instructor OK ______________ Score __________

ASE Lab Preparation Worksheet #2-9

REPLACE A WIPER BLADE

Name___________________________ Class ___________________

OBJECTIVE:

Upon completion of this assignment, you should be able to replace a vehicle's wiper blades.

DIRECTIONS:

Before beginning this lab assignment, review the worksheet completely. Fill in the information in the spaces provided as you complete each task.

TOOLS AND EQUIPMENT REQUIRED:

Safety glasses, fender covers

PARTS AND SUPPLIES:

Wiper blades

PROCEDURE:

Vehicle year ___________ Make _________________ Model __________________

Repair Order # ________ Engine size ____________ # of Cylinders ___________

1. Measure the length of the old wiper blade. 12″ ____ 13″ ____ 14″ ____ 15″ ____ Other ____
2. The wiper blade assembly is: Refillable ____ Non-refillable ____
3. If the wiper blade is of the refillable type, what type of locking mechanism does it use?

   Plastic button ____

   Metal end clip ____

   Notched ____

Push Button Refill

End Clip Refill

Notched Flexor Refill Coin Removal

4. Obtain the two new replacement wiper blade assemblies or refills.
5. Compare the new parts with the old wiper blades. Do they look like they are the right replacement parts?

   Yes _____No _____

   If there is any doubt consult your instructor.
6. Carefully read and follow the instructions that come with the replacement parts. Did you read the instructions before you started to replace the wiper blades?

   Yes _____No _____
7. Be especially careful not to scratch the windshield glass or paint. Remove the old blade assembly from the wiper arm.

   Note: Some shops place a piece of cardboard over the windshield to avoid damaging it during a wiper blade replacement.

8. Install the new wiper blade assemblies or refills. Be certain that they fit properly.

   Note: Failure to properly install the wiper blade can result in a scratched windshield, which is costly to replace.

9. Check the operation of the windshield wipers. Do the windshield wipers work?

   Yes ____ No ____

   If the wipers did not work properly, what is the problem?

   ____

10. Before completing the paperwork, clean your work area, put the tools in their proper places, and wash your hands.

11. Were there any other recommendations for needed service or unusual conditions that you noticed while replacing the wiper blades?

    Yes ____ No ____

12. Record your recommendations for additional service or repairs on the repair order.

13. Complete the repair order (R.O.).

STOP

Instructor OK ________________ Score ____________

ASE Lab Preparation Worksheet #2-10

ON-THE-GROUND SAFETY CHECKLIST

Name______________________________ Class ____________________

OBJECTIVE:

Upon completion of this assignment, you should be able to inspect a vehicle's safety features that are accessible without raising the vehicle.

DIRECTIONS:

Before beginning this lab assignment, review the worksheet completely. Fill in the information in the spaces provided as you complete each task.

TOOLS AND EQUIPMENT REQUIRED:

Safety glasses, fender covers, shop towel

PROCEDURE:

Vehicle year ____________ Make __________________ Model ____________________

Repair Order # ________ Engine size _____________# of Cylinders ____________

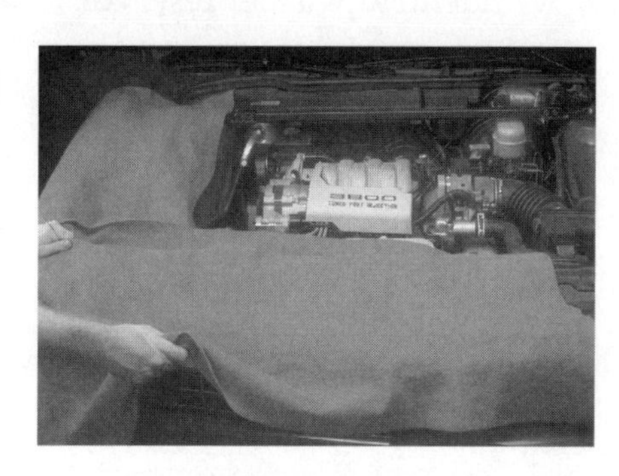

1. Open the hood and place fender covers on the fenders and over the front body parts.
2. Inspect the condition of the following:

	OK	Needs Attention
Lights:		
Brake	________	________
Tail	________	________
Headlights	________	________
Other	________	________
Wiper blade condition	________	________
Horn	________	________
Brake pedal travel	________	________
Emergency brake adjustment	________	________
Seat belts (not loose or damaged)	________	________
Windows (cracked or broken)	________	________
Mirrors (cracked, loose, or broken)	________	________
Door latches (lube)	________	________
Exhaust leaks (listen)	________	________
Tire pressure check (cold)	________	________

	OK	Needs Attention
Check tires for nails	______	______
Tire condition	______	______
Lug nut torque check (Spec. ______ ft.-lb)	______	______
V-belt tension and condition	______	______
Hood latch	______	______
Brake fluid level	______	______
Battery water check and fill	______	______
Electrical wiring (inspect)	______	______
Hose condition:		
Radiator hoses	______	______
Heater hoses	______	______
Power steering hoses	______	______
Fuel hoses	______	______
Windshield washer reservoir	______	______

Describe any other problems noticed during the vehicle inspection.

____

____

3. Before completing the paperwork, clean your work area, put the tools in their proper places, and wash your hands.
4. List any other recommendations for needed service or unusual conditions that you noticed while completing the safety inspection.

____

____

5. Record your recommendations for additional service or repairs on the repair order.
6. Complete the repair order (R.O.).

Instructor OK ______________ Score __________

ASE Lab Preparation Worksheet #2-11

CHECK AUTOMATIC TRANSMISSION FLUID (ATF) LEVEL

Name_____________________________ Class ____________________

OBJECTIVE:

Upon completion of this assignment, you should be able to check an automatic transmission for the correct fluid level. This task will help prepare you to pass the ASE certification examination in automatic transmissions.

DIRECTIONS:

Before beginning this lab assignment, review the worksheet completely. Fill in the information in the spaces provided as you complete each task.

TOOLS AND EQUIPMENT REQUIRED:

Safety glasses, fender covers

PROCEDURE:

Vehicle year ___________ Make __________________ Model ___________________

1. What type of ATF is specified for this vehicle?

   Dexron ____ Dexron II ____ Type F ____ CJ ____ Mercon ____ Type T ____ Other ____

2. To obtain an accurate fluid level reading, the fluid must be at normal operating temperature.

   Normal operating temperature? Yes ____ No ____

   Note: Cold fluid will be approximately one pint lower than warm fluid.

Cool (65°–85°F) (18°–30°C)

Hot (190°–200°F) (88°–93°C)

Add .5 liter (1 pt.)

Full hot

Warm

Note: Do not overfill. It takes only one pint to raise the level from "add" to "full" with a hot transmission.

3. The vehicle must be on level ground.
4. Locate and remove the transmission dipstick.
5. Wipe the dipstick with a clean shop towel.
6. Does the fluid wipe off easily?

   Yes ____No ____

   Note: If the fluid has become excessively hot, it can become sticky. The transmission will require rebuilding.

7. Are fluid checking instructions stamped on the dipstick?

   Yes ____No ____

8. What is the specified gearshift selector position during the fluid level check?

   P____ R____ N____ D____ L____

9. Be certain the parking brake is set.
10. Start the engine. Insert the dipstick until it seats firmly on the dipstick tube.

TURN

11. Remove the dipstick and note the reading.

    Full ____ Low ____

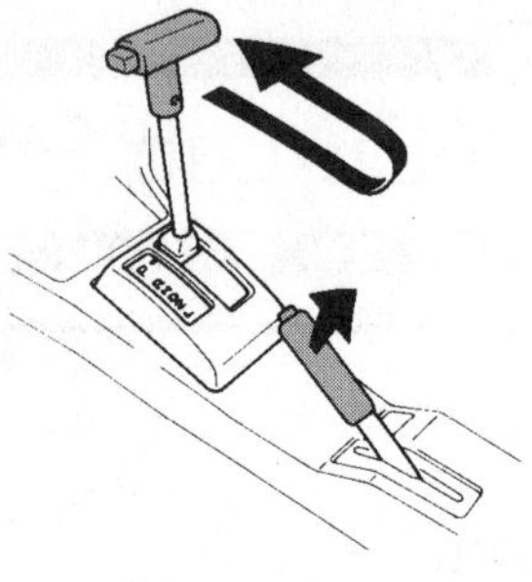

12. What color is the fluid?

    Red ____ Brown ____ Pink ____

13. Does the fluid smell burnt?

    Yes ____No ____

14. Does any fluid need to be added to the transmission?

    Yes ____No ____

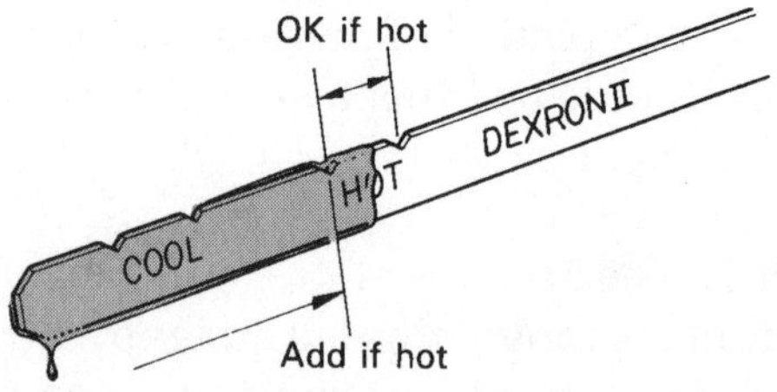

15. Was any ATF added to the transmission?

    Yes ____No ____

    What type?

    Dexron ____ Dexron II ____ Type F ____ CJ ____ Mercon ____ Type T ____ Other ____

16. Before completing the paperwork, clean your work area, put the tools in their proper places, and wash your hands.

17. Were there any other recommendations for needed service or unusual conditions that you noticed while checking the transmission fluid level?

    Yes ____No ____

18. Record your recommendations for additional service or repairs on the repair order.

19. Complete the repair order (R.O.).

STOP

Instructor OK ______________ Score __________

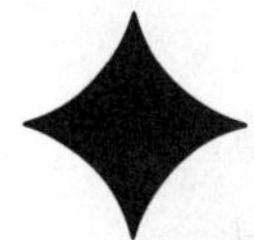

ASE Lab Preparation Worksheet #2-12

INSPECT SHOCK ABSORBERS

Name______________________________ Class __________________

OBJECTIVE:

Upon completion of this assignment, you should be able to inspect a vehicle's shock absorbers. This task will help prepare you to pass the ASE certification examination in suspension and steering.

DIRECTIONS:

Before beginning this lab assignment, review the worksheet completely. Fill in the information in the spaces provided as you complete each task.

TOOLS AND EQUIPMENT REQUIRED:

Safety glasses, shop towel

PROCEDURE:

Vehicle year __________ Make ________________ Model _________________

1. Perform a bounce test.

   a. Push down hard two or three times on the bumper at each corner of the car. If the shocks are operating properly, the spring should oscillate about 1.5 cycles then stop. Resistance should be equal from side to side.

Bounce vehicle body

Bounce vehicle body once or twice.
Bouncing should be damped quickly.
Vehicle body movement
Stop bouncing

Note: Do not compare front operation to rear operation.

Results: Shock resists spring oscillation __________

Right to left resistance equal __________

   b. Is there unusual noise during the bounce test? Yes ____ No ____

Note: The sound of fluid being forced through the valves in the shock is normal.

2. Inspect tires for "cupping" wear.

   Cupping wear: Yes ____ No ____

Note: Cupping indicates that the tire has been hopping. Tire imbalance could also be a contributing factor.

3. Perform a visual inspection.

   a. Condition of the shock mounts and rubber cushions.

      Good ____ Bad ____

   b. Has any fluid leaked out of a shock?

      Dry ____ Leaking ____

Note: It is normal for a slight amount of moisture to be on the seal.

4. Is the outside of the shock body damaged? Yes ____ No ____

Note: Shock absorbers are always replaced in pairs (front and/or rear).

5. Before completing the paperwork, clean your work area, put the tools in their proper places, and wash your hands.
6. Record any other recommendations for needed service or unusual conditions that you noticed while testing the shock absorbers.

Recommendations: ____

____

____

____

Part II

ASE Lab Preparation Worksheets

Service Area 3

Under Vehicle Services

ASE Lab Preparation Worksheet #3-1

TIRE INSPECTION

Name___________________________ Class ____________________

OBJECTIVE:

Upon completion of this assignment, you should be able to inspect the condition of the tires. This task will help prepare you to pass the ASE certification examination in suspension and steering.

DIRECTIONS:

Before beginning this lab assignment, review the worksheet completely. Fill in the information in the spaces provided as you complete each task.

TOOLS AND EQUIPMENT REQUIRED:

Safety glasses, jack stands or vehicle lift, shop towel

PROCEDURE:

Vehicle year ___________ Make _________________ Model __________________

Tire Size ____________ Tire Manufacturer ______________________________

1. Raising the vehicle onto jack stands or a lift allows the tires to be more easily inspected.
2. Visually inspect the tires. Check any of the following that apply.

   Nails ____ Glass ____

   Screws ____ Cracks ____

   Tears ____ Bulges ____ Other ____
3. Is there any sign that the tread is separating from the tire casing?

   Yes ____ No ____
4. Tire sidewall condition:

   Good condition? ____

   Sun cracks? ____

   Whitewall scuffed? ____

   Bubble or separation? ____

   Cut? ____
5. Spin the wheels. Do the tires appear to spin "true"?

   Left front: Yes ____ No ____ Right front: Yes ____ No ____

   Left rear: Yes ____ No ____ Right rear: Yes ____ No ____

6. Does the tire show signs of underinflation or overinflation wear?

   Yes ____ No ____

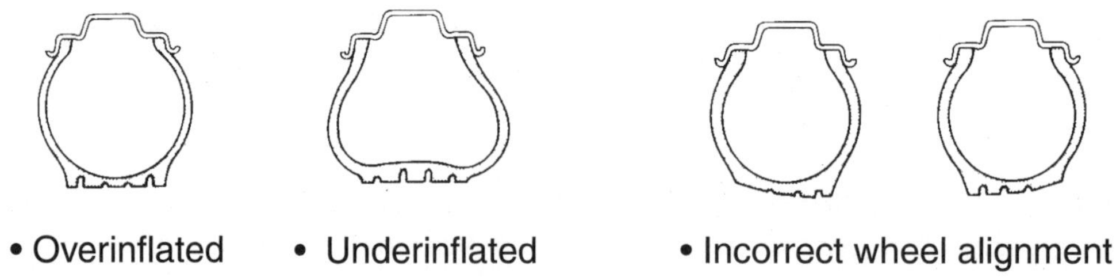

• Overinflated • Underinflated • Incorrect wheel alignment

7. Measure the depth of the tread.

   Front: Left ____ /32″

   Right ____ /32″

   Rear: Left ____ /32″

   Right ____ /32″

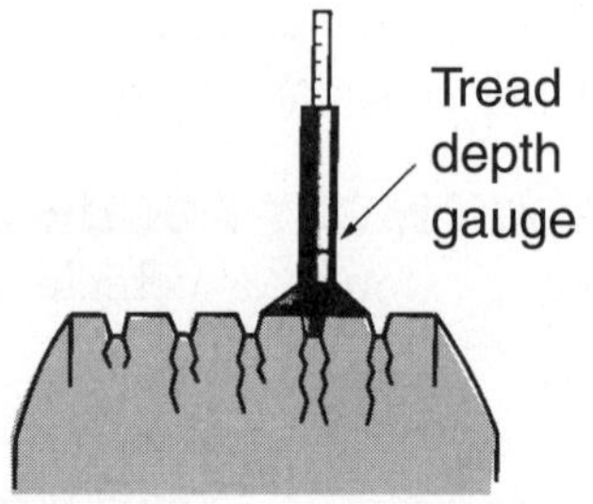

Tread depth gauge

8. Are the tread wear bars even with the surface of the tread (less than 1/16″ of tread remaining)?

   Yes ____ No ____

9. Are there signs of unusual tread wear? Yes ____ No ____

10. Are any of the tires worn more than the others?

    Yes ____ No ____

11. Should a tire rotation be recommended?

    Yes ____ No ____

12. Are all of the lug nuts on each wheel?

    Yes ____ No ____

13. Are any of the wheel covers missing?

    Yes ____ No ____

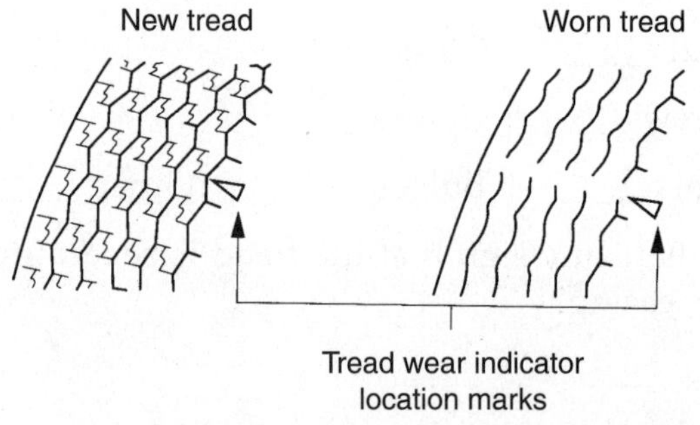

New tread

Worn tread

Tread wear indicator location marks

14. Before completing the paperwork, clean your work area, put the tools in their proper places, and wash your hands.

15. Record any other recommendations for needed service or unusual conditions that you noticed while inspecting the tires.

Recommendations: ____

____

____

____

STOP

ASE Lab Preparation Worksheet #3-2

ADJUST TIRE PRESSURES

Name____________________________ Class ________________

OBJECTIVE:

Upon completion of this assignment, you should be able to check tire pressures and correct them to the vehicle manufacturer's specification. This task will help prepare you to pass the ASE certification examination in suspension and steering.

DIRECTIONS:

Before beginning this lab assignment, review the worksheet completely. Fill in the information in the spaces provided as you complete each task.

TOOLS AND EQUIPMENT REQUIRED:

Safety glasses, tire pressure gauge, air chuck, shop towel

PROCEDURE:

Vehicle year ___________ Make _________________ Model __________________

1. Locate the tire pressure specifications.

   Front ____ psi Rear ____ psi

2. Tire pressures should be checked when the tires are:

   Hot ____ Cold ____

3. Where was the tire pressure specification located?

   Driver's door or door post ____ Owner's manual ____

   Glove compartment label ____ Service manual ____

   If a service manual was used to find the specifications, which one was used and what page number was the specification on?

   Service manual ___________________________ Page ______ N/A _________

4. Remove the valve stem caps.

5. Check the condition of the valve stems by bending each of them back and forth.

   Do any of the valve stems need to be replaced?

   Yes ____ No ____

   Which, if any, of the valve stems need to be replaced?

   Right front ____ Right rear ____

   Left front ____ Left rear ____

6. Use a tire pressure gauge to check tire pressures.

7. Record tire pressures below.

	LF	RF	LR	RR
Before adjusting air pressure	____	____	____	____
After adjusting air pressure	____	____	____	____

8. Reinstall valve stem caps.
9. Check the general condition of the tires as you check the air pressure.

	LF	RF	LR	RR
Good	____	____	____	____
Fair	____	____	____	____
Unsafe	____	____	____	____

Describe the problem with any unsafe tires below.

____

____

____

Scale extension
Read pressure here
Air check

10. Check the pressure and the condition of the *spare tire.*

Type of spare tire: Standard tire ____ Compact spare ____ No spare ____

Does it have the proper air pressure? Yes ____ No ____

Spare tire condition: Good ____ Fair ____ Needs to be replaced ____

11. Before completing the paperwork, clean your work area, put the tools in their proper places, and wash your hands.
12. Record any other recommendations for needed service or unusual conditions that you noticed while checking the tire pressure.

Recommendations: ____

____

____

____

STOP

Instructor OK ______________ Score __________

ASE Lab Preparation Worksheet #3-3

TIRE WEAR INSPECTION

Name______________________________ Class __________________

OBJECTIVE:

Upon completion of this assignment, you should be able to inspect the condition of the tires for serviceability and abnormal wear. This task will help prepare you to pass the ASE certification examination in suspension and steering.

DIRECTIONS:

Before beginning this lab assignment, review the worksheet completely. Fill in the information in the spaces provided as you complete each task.

TOOLS AND EQUIPMENT REQUIRED:

Safety glasses, tire depth gauge

PROCEDURE:

Vehicle year __________ Make ________________ Model _________________

Inspect the vehicle's tires and record the results below. Indicate a problem tire using the following abbreviations. Right front (RF), Left front (LF), Right rear (RR), Left rear (LR), and Spare (S).

1. Do all of the tires on the vehicle have the same tread design? Yes ____ No ____

   If yes, which ones?

   RF ____ LF ____ RR ____ LR ____ S ____

2. Are any of the tires worn to the wear bars?

   Yes ____ No ____

   If yes, which ones?

   RF ____ LF ____ RR ____ LR ____ S ____

3. Measure the tread depth for each of the tires.

   RF ____ LF ____ RR ____ LR ____ S ____

4. Inspect each of the tires for unusual wear and indicate problem tire(s).

   Underinflation wear RF ____ LF ____ RR ____ LR ____ S ____

   Overinflation wear RF ____ LF ____ RR ____ LR ____ S ____

   Camber wear (one side of the tread)?

   RF ____ LF ____ RR ____ LR ____ S ____

   Toe wear (feathered edge across the tread)?

   RF ____ LF ____ RR ____ LR ____ S ____

   Cupped wear?

   RF ____ LF ____ RR ____ LR ____ S ____

Cuts in tire sidewall?

RF ____ LF ____ RR ____ LR ____ S ____

5. List any tires that should be replaced.

RF ____ LF ____ RR ____ LR ____ S ____

6. Abnormal tire wear can be due to other problems with the vehicle, driver habits, or road conditions. Which of the following could cause abnormal tire wear? Check all that apply.

   a. High-speed driving ______

   b. Bad shocks ______

   c. Bad transmission ______

   d. Mountain driving ______

   e. Loud stereo ______

7. Before completing the paperwork, clean your work area, put the tools in their proper places and wash your hands.

8. Record any other recommendations for needed service or unusual conditions that you noticed while inspecting the tires for wear.

Recommendations: ____

____

____

____

STOP

Instructor OK ______________ Score __________

ASE Lab Preparation Worksheet #3-4

EXHAUST SYSTEM INSPECTION

Name_________________________ Class ________________

OBJECTIVE:

Upon completion of this assignment, you should be able to inspect a vehicle's exhaust system for leaks and damage. This task will help prepare you to pass the ASE certification examination in engine performance.

DIRECTIONS:

Before beginning this lab assignment, review the worksheet completely. Fill in the information in the spaces provided as you complete each task.

TOOLS AND EQUIPMENT REQUIRED:

Safety glasses, jack stands or vehicle lift, shop towel

PROCEDURE:

Vehicle year __________ Make ________________ Model _________________

1. Raise the vehicle to gain easy access to the exhaust system.
2. Identify the components of the exhaust system on the vehicle that you are inspecting.

   Check all that apply.

   Single exhaust ____

   Dual exhaust ____

   Catalytic converter ____

   Muffler ____

   Resonator ____

3. How is the vehicle you are inspecting equipped?

   Check all that apply.

   Stock exhaust system? Yes ____ No ____

   Modified exhaust system? Yes ____ No ____

   Tailpipe after rear axle? Yes ____ No ____

   Pipe exiting in front of rear wheels?

   Yes ____ No ____

4. Visually inspect the components of the exhaust system.

   a. Condition of muffler hangers/supports.

   OK ____ Bad ____

   b. Holes in the muffler? Yes ____ No ____

   c. Holes in any of the pipes? Yes ____ No ____

d. Do the clamps appear to be tight? Yes ____ No ____

e. Catalytic converter damaged? Yes ____ No ____

f. Exhaust system shields in place? Yes ____ No ____

5. Start the engine and check for leaks in each of these components:

	OK	Leaks	Parts needed
Exhaust manifold	____	____	____
Exhaust manifold gasket	____	____	____
Header pipe	____	____	____
Catalytic converter	____	____	____
Muffler	____	____	____
Exhaust pipe	____	____	____
Resonator	____	____	____
Tailpipe	____	____	____

Other ____

6. Check the condition of the exhaust system clamps and hangers. OK ____ Parts needed ____

7. Before completing the paperwork, clean your work area, put the tools in their proper places, and wash your hands.

8. Record any other recommendations for needed service or unusual conditions that you noticed while checking the exhaust system.

Recommendations: ____

____

____

____

Instructor OK ______________ Score __________

ASE Lab Preparation Worksheet #3-5

INSPECT SUSPENSION AND STEERING LINKAGE

Name______________________________ Class __________________

OBJECTIVE:

Upon completion of this assignment, you should be able to inspect front suspension and steering linkage. This task will help prepare you to pass the ASE certification examination in suspension and steering.

DIRECTIONS:

Before beginning this lab assignment, review the worksheet completely. Fill in the information in the spaces provided as you complete each task.

TOOLS AND EQUIPMENT REQUIRED:

Safety glasses, jack stands or vehicle lift, shop towel

PROCEDURE:

Vehicle year __________ Make ________________ Model _________________

Visual Inspection

1. Check the steering system for looseness (dry park check).

   With the wheels on the ground, have an assistant turn the steering wheel back and forth a short distance while you look for looseness in steering linkage. A power steering car must have the engine running for this test.

   Loose ____ Tight ____

2. Raise the vehicle.

3. Inspect suspension bushings visually and with a prybar. Record results below.

	Good	Need Repair	N/A
Upper control arm bushings	____	____	____
Lower control arm bushings	____	____	____
Sway bar bushings	____	____	____
Strut rod bushing	____	____	____

4. Inspect steering linkage pivot connections. Firmly grasp the part and rock it to check for looseness.

	Good	Need Repair	N/A
Tie-rod ends	____	____	____
Idler arm	____	____	____
Pitman arm	____	____	____
Center link	____	____	____
Rack bushings	____	____	____

TURN

5. Check wheel bearing adjustment. Refer to Worksheet 9-9, Adjust a Tapered Roller Wheel Bearing.

   OK ____ Loose ____

6. Check ball joints for looseness. Refer to Worksheet 9-16, Check Ball Joint Wear.

   OK ____ Loose ____

7. Inspect rubber grease boots on tie-rods and ball joints.

	Good	Damaged
Tie-rod end boots	____	____
Ball joint boots	____	____

8. Inspect the shock absorbers. Refer to Worksheet 2-12, Inspect Shock Absorbers.

   OK ____ Dented ____ Leaking ____

9. Inspect the steering gear.

   OK ____ Leaking ____

   What type of steering gear assembly is used on this vehicle?

   Rack and pinion ____

   Recirculating ball and nut ____

   Other ____________________

10. Lower the vehicle, open the hood, and place fender covers on the fenders and front body parts.

11. Inspect the power steering system.

	Good	Needs Attention	N/A
Fluid level	____	____	____
Leakage	____	____	____
Drive belt condition	____	____	____
Drive belt tension	____	____	____

12. Before completing the paperwork, clean your work area, put the tools in their proper places, and wash your hands.

13. Record any other recommendations for needed service or unusual conditions that you noticed while inspecting the suspension and steering.

Recommendations: ____

____

____

____

STOP

ASE Lab Preparation Worksheet #3-6

CHASSIS LUBRICATION

Name______________________________ Class ______________

OBJECTIVE:

Upon completion of this assignment, you should be able to lubricate steering and suspension components. This task will help prepare you to pass the ASE certification examination in suspension and steering.

DIRECTIONS:

Before beginning this lab assignment, review the worksheet completely. Fill in the information in the spaces provided as you complete each task.

TOOLS AND EQUIPMENT REQUIRED:

Safety glasses, fender covers, jack stands or vehicle lift, grease gun, shop towel, service manual

PARTS AND SUPPLIES:

Grease, hinge lubricant, door latch lubricant

PROCEDURE:

Vehicle year __________ Make ________________ Model __________________

Repair Order # ________ Engine ________________ # of Cylinders ___________

1. Refer to a service manual for the number and location of the lubrication points.

   Number of fittings ____ N/A ____

   Name of service manual used __________________________________ Page # ______

2. Open the hood and place fender covers on the fenders and front body parts.
3. Raise the vehicle on a hoist or with a jack. Be sure to support it with jack stands.
4. Locate all of the fittings to be greased and wipe them clean with a shop towel.

   Number of fittings located ______

   Cleaned? Yes ____ No ____

   The vehicle is equipped with:

   Plugs ____ Zerk fittings ____

   Note: Some vehicles have plugs installed in place of the grease fittings. These will need to be removed and a fitting temporarily installed. Future grease jobs are easier if the plugs are replaced with standard zerk grease fittings.

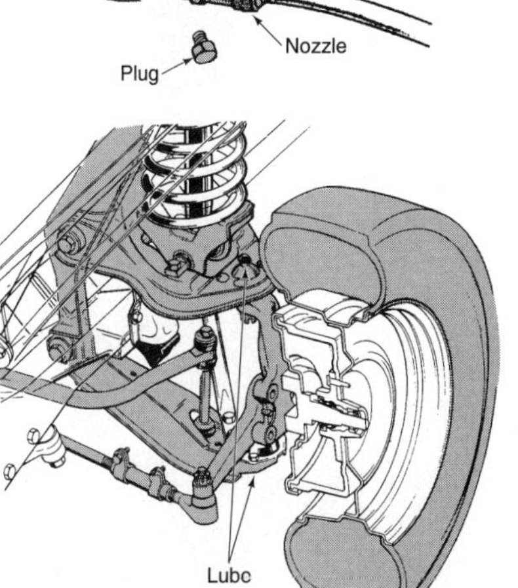

Note: Pumping too much grease into the fitting can damage the seals. There are two types of joint seals.

❏ *Sealed boot.* Apply only enough grease to slightly bulge the boot.

❏ *Non-sealed.* Apply grease until the old grease has been flushed from the joint.

5. Push the grease gun straight onto the fitting. Hold it firmly in place and pump the grease into the joint slowly to prevent damage to the grease seals.

   Type of grease gun used:

   Hand pump ____ Air operated ____

6. Wipe excess grease from the fittings.

7. Lower the vehicle. Lubricate the moving parts of the hinges with a small amount of oil or lubricant.

   Door hinges _________

   Trunk hinges _________

   Hood hinges _________

8. Lubricate the hood and door latches.

   Note: The door latches are located where people rub against them when entering and exiting the vehicle. To prevent damage to clothing, use a nonstaining lubricant that does not attract dirt.

   Door latches ____ Trunk latch ____ Hood latch ____

9. Before completing the paperwork, clean your work area, put the tools in their proper places, and wash your hands.

10. Record any other recommendations for needed service or unusual conditions that you noticed while lubricating the chassis.

11. Complete the repair order (R.O.).

Instructor OK ______________ Score __________

ASE Lab Preparation Worksheet #3-7

CHECK THE FLUID LEVEL IN A MANUAL TRANSMISSION

Name______________________________ Class __________________

OBJECTIVE:

Upon completion of this assignment, you should be able to check the fluid level in a manual transmission. This task will help prepare you to pass the ASE certification examination in manual drivetrain and axles.

DIRECTIONS:

Before beginning this lab assignment, review the worksheet completely. Fill in the information in the spaces provided as you complete each task.

PROCEDURE:

Vehicle year __________ Make ________________ Model ________________

Repair Order # ________ Engine size ____________ # of Cylinders ___________

1. The vehicle is equipped with a:

   RWD manual transmission ____ FWD manual transaxle ____

2. What type of fluid is required? ____________________
3. Raise the vehicle on a lift.
4. Locate the fill plug on the side of the transmission or transaxle.

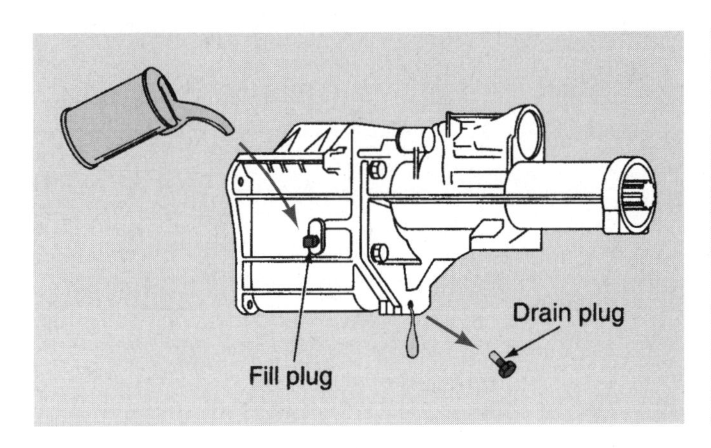

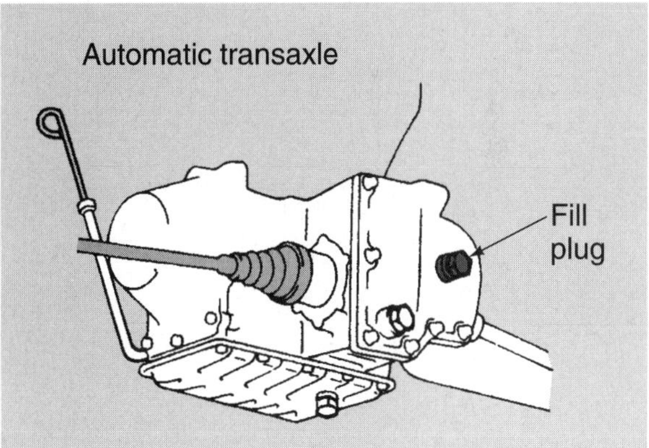

5. Remove the plug and check the fluid level. The fluid should be level with the bottom of the fill plug hole.

   OK ____ Low ____ N/A ____

6. Add the correct fluid as required.
7. Install the fill plug.

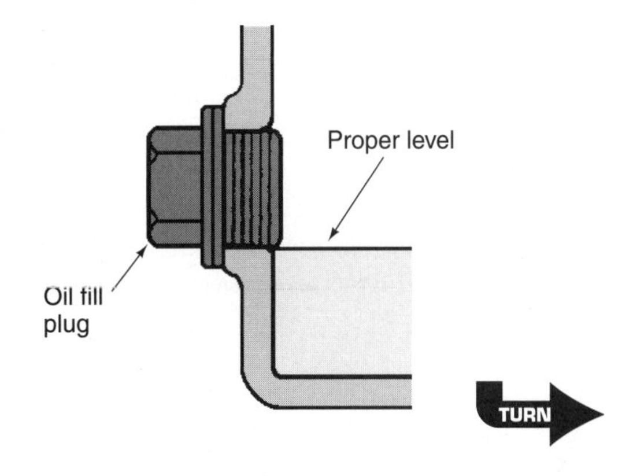

TURN

Do not overtighten the fill plug. The thread is usually tapered (NPT). An overtightened fill plug can cause the transmission case to crack.

8. Lower the vehicle.
9. Before completing the paperwork, clean your work area, put the tools in their proper places, and wash your hands.
10. List any recommendations for unusual conditions or needed service.

    ____

    ____

    ____

11. Record your recommendations for needed service or additional repairs on the repair order.
12. Complete the repair order (R.O.).

Instructor OK ______________ Score __________

ASE Lab Preparation Worksheet #3-8

CHECK THE FLUID LEVEL IN A DIFFERENTIAL

Name______________________________ Class ____________________

OBJECTIVE:

Upon completion of this assignment, you should be able to check the fluid level in a differential. This task will help prepare you to pass the ASE certification examination in manual drivetrain and axles.

DIRECTIONS:

Before beginning this lab assignment, review the worksheet completely. Fill in the information in the spaces provided as you complete each task.

PROCEDURE:

Vehicle year ____________ Make __________________ Model ____________________

Repair Order # ________ Engine size _____________ # of Cylinders ____________

1. What type of fluid is required? ______________________
2. Raise the vehicle on a lift.
3. Locate the fill plug on the differential.
4. Remove the plug and check the fluid level. The fluid should be level with the bottom of the fill plug hole.

   OK ____ Low ____ N/A ____
5. Add the correct fluid as required.
6. Install the fill plug.

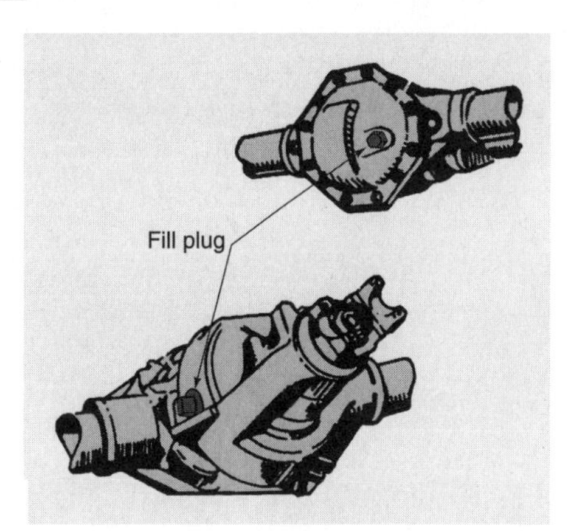

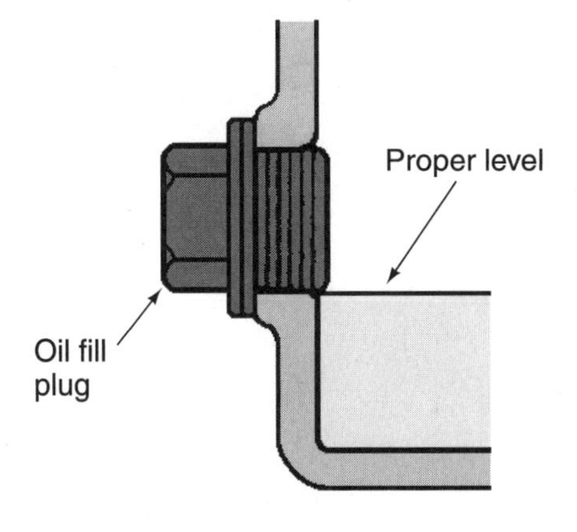

Do not overtighten the fill plug. The thread is usually tapered (NPT). An overtightened fill plug can damage the differential.

7. Lower the vehicle.
8. Before completing the paperwork, clean your work area, put the tools in their proper places, and wash your hands.
9. List any unusual conditions or recommendations for needed service.

    ____

    ____

    ____

10. Record your recommendations for needed service or additional repairs on the repair order.
11. Complete the repair order (R.O.).

Instructor OK ______________ Score __________

ASE Lab Preparation Worksheet #3-9

COMPLETE MAINTENANCE AND INSPECTION SERVICE

Name______________________________ Class ____________________

OBJECTIVE:

Upon completion of this assignment, you should be able to do a complete lubrication service and comprehensive safety inspection on a vehicle.

DIRECTIONS:

Before beginning this lab assignment, review the worksheet completely. Fill in the information in the spaces provided as you complete each task.

TOOLS AND EQUIPMENT REQUIRED:

Safety glasses, fender covers, jack and jack stands or vehicle lift, drain pan, filter wrench, shop towels

PARTS AND SUPPLIES:

Oil, oil filter

RELATED INFORMATION:

Lubrication/safety service is very important, both to the customer and the service technician. Potential component failures and safety problems can be identified during the inspection. Getting the problems repaired before there is a breakdown can save the customer from the inconvenience of being without his or her vehicle, not to mention the unexpected expense. A considerable amount of service and repair work can be identified from a properly performed inspection.

During a vehicle inspection, items in need of repairs are located and documented. A lubrication/safety service includes underhood and underbody inspections. The underhood and body inspection is done while the car is on the ground. After completing inspection and maintenance under the hood, the vehicle is raised in the air to perform an undercar inspection. The oil and filter are usually changed while the vehicle is in the air. Position the vehicle correctly on the lift using the specified lift contact points.

When performing undercar service, practice developing an efficient routine. After raising the car on the lift, start at the front on the passenger side and work around the car, finishing up at the front again. Some technicians prefer to start with undercar service, including draining the oil and changing the filter. The technician would then complete underhood services while refilling the crankcase.

PROCEDURE:

Vehicle year ___________ Make _________________ Model __________________

1. Fill in the Car Care Service form at the end of this assignment as you complete the lubrication/safety service.
2. Practice developing an efficient routine.

   Where did you start the job?

   Underhood ____ Undercar ____
3. Before working under the hood, place fender covers on the fenders and front body parts.
4. Before starting the undercar service, raise the vehicle on a lift or place it on jack stands.

TURN

5. After finishing the service but before completing the paperwork, clean your work area, put the tools in their proper places, and wash your hands.
6. Record any recommendations for needed service or unusual conditions that you noticed while doing the lubrication/safety service.

Recommendations: ____

____

____

____

____

____

____

____

____

____

____

STOP

CERTIFIED CAR CARE SERVICE

Customer Name ________________

Address ________________ City ________ Zip Code ________ Phone ________

Date ________________ Time ________

Vehicle ________ Year ________ Model ________ License Number ________ Odometer Reading ________

ELECTRICAL SYSTEM CHECKS

________Wiring Visual Inspection
Battery
________Top Off Water Level
Posts and Cables
________Clean ________Corroded
________Damaged
Battery Condition
________Good ________Replace
________Recharge

LIGHTS

________Park ________Brake
________Signal ________Emergency
________Dash Lights Back-up
Headlight Operation
________High Left
________Low Left
________High Right
________Low Left
________Horn Operation

FUEL SYSTEM CHECKS

________Condition of Hoses
________Gas Cap Condition
________Air Cleaner
________Crankcase Vent Filter
________Fuel Filter (miles until change suggested) ______

COOLING SYSTEM CHECKS

________Level
________Strength of Coolant
(Protection to________________°)
________No Leaks

Condition of Hoses
________Radiator
________Heater
________Thermostat Bypass
Hose (if so equipped)

Pressure Test
________Radiator
________Cap
________Condition of Coolant
Pump Belt

BRAKE INSPECTION

________Pedal Travel
________Emergency Brake
________Brake Hoses and Lines
________Master Brake Cylinder-
Fluid Level and Condition

ON-GROUND STEERING, SUSPENSION, DRIVE LINE CHECKS

________ Steering Wheel Freeplay
________ Power Steering Fluid Level
________ Shock Absorber Bounce Test

	Good	Unsafe
Front	______	______
Rear	______	______

________No Squeaks
________Ride Height Check
________Check ATF Level

VISIBILITY

________Mirrors
________Wiper Blades
Wiper Operation
______fast ______slow
________Washer Fluid and Pump
________Clean and Inspect all Glass

UNDERCAR SERVICE

________Drain Crankcase (if ordered)
________Remove and replace oil filter
________Inspect Under Car for
Oil, Gasoline, and Coolant Leaks
________Check Crankcase Oil Level
________Check Oil Filter for Leaks

INFLATE AND CHECK TIRES

Inflate to __________ lb.

Tire Condition:

	Good	Fair	Unsafe
RF	________	________	________
LF	________	________	________
RR	________	________	________
LR	________	________	________

Inflate and Check Spare
Good ______ Fair ______ Unsafe ______

SUSPENSION AND STEERING

________Inspect Steering Linkage
________Inspect Shock Absorbers
________Inspect Suspension
Bushings
________Clean Lubrication Fittings
________Lubricate Fittings
________Ball Joints
________Inspect Ball Joint Seals
________Ball Joint Wear
Inspection
________Inspect Ride Height

UNDERCAR FUEL SYSTEM CHECKS

________Condition of Fuel Hoses
________Condition of Fuel Tank

DRIVE LINE CHECKS

________Check Universal
or CV Joints
Inspect Gear Cases
________Transmission
________Differential
Replace Drain Plugs
________Inspect Motor Mounts
________Lubricate Door and Hood Hinge
and Latches

EXHAUST SYSTEM CHECKS

________Mufflers and Pipes
________Pipe Hangers
________Exhaust Leaks
________Heat Riser

FINAL VEHICLE PREPARATION

________Replace Crankcase Oil
________Clean Windows, Vacuum Interior
________Fill Out and Affix Door Jamb
Record to Door Post
________Complete a Repair Order

NOTES (specs, procedures, additional service or repair information):

ADDITIONAL RECOMMENDATIONS FOR SERVICE OR REPAIRS:

CERTIFIED CAR CARE SERVICE

Customer Name ____________________

Address ____________________ City __________ Zip Code __________ Phone __________

Date ____________________ Time __________

Vehicle __________ Year __________ Model __________ License Number __________ Odometer Reading __________

ELECTRICAL SYSTEM CHECKS

_____ Wiring Visual Inspection

Battery

_____ Top Off Water Level

Posts and Cables

_____ Clean _____ Corroded

_____ Damaged

Battery Condition

_____ Good _____ Replace

_____ Recharge

LIGHTS

_____ Park _____ Brake

_____ Signal _____ Emergency

_____ Dash Lights Back-up

Headlight Operation

_____ High Left

_____ Low Left

_____ High Right

_____ Low Left

_____ Horn Operation

FUEL SYSTEM CHECKS

_____ Condition of Hoses

_____ Gas Cap Condition

_____ Air Cleaner

_____ Crankcase Vent Filter

_____ Fuel Filter (miles until change suggested) _____

COOLING SYSTEM CHECKS

_____ Level

_____ Strength of Coolant

(Protection to __________ °)

_____ No Leaks

Condition of Hoses

_____ Radiator

_____ Heater

_____ Thermostat Bypass

Hose (if so equipped)

Pressure Test

_____ Radiator

_____ Cap

_____ Condition of Coolant

Pump Belt

BRAKE INSPECTION

_____ Pedal Travel

_____ Emergency Brake

_____ Brake Hoses and Lines

_____ Master Brake Cylinder-

Fluid Level and Condition

ON-GROUND STEERING, SUSPENSION, DRIVE LINE CHECKS

_____ Steering Wheel Freeplay

_____ Power Steering Fluid Level

_____ Shock Absorber Bounce Test

	Good	Unsafe
Front	_____	_____
Rear	_____	_____

_____ No Squeaks

_____ Ride Height Check

_____ Check ATF Level

VISIBILITY

_____ Mirrors

_____ Wiper Blades

Wiper Operation

_____ fast _____ slow

_____ Washer Fluid and Pump

_____ Clean and Inspect all Glass

UNDERCAR SERVICE

_____ Drain Crankcase (if ordered)

_____ Remove and replace oil filter

_____ Inspect Under Car for

Oil, Gasoline, and Coolant Leaks

_____ Check Crankcase Oil Level

_____ Check Oil Filter for Leaks

INFLATE AND CHECK TIRES

Inflate to _____ lb.

Tire Condition:

	Good	Fair	Unsafe
RF	_____	_____	_____
LF	_____	_____	_____
RR	_____	_____	_____
LR	_____	_____	_____

Inflate and Check Spare

Good _____ Fair _____ Unsafe _____

SUSPENSION AND STEERING

_____ Inspect Steering Linkage

_____ Inspect Shock Absorbers

_____ Inspect Suspension

Bushings

_____ Clean Lubrication Fittings

_____ Lubricate Fittings

_____ Ball Joints

_____ Inspect Ball Joint Seals

_____ Ball Joint Wear

Inspection

_____ Inspect Ride Height

UNDERCAR FUEL SYSTEM CHECKS

_____ Condition of Fuel Hoses

_____ Condition of Fuel Tank

DRIVE LINE CHECKS

_____ Check Universal

or CV Joints

Inspect Gear Cases

_____ Transmission

_____ Differential

Replace Drain Plugs

_____ Inspect Motor Mounts

_____ Lubricate Door and Hood Hinge

and Latches

EXHAUST SYSTEM CHECKS

_____ Mufflers and Pipes

_____ Pipe Hangers

_____ Exhaust Leaks

_____ Heat Riser

FINAL VEHICLE PREPARATION

_____ Replace Crankcase Oil

_____ Clean Windows, Vacuum Interior

_____ Fill Out and Affix Door Jamb

Record to Door Post

_____ Complete a Repair Order

NOTES (specs, procedures, additional service or repair information):

ADDITIONAL RECOMMENDATIONS FOR SERVICE OR REPAIRS:

STOP

Part II

ASE Lab Preparation Worksheets

Service Area 4

Tire and Wheel Service

ASE Lab Preparation Worksheet #4-1

TIRE IDENTIFICATION

Name______________________________ Class ________________

OBJECTIVE:

Upon completion of this assignment, you should be able to read and understand the tire sidewall markings. This task will help prepare you to pass the ASE certification examination in suspension and steering.

DIRECTIONS:

Before beginning this lab assignment, review the worksheet completely. Fill in the information in the spaces provided as you complete each task.

TOOLS AND EQUIPMENT REQUIRED:

Safety glasses

PROCEDURE:

Vehicle year __________ Make ________________ Model _________________

Repair Order # ________

Inspect the right front tire for the assigned vehicle and record the following information:

Code number (who made it, where, and when)
Size
Maximum cold air pressure
DOT
Pressure load maximum
UTQG rating
NAME OF TIRE
P205/60HR15 90H
DOT XXXX XXXX • TREAD 4 PLIES • 2 XXXX CORD • 2 XXXX CORD
MAX LOAD 590 KG (1301 LBS) @ 240 KPA (35PSI) MAX PRES
TUBELESS RADIA
SIDEWALL 2 PLIES XXXX CORD
TREADWEAR 160 TRACTION B TEMPERATURE A
MANUFACTURER'S NAME

1. Identify the type of tire construction.

   Radial ____ Belted Bias ____

   Blackwall ____ Whitewall ____

2. Number of sidewall plies:

   1 ____ 2 ____ 3 ____ Other ____

3. Number of tread plies:

   2 ____ 3 ____ 4 ____ Other ____

4. If the tire has belts, how many are there? ________

5. What material(s) are they made of (e.g., steel, rayon, nylon)?

   ____

6. What is the DOT Manufacturer's Code Number?

   ____

7. What brand name is on the tire?

   ____

8. What size is the tire?

   ____

9. What is the wheel rim diameter?

   13″ ____ 14″ ____ 15″ ____ 16″ ____ 17″ ____ 18″ ____ Other ____

10. What is the aspect ratio of the tire? ______________________________

11. What is the maximum air pressure for the tire? ____ psi

12. List the UTQG rating from the tire sidewall.

    Treadwear __________

    Traction __________

    Temperature __________

13. Is this an appropriate tire for the vehicle being inspected?

    Yes ____ No ____

14. Is the tire listed as M & S? Yes ____ No ____

15. What does M & S mean?

    ____

16. What is the tire's speed rating number? ____________

17. What maximum speed is the tire rated for?

    __________ mph

18. What is the maximum load rating for the tire?

    ____ lb @ ____ psi air pressure

19. Before completing the paperwork, clean your work area, put the tools in their proper places, and wash your hands.

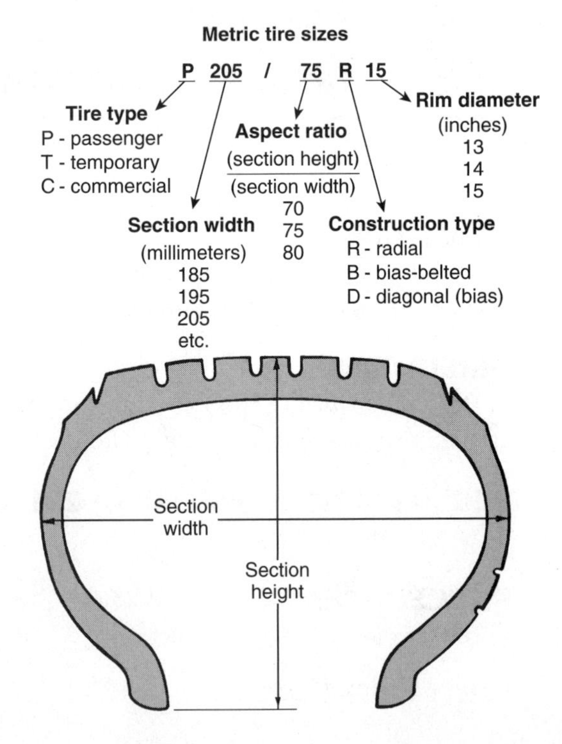

Instructor OK ______________ Score ______________

ASE Lab Preparation Worksheet #4-2

TIRE MAINTENANCE

Name______________________________ Class ____________________

OBJECTIVE:

Upon completion of this assignment, you should be able to rotate the tires. This task will help prepare you to pass the ASE certification examination in suspension and steering.

DIRECTIONS:

Before beginning this lab assignment, review the worksheet completely. Fill in the information in the spaces provided as you complete each task.

TOOLS AND EQUIPMENT REQUIRED:

Safety glasses, jack stands or vehicle lift, ratchet and sockets, torque wrench, shop towel, air impact wrench, impact sockets, tire gauge, air chuck

PROCEDURE:

Vehicle year ____________ Make __________________ Model ___________________

Repair Order # _________ Tire size ____________ Tire Manufacturer ______________________________

1. Check and adjust tire pressures. (Review Worksheet 3-2.)

   When checking the air pressure, tires should be: Hot ____ Cold ____

Record Pressures:	Before	After
Left front	____ psi	____ psi
Right front	____ psi	____ psi
Left rear	____ psi	____ psi
Right rear	____ psi	____ psi

2. Determine proper rotation pattern. Use numbers, lines, and arrows to indicate the rotation pattern to be used:

Prior Positions

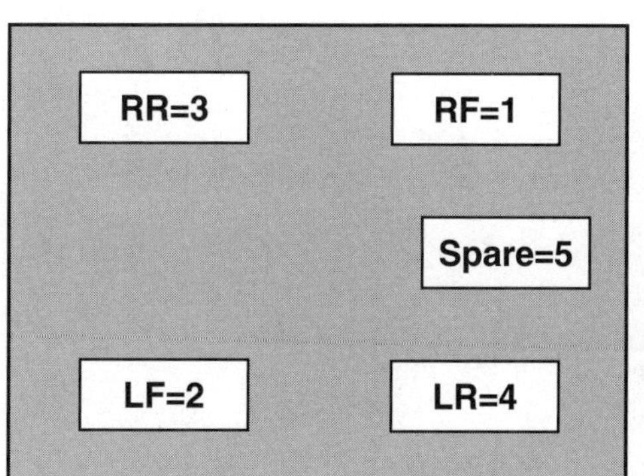

Rotated Positions
(enter in box below)

<--Front

TURN

3. How many tires will be rotated: 2 ____ 4 ____ 5 ____ (spare)
4. What is the torque specification for the lug nuts? ____ ft.-lb
5. Where did you locate the lug nut torque specification? ________________________________

Rotate the Tires

6. Remove the wheels.
   a. Loosen each lug nut about 1/4 turn before raising wheels off the ground.

   Note: With the air impact wrench, you do *not* need to loosen the lug nuts prior to lifting the vehicle off the ground.

   b. Which direction did you turn the lug nuts to loosen them?

   Clockwise ____ Counterclockwise ____

   c. When did you loosen the lugs?

   Before raising the vehicle ____ After raising the vehicle ____

Reinstalling Wheels

7. Check the condition of all tubeless valve stems.

   Does any need to be replaced? Yes ____ No ____

   If yes, which one(s)? LF ____ RF ____ RR ____ LR ____ Spare ____
8. Rotate wheels to desired positions.
9. Lift the wheel assembly onto the lug bolts.

CAUTION When lifting heavy objects, remember to lift with your legs, not your back.

10. Start the lug nuts onto the bolt threads.
11. Turn each lug nut by hand for at *least* one turn.
12. Which way does the tapered side of the lug nut face?

    Toward the rim ____ Away from the rim ____
13. Tighten lug nuts only until they are "snug", using the correct pattern.
14. Sketch the tightening pattern below.

15. Lower the vehicle to the ground.
16. Use a torque wrench to tighten each lug nut to specifications. ____ ft.-lb

CONTINUED

Note: Failure to tighten a wheel to the proper torque could result in a failing grade for the course!

17. Reinstall the wheel covers.

    a. Did you use a rubber mallet to install the wheel covers? Yes ____ No ____

    b. Are all wheel covers clean and in good condition? Yes ____ No ____

18. Before completing the paperwork, clean your work area, put the tools in their proper places, and wash your hands.

19. Record your recommendations for additional service or repairs on the repair order.

STOP

Instructor OK ____________ Score ________

ASE Lab Preparation Worksheet #4-3

REPLACE A RUBBER VALVE STEM

Name____________________ Class ____________

OBJECTIVE:

Upon completion of this assignment, you should be able to replace the valve stem in a tubeless tire. This task will help prepare you to pass the ASE certification examination in suspension and steering.

DIRECTIONS:

Before beginning this lab assignment, review the worksheet completely. Fill in the information in the spaces provided as you complete each task.

TOOLS AND EQUIPMENT REQUIRED:

Safety glasses, tire changer, valve stem installation tool, air chuck, tire gauge, shop towel

PROCEDURE:

Vehicle year __________ Make ______________ Model ______________

Repair Order # ________ Tire size __________ Tire Manufacturer ____________________

1. Locate a replacement valve stem of the proper length and diameter.

   The wheel is used with a:

   Wheel cover (long stem) ____

   Hub cap (short stem) ____

2. Install the wheel on the tire machine with the valve stem facing the bead breaker.

Short Long Large diameter

Note: Before using the tire machine check with your instructor.

3. Break down the outer bead.
4. Put rubber lube on the part of the valve stem that is inside the rim. Use the installation tool to remove the stem.
5. If the valve stem will not come out, use diagonal cutters or another suitable tool to cut the valve stem from outside of the wheel. Hold onto the bottom end of the valve stem while cutting so it does not drop into the tire.
6. Lube the replacement valve stem with rubber lube.
7. Use the installation tool to pull the new stem into place.
8. Reinflate the tire to the proper pressure.
9. What pressure did you inflate the tire to? ____ psi
10. Before completing the paperwork, clean your work area, put the tools in their proper places, and wash your hands.

11. Were there any other recommendations for needed service or unusual conditions that you noticed?

    Yes _____ No _____

12. Record your recommendations for additional service or repairs on the repair order.
13. Complete the repair order (R.O.).

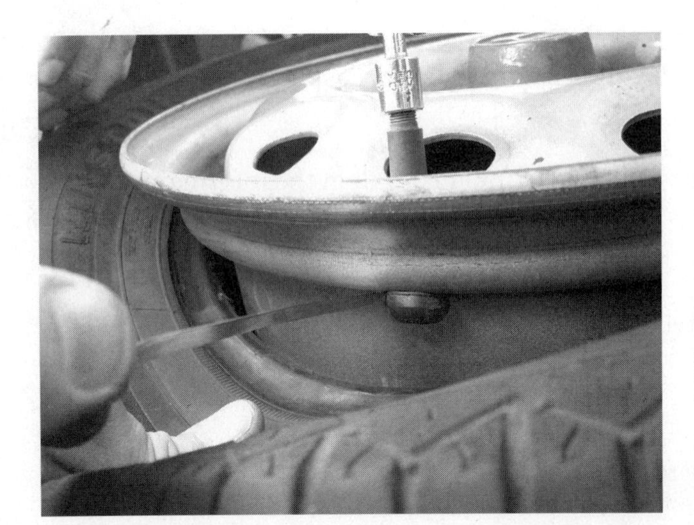

STOP

Instructor OK ______________ Score __________

ASE Lab Preparation Worksheet #4-4

DISMOUNT AND MOUNT TIRES (CENTER POST TIRE CHANGER)

Name______________________________ Class ____________________

OBJECTIVE:

Upon completion of this assignment, you should be able to mount and dismount a tire using the center post tire changer. This task will help prepare you to pass the ASE certification examination in suspension and steering.

DIRECTIONS:

Before beginning this lab assignment, review the worksheet completely. Fill in the information in the spaces provided as you complete each task.

TOOLS AND EQUIPMENT REQUIRED:

Safety glasses, jack stands or vehicle lift, center post tire changer valve, stem tool, air chuck, tire gauge

Note: If this is the first time you are doing this job, you must have supervision.

PROCEDURE:

Vehicle year ___________ Make ________________ Model _________________

Repair Order # ________ Tire size ___________ Tire Manufacturer ___________________________

A. Dismounting a Tire

1. Raise the vehicle and remove the wheel and tire assembly.
2. Unscrew the valve core from inside the valve stem.
3. Mount the wheel on the tire changer with the "drop center" offset facing up.

   Is this the valve stem side of the rim?

   Yes ____ No ____

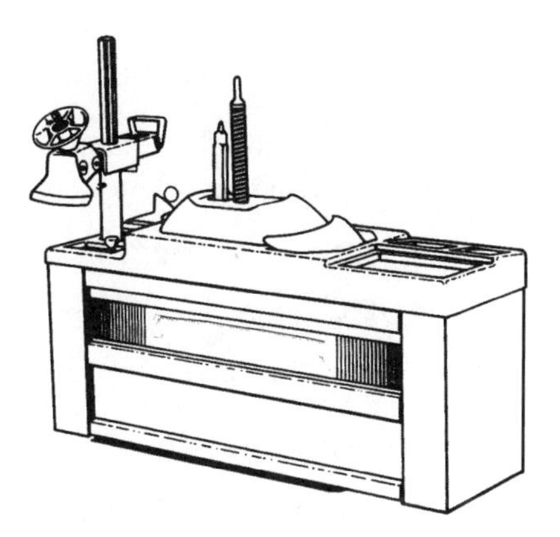

4. Align the positioning pin in one of the wheel lug holes and tighten the *holddown cone* hand tight.
5. Adjust the detent control to the proper rim size (if applicable).
6. Set the bead breaker into position.
7. Step on the air control valve to break the top and bottom beads free from the rim.
8. Apply rubber lube to the bead surfaces of the tire.
9. Insert the combination tool under the bead of the tire and position the tool on the center post.
10. Hold the tool in place as you step on the air control valve.

11. Remove the top bead. Push the bead of the tire opposite the tire tool down into the drop center of the rim for easier removal.

Watch your fingers!

12. Remove the tire from the rim by repeating steps 9 through 11 to the lower bead.
13. Inspect the bead seat area of the rim for rust and dirt. Clean as necessary.

    OK ____ Needs cleaning ____

B. Mounting a Tire

1. Apply rubber lube or soap to both tire beads.
2. Position the installation tool on the center post of the tire machine and rotate the tire clockwise until the lower bead holds the tire tool in place.
3. Slide the tire clockwise until part of its lower bead falls into the rim drop center.

CAUTION Keep hands out of the way!

4. Step on the foot pedal and follow the tool with your *right* hand. Keep the installed part of the bead pushed into the drop center so the rest of the bead can be installed without damaging the tire.
5. Mount the upper bead in the same manner.

C. Inflating a Tire

1. Coat the tire beads with rubber lube.
2. If the tire changer has an "air ring," leave the tire mounted on the tire changer while inflating the tire.
3. *Stand to the side when inflating.*
4. *Keep your hands clear of the tire and the tire changer!*
5. Begin to inflate the tire. As soon as it begins to inflate, loosen the holddown cone one turn.
6. Was there a loud "pop" during inflation? Yes ____ No ____
7. If more than 30 psi is required to seat the beads completely, call your instructor. Was more than 30 psi required?

   Yes ____ No ____

Note: Never inflate the tire to more than the maximum pressure allowed for the tire being inflated.

8. When the beads are seated, install the valve core and inflate the tire to the vehicle manufacturer's specifications.

   What is the recommended air pressure for the vehicle that the tire will be installed on?

   Front __________

   Rear __________

9. Before completing the paperwork, clean your work area, put the tools in their proper places and wash your hands.
10. Were there any other recommendations for needed service or unusual conditions that you noticed?

    Yes _____ No _____
11. Record your recommendations for additional service or repairs on the repair order.
12. Complete the repair order (R.O.).

Instructor OK ______________ Score __________

ASE Lab Preparation Worksheet #4-5

DISMOUNT AND MOUNT TIRES (RIM CLAMP-STYLE TIRE CHANGER)

Name______________________________ Class ____________________

OBJECTIVE:

Upon completion of this assignment, you should be able to use a European-style tire changer to dismount and mount a tire. This task will help prepare you to pass the ASE certification examination in suspension and steering.

DIRECTIONS:

Before beginning this lab assignment, review the worksheet completely. Fill in the information in the spaces provided as you complete each task.

TOOLS AND EQUIPMENT REQUIRED:

Safety glasses, shop towel, European-style tire changer, valve core tool

PROCEDURE:

Vehicle year ___________ Make ________________ Model __________________

Repair Order # ________ Tire size ___________ Tire Manufacturer ____________________________

Note: If this is the first time you are doing this job, you must have supervision.

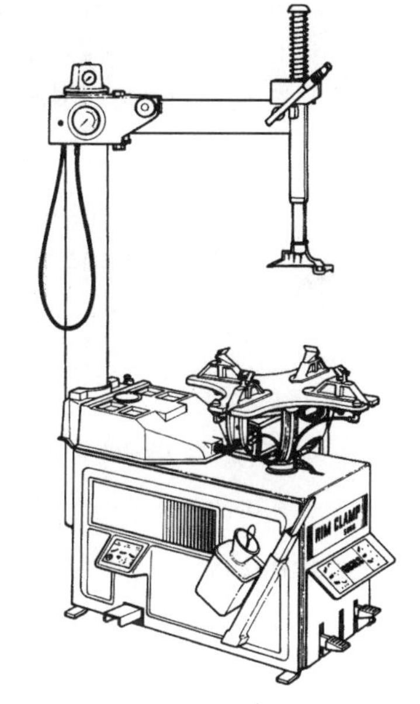

1. Dismounting the Tire:
   a. Raise the vehicle and remove the wheel and tire assembly from the car.
   b. Unscrew the tire's valve core to allow the air to escape.
   c. Place the rim under the bead breaker attachment with the "drop center" offset toward the bead breaker. Is this the valve stem side of the rim?

      Yes ____ No ____
   d. Force the tire bead away from the safety ledges on both sides of the wheel.
   e. Install the tire and wheel assembly on the top of the tire changer. Push on the air control to clamp the wheel to the machine.
   f. Apply rubber lube to the bead surfaces of the tire.
   g. Adjust the top arm of the tire changer so that it almost contacts the edge of the rim.
   h. Use the tire iron to pry the bead over the edge of the rim.
   i. Hold the tire iron in place as you step on the foot pedal, which turns the wheel against the tool.

j. Remove the top bead. Push the bead of the tire opposite the tire iron down into the drop center of the rim for easier removal.

Do not allow your fingers to become trapped in between the tire bead and the rim. Serious injury could result!

k. Remove the lower bead in the same manner.

l. Inspect the bead seat area of the rim for rust and dirt.

OK ____ Needs cleaning ____

2. Remounting the Tire on the Wheel:

a. Apply rubber lube to tire beads.

b. Position the tire on the wheel and slide the lower bead over the rim by rotating it clockwise. When the tire cannot be installed further, its bead should be wedged against the top post of the tire machine. (This is a curved area that looks like a tool.)

c. Step on the air control to rotate the tire clockwise. This will force the remaining part of the bead over the top edge of the rim. Be sure the part of the tire's bead that is already over the edge of the rim remains in the drop center of the rim.

CAUTION Do not allow your fingers to become trapped in between the tire bead and the rim. Serious injury could result!

d. *Keep hands out of the way!* Step on the foot pedal and follow the tool with your *right* hand, keeping the bead pushed into the drop center. The lower bead is now installed.

e. Mount the upper bead in the same manner.

3. Inflating:

a. Be sure beads are coated with rubber lube.

b. The tire changer has an "air ring" that is used to fill the lower bead air gap while inflating the tire. Apply the correct foot pedal to inflate the tire. The pedal has two positions. Pushing on it all of the way forces air into the air ring during initial tire inflation.

c. *Stand to the side when inflating.* There is a connector that attaches the inflation hose to the tire valve so you do not have to hold it while inflating the tire.

Keep your hands out of the way!!

d. Begin to inflate the tire. As soon as it begins to inflate, loosen the holddown cone one turn.

e. Was there a loud "pop"?

Yes ____ No ____

f. If more than 30 psi is required to seat the beads completely, call your instructor. More than 30 psi required?

Yes ____ No ____

g. When the beads are seated, install the valve core and inflate to the manufacturer's specifications.

h. Before completing the paperwork, clean your work area, put the tools in their proper places, and wash your hands.

4. Were there any other recommendations for needed service or unusual conditions that you noticed while dismounting and mounting the tire?

Yes ____ No ____

5. Record your recommendations for additional service or repairs on the repair order. Complete the repair order (R.O.).

STOP

ASE Lab Preparation Worksheet #4-6

REPAIR A TIRE PUNCTURE

Name_____________________________ Class ____________________

OBJECTIVE:

Upon completion of this assignment, you should be able to repair a tire that is leaking. This task will help prepare you to pass the ASE certification examination in suspension and steering.

DIRECTIONS:

Before beginning this lab assignment, review the worksheet completely. Fill in the information in the spaces provided as you complete each task.

TOOLS AND EQUIPMENT REQUIRED:

Safety glasses, tire soak tank, valve stem installation tool, air chuck, tire gauge, pliers, burr tool, tire probe, vulcanizing cement, shop towel, tire spreading fixture, vacuum cleaner, patch stitcher, tire patches

PROCEDURE:

Vehicle year ___________ Make _________________ Model __________________

Repair Order # ________ Tire size ____________ Tire Manufacturer ____________________________

A. Locate the Leak in the Tire:

1. Inflate the tire to the maximum pressure listed on the tire's sidewall.

   What is the maximum pressure? ____ psi

2. Submerge the tire in the soak tank.
3. Rotate the tire, watching for bubbles as the tread area leaves the water.

   Any leaks in the tread area? Yes ____ No ____

4. Inspect the bead area for bubbles in the same manner.

   Any leaks from the bead area? Yes ____ No ____

Note: If the leak is at the bead area, the tire will need to be dismounted and the bead area of the tire and rim cleaned and inspected.

5. While the valve stem is under water, push on it while looking for bubbles.

   Any leaks from the valve stem? Yes ____ No ____

   If the valve stem is leaking, replace the valve stem. Refer to Worksheet 4-3.

   If the valve core is leaking, tighten or replace the valve core.

6. Use a marking crayon to mark the location of any leaks.

B. Repair a Punctured Tire:

1. Remove the tire from the rim. Refer to Worksheet 4-4 or 4-5.
2. Mount the tire on a tire spreading fixture.
3. Is the item that punctured the tire still present?
   Yes ____ No ____ If it is, remove it.
4. Probe the hole gently in the same direction as the nail or screw entered the tire.
5. Clean the area to be repaired.

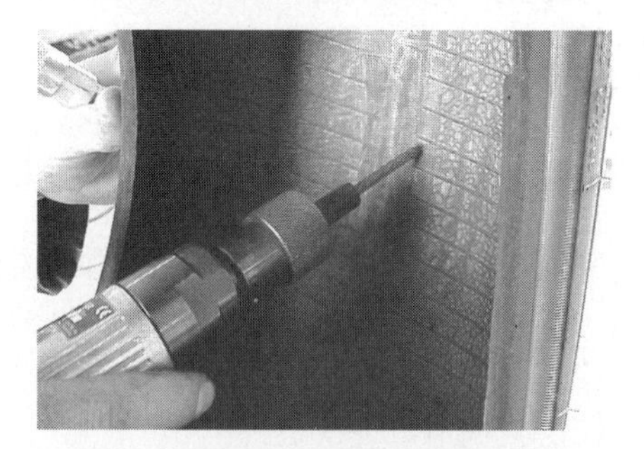

6. Ream the hole with a burr tool.
7. Install vulcanizing cement on the probe and probe the hole.
8. Install a tire plug into the hole until it extends slightly from both the inner and outer surfaces of the hole.
9. Cut off the tire plug and grind it down until it is *almost* flush. Be careful not to grind into the inner surface of the tire.
10. Select the correct type of patch.

    Radial ____ Bias ____ Universal ____

11. Use the patch and a marking crayon to outline the repair area (slightly larger than the patch).
12. Use liquid buffer and a scraper to clean the area to be patched.
13. Vacuum any debris from inside the tire.
14. Coat the repair area with vulcanizing cement.
15. Allow the cement to dry **completely**.

16. Apply the patch to the cemented area. Use a stitcher to seat it. Remove the plastic from the back of the patch.
17. Remount the tire, inflate, and test for leaks.
18. Before completing the paperwork, clean your work area, put the tools in their proper places, and wash your hands.
19. Were there any other recommendations for needed service or unusual conditions that you noticed while repairing the tire?

    Yes ____ No ____

    Record your recommendations for additional service or repairs on the repair order.

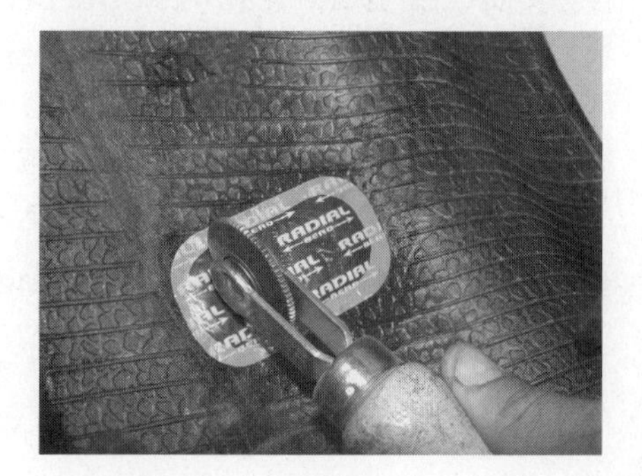

20. Complete the repair order (R.O.).

Instructor OK ______________ Score __________

ASE Lab Preparation Worksheet #4-7

COMPUTER TIRE BALANCE

Name____________________________ Class __________________

OBJECTIVE:

Upon completion of this assignment, you should be able to balance a tire using a computer wheel balancer. This task will help prepare you to pass the ASE certification examination in suspension and steering.

DIRECTIONS:

Before beginning this lab assignment, review the worksheet completely. Fill in the information in the spaces provided as you complete each task.

TOOLS AND EQUIPMENT REQUIRED:

Safety glasses, computer wheel balancer, wheel weights, weight hammer

PROCEDURE:

Vehicle year ___________ Make _________________ Model _________________

Repair Order # ________ Tire size ___________ Tire Manufacturer ____________________________

1. Remove all rocks and mud from the tire and rim.
2. Mount the wheel on the balancer.
3. Enter the required information into the computer as necessary.
   a. Enter wheel rim size.

      13″ ____ 15″ ____

      14″ ____ 16″ ____
   b. Enter the distance from the balancer to the edge of the wheel rim.

      Distance ______________
   c. Measure the width of wheel rim from bead to bead with the special caliper.

      Width _____ ″
4. Drop the protective cover over the wheel.
5. Depress the switch to spin the wheel and tire assembly.
6. When the wheel stops spinning, raise the cover and rotate the wheel until it is at the indicated position for the left side of the tire.

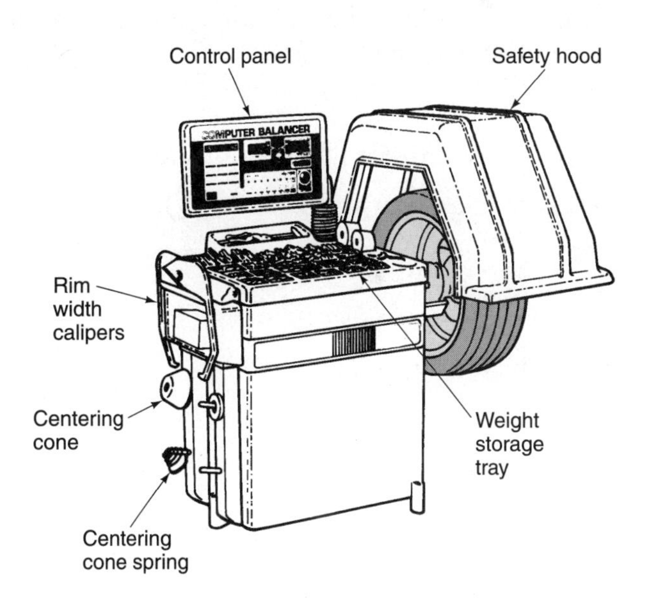

7. Install the specified weight onto the wheel rim in line with the weight line on the wheel balancer.

   How much weight was added to the rim at this point? _______ oz.

8. Position the wheel at the indicated position for the right side and install proper weight.

   How much weight was added to the rim at this point? _______ oz.

9. Spin the wheel once again to check the accuracy of the weight installation.

   OK ____ Rebalance ____

   How much total weight was added to the rim? _______ oz.

Note: If the tire is still not correctly balanced when rechecked, do not add another weight to correct the imbalance. Move the weight halfway to the new indicated position or start the balance procedure over.

10. Before completing the paperwork, clean your work area, put the tools in their proper places, and wash your hands.

11. Were there any other recommendations for needed service or unusual conditions that you noticed while balancing the tire?

    Yes ____ No ____

12. Record your recommendations for additional service or repairs on the repair order.

13. Complete the repair order (R.O.).

Part II

ASE Lab Preparation Worksheets

Service Area 5

Service Information

Instructor OK _______________ Score __________

ASE Lab Preparation Worksheet #5-1

UNDERHOOD LABEL WORKSHEET

Name_____________________________ Class ____________________

OBJECTIVE:

Completion of this assignment should prepare you to obtain important vehicle information from an underhood emission label. This task will help prepare you to pass the ASE certification examination in engine performance.

DIRECTIONS:

Complete this worksheet on a vehicle assigned by your instructor. All of the requested information may not be found on the underhood label of the vehicle that you are inspecting. If the information is not on the label, write N/A (not available) in the answer space. Before beginning this lab task, review the worksheet completely. Fill in the information in the spaces provided as you complete this assignment.

TOOLS AND EQUIPMENT REQUIRED:

Safety glasses, fender covers

PROCEDURE:

Vehicle year ___________ Make __________________ Model ___________________

1. Open the vehicle's hood and put fender covers over the fenders.
2. Are there any other underhood labels in the engine compartment?

   Air conditioning ____ Belt Routing ____

   Vacuum hose routing ____

   Emission updates ____

   List any others ______________________________

3. Locate the underhood **emission** label. Where under the hood is the label located?

   Hood ____

   Radiator support ____

   Inner fender ____

   Engine ____

4. Record the following information from the label.

VEHICLE EMISSION CONTROL INFORMATION · ACURA NSX
ENGINE FAMILY MHN3.0V5F3AX
EVAPORATIVE FAMILY 91FK
DISPLACEMENT 2977cm³
182CID
TWC HO2S
EGR SMPI
CATALYST

TUNE UP SPECIFICATIONS
TUNE UP CONDITIONS:
ENGINE AT NORMAL OPERATING TEMPERATURE
ALL ACCESSORIES TURNED OFF, COOLING FAN OFF
TRANSMISSION IN NEUTRAL
ADJUSTMENTS TO BE MADE IN ACCORDANCE WITH INDICATIONS GIVEN IN SHOP MANUAL.

IDLE SPEED	5 SPEED TRANSMISSION	800±50 rpm
	AUTOMATIC	750±50 rpm
IGNITION TIMING AT IDLE		15±2° BTDC
VALVE LASH	SETTING POINTS BETWEEN CAMSHAFT AND ROCKER ARM	IN. 0.17±0.02mm COLD
		EX. 0.19±0.02mm COLD

NO OTHER ADJUSTMENTS NEEDED.

THIS VEHICLE CONFORMS TO U.S. EPA AND STATE OF CALIFORNIA REGULATIONS APPLICABLE TO 1991 MODEL YEAR NEW MOTOR VEHICLES.

UNDERHOOD LABEL

Note: The engine idle speed and/or the ignition timing are not adjustable on all vehicles. Also, not all underhood labels have all of the information that is requested here. If the information is not given on the underhood label or if adjustments are not necessary, indicate that by entering "N/A" for that item.

Transmission type:

Automatic ____ Manual ____

Engine idle speed ____ rpm

Engine idle mixture ______________________________

Ignition timing ______________________________

Idle speed adjustment procedure ____

____

Engine timing adjustment procedure ____

____

Idle mixture adjustment procedure ____

____

Engine oil API classification:

SA ____ SB ____ SE ____ SF ____ SG ____ SH ____ SJ ____ SK ____

Emission certification year ________

The vehicle is certified for sale by: EPA ____ California ____ Other ____

Recommended spark plug ______________________________

Spark plug gap 0. ____ "

Does the engine have adjustable valves? Yes ____ No ____

If the valves are adjustable, what is the lash (clearance) specification?

Intake 0. ____ " Exhaust 0. ____ "

Is a vacuum hose routing diagram included on the label? Yes ____ No ____

5. When finished, remove the fender covers, close the hood, and complete the paperwork.

STOP

Instructor OK ____________ Score ____________

ASE Lab Preparation Worksheet #5-2

MOTOR SERVICE MANUAL WORKSHEET—ENGINE SPECIFICATIONS

Name____________________ Class ____________

OBJECTIVE:

Upon completion of this assignment, you should be able to use a Motor Service Manual to locate specifications required to service and repair a vehicle. This task will help prepare you to pass the ASE certification examination in engine repair and engine performance.

DIRECTIONS:

Use a Motor Service Manual and your own vehicle or one assigned by your instructor to locate the information requested. Most vehicles are equipped differently. If an item does not apply, put N/A in the answer blank.

TOOLS AND EQUIPMENT REQUIRED:

Motor Service Manual

PROCEDURE:

Vehicle year ____________ Make ________________ Model ________________

Vehicle Identification Number (VIN) ____________________________

Engine Type: V8 ____ V6 ____ In-line-6 ____ In-line-4 ____ Other ____

Engine Size (C.I. or liters) ____

Fuel System Type: Carburetor ____ Fuel Injection ____

Transmission: Automatic ____ Manual ____

Drive Wheels:

Front-wheel drive ____ Rear-wheel drive ____

4-wheel drive ____ All-wheel drive ____

Use the service manual that you have selected and answer the following questions:

1. What manual are you using? ____________________
2. What model years are included? ____________________
3. Are imported vehicles included in this service manual? Yes ____ No ____
4. Are light-duty trucks included? Yes ____ No ____
5. Look at the table of contents. Into how many sections is the manual divided? ____
6. In what section would information on how to overhaul a differential be located?

____

Locate the following information for the selected vehicle:

General Engine Specifications

7. Which digit of the VIN code describes the engine for this vehicle? Circle one:

   1st 2nd 3rd 4th 5th 6th 7th 8th 9th 10th 11th 12th

8. Use the VIN code to identify the engine ______________________________

9. Standard engine bore (STD) _______

10. Engine stroke _______

11. Compression ratio ___ :1

12. Maximum brake horsepower _________ at _________ rpm

13. Maximum torque _________ ft.-lb at _________ rpm

14. Normal oil pressure _________ psi at _________ rpm

Piston, Crankshaft, and Bearing Specifications

15. Crankshaft main journal diameter . _______ "

16. Main bearing oil clearance 0. _______ "

17. Crankshaft rod journal diameter . _______ "

18. Rod bearing oil clearance 0. _______ "

19. Piston clearance 0. _______ "

20. Piston ring end gap (compression ring) 0. _______ "

Valve Specifications

21. Valve stem diameter: Intake 0. _______ " Exhaust 0. _______ "

22. Valve seat angle: Intake _______ ° Exhaust _______ °

23. How are the valves arranged? (example, EIIEEIIE) ______________________________

24. Valve lash (clearance) Intake 0. _______ " Exhaust 0. _______ "

    Considerations when checking or adjusting the valve clearance:

    ____

Tightening Specifications

25. Main bolt (cap) torque ____ ft.-lb

26. Rod bolt torque ____ ft.-lb

27. Head bolt torque ____ ft.-lb

Instructor OK ________________ Score ____________

ASE Lab Preparation Worksheet #5-3

MOTOR SERVICE MANUAL WORKSHEET—MAINTENANCE SPECIFICATIONS

Name______________________________ Class ____________________

OBJECTIVE:

Upon completion of this assignment, you should be able to use a Motor Service Manual to locate specifications required to service and repair a vehicle. This task will help prepare you to pass the ASE certification examination in engine repair and engine performance.

DIRECTIONS:

Use a Motor Service Manual and your own vehicle or one assigned by your instructor to locate the information requested. Most vehicles are equipped differently. If an item does not apply, put N/A in the answer blank.

TOOLS AND EQUIPMENT REQUIRED:

Motor Service Manual

PROCEDURE:

Vehicle year ___________ Make __________________ Model ___________________

Vehicle Identification Number (VIN) ____

Engine Type: V8 ____ V6 ____ In-line-6 ____ In-line-4 ____ Other ____

Engine Size (C.I. or liters) ____

Fuel System Type: Carburetor ____ Fuel Injection ____

Transmission: Automatic ____ Manual ____

Drive Wheels:

Front-wheel drive ____ Rear-wheel drive ____

4-wheel drive ____ All-wheel drive ____

VIN plate location

Instrument panel

Use the service manual that you have selected and answer the following questions:

1. What manual are you using? ________________________________
2. What model years are included? ________________________________
3. Are imported vehicles included in this service manual? Yes ____ No ____
4. Are light-duty trucks included? Yes ____ No ____
5. Look at the table of contents. Into how many sections is the manual divided? ____

Tune-up Specifications

6. Spark plug gap 0. _______ "

7. Firing order ______________________________

8. Which direction does the distributor rotate? If not equipped with a distributor, describe the timing of the firing order and special procedure. Clockwise ____ Counterclockwise ____

9. In the spaces below, draw a sketch similar to the ones in the service manual that shows the firing order, cylinder numbering sequence, and the direction of distributor rotation.

10. Idle speed Curb idle ____ Fast idle ____

    Consideration when checking or adjusting the idle speed:

    ____

11. Point gap 0. _______ " N/A ____

12. Dwell ____ ° N/A ____

13. Ignition timing ____ degrees _________ TDC at _________ rpm.

    Consideration when checking or adjusting the timing:

    ____

    Draw a sketch below that represents the ignition timing marks.

Capacities

14. Cooling system capacity ____ quarts with A/C ____ quarts without A/C

15. Radiator cap pressure ____ psi ____ bar

16. Fuel tank ____ gallons to fill

17. Engine oil refill ____ quarts without filter ____ quarts with filter

Tightening Specifications

18. Spark plug torque _____ ft.-lb
19. Intake manifold bolt torque _____ ft.-lb
20. Draw a sketch below of the cylinder head tightening sequence.

STOP

ASE Lab Preparation Worksheet #5-4

MITCHELL SERVICE MANUAL WORKSHEET—ENGINE SPECIFICATIONS

Name__________________________ Class ________________

OBJECTIVE:

Upon completion of this assignment, you should be able to use a Mitchell's Service and Repair Manual to locate specifications required to service and repair a vehicle. This task will help prepare you to pass the ASE certification examination in engine repair and engine performance.

DIRECTIONS:

Use a Mitchell's Service and Repair Manual and your own vehicle or one assigned by your instructor to locate the information requested. Most vehicles are equipped differently. If an item does not apply, put N/A in the answer blank.

TOOLS AND EQUIPMENT REQUIRED:

Mitchell's Service and Repair Manual

PROCEDURE:

Vehicle year __________ Make ________________ Model ________________

Vehicle Identification Number (VIN) ______________________________

Engine Type: V8 ____ V6 ____ In-line-6 ____ In-line-4 ____ Other ____

Engine Size (C.I. or liters) ______________

Fuel System Type: Carburetor ____ Fuel Injection ____

Transmission: Automatic ____ Manual ____

Drive Wheels:

Front-wheel drive ____ Rear-wheel drive ____

4-Wheel drive ____ All-wheel drive ____

VIN plate location

Instrument panel

Use the service manual that you have selected and answer the following questions:

1. What manual are you using?

____

2. What model years does it cover?

____

3. Are imported vehicles included in this service manual? Yes ____ No ____
4. Are light-duty trucks included? Yes ____ No ____
5. Look at the table of contents. Into how many sections is the manual divided? ________
6. In what section would information on how to overhaul a differential be located?

____

TURN

Locate the following information for the vehicle selected:

General Engine Specifications

7. Which digit of the VIN code identifies the engine for this vehicle? Circle one:

   1st 2nd 3rd 4th 5th 6th 7th 8th 9th 10th 11th 12th

8. Using the VIN code, which engine is in the vehicle? ______________________
9. Standard engine bore (STD) __________
10. Engine stroke __________
11. Compression ratio __________
12. Maximum brake horsepower ______ at ______ rpm
13. Maximum torque ______ ft.-lb at ______ rpm
14. Normal oil pressure ______ psi at ______ rpm

Piston, Crankshaft, and Bearing Specifications

15. Crankshaft main journal diameter . ______ "
16. Main bearing oil clearance 0. ______ "
17. Crankshaft rod journal diameter . ______ "
18. Rod bearing oil clearance 0. ______ "
19. Piston clearance 0. ______ "
20. Piston ring end gap (compression ring) 0. ______ "

Valve Specifications

21. Valve stem diameter

    Intake 0. ______ " Exhaust 0. ______ "

22. Valve seat angle

    Intake ______ ° Exhaust ______ °

23. How are the valves arranged? (example, EIIEEIIE) ______________________
24. Valve lash (clearance) Intake 0. ______ " Exhaust 0. ______ "

Considerations when checking or adjusting the valve clearance:

____

Tightening Specifications

25. Main bolt (cap) ____ ft.-lb
26. Rod bolt (cap) ____ ft.-lb
27. Head bolt (cap) ____ ft.-lb

Instructor OK ______________ Score __________

ASE Lab Preparation Worksheet #5-5

MITCHELL SERVICE MANUAL WORKSHEET—MAINTENANCE SPECIFICATIONS

Name_____________________________ Class ____________________

OBJECTIVE:

Upon completion of this assignment, you should be able to use a Mitchell's Service and Repair Manual to locate specifications required to service and repair a vehicle. This task will help prepare you to pass the ASE certification examination in engine repair and engine performance.

DIRECTIONS:

Use a Mitchell's Service and Repair Manual and your own vehicle or one assigned by your instructor to locate the information requested. Most vehicles are equipped differently. If an item does not apply, put N/A in the answer blank.

TOOLS AND EQUIPMENT REQUIRED:

Mitchell's Service and Repair Manual

PROCEDURE:

Vehicle year ___________ Make _________________ Model __________________

Vehicle Identification Number (VIN) ______________________________________

Engine Type: V8 ____ V6 ____ In-line-6 ____ In-line-4 ____ Other ____

Engine Size (C.I. or liters) ______________

Fuel System Type: Carburetor ____ Fuel Injection ____

Transmission: Automatic ____ Manual ____

Drive Wheels:

Front-wheel drive ____ Rear-wheel drive ____

4-Wheel drive ____ All-wheel drive ____

Use the service manual that you have selected and answer the following questions:

1. What manual are you using?

____

2. What model years does it cover?

____

3. Are imported vehicles included in this service manual? Yes ____ No ____
4. Are light-duty trucks included? Yes ____ No ____

Tune-up Specifications

5. Spark plug gap 0. ____ "
6. Firing order ________________________________
7. Which direction does the distributor rotate? Clockwise _______ Counterclockwise _______ N/A _______
8. In the spaces below, draw a sketch of the distributor and engine that shows the firing order, cylinder numbering sequence, and the direction of distributor rotation.

9. Idle speed Curb idle _______ Fast idle _______

   Consideration when checking or adjusting the idle speed:

   ____

10. Ignition timing _______ degrees _______ TDC at _______ rpm _______ N/A

    Consideration when checking or adjusting the timing:

    ____

11. Draw a sketch below that represents the ignition timing marks. If timing is not adjustable, mark "N/A" in the box.

Capacities

12. Cooling system capacity ____ quarts with A/C ____ quarts without A/C
13. Radiator cap pressure ____ psi ____ bar
14. Fuel tank ____ gallons to fill
15. Engine oil refill ____ quarts without filter ____ quarts with filter

Tightening Specifications

16. Spark plug torque ____ ft.-lb
17. Intake manifold bolt torque ____ ft.-lb

STOP

Instructor OK ______________ Score __________

ASE Lab Preparation Worksheet #5-6

COMPUTERIZED SERVICE INFORMATION

Name____________________________ Class ____________________

OBJECTIVE:

Upon completion of this assignment, you should be able to locate service specifications using a computerized service information system. This task will help prepare you to pass the ASE certification examination in all service areas.

DIRECTIONS:

Before beginning this lab assignment, review the worksheet completely. Computerized service information can be found on a computer hard drive, a CD-ROM, a DVD, or on the Internet. Fill in the information in the spaces provided as you complete each task.

TOOLS AND EQUIPMENT REQUIRED:

Computer station, Mitchell On-Demand, or Alldata information systems

PROCEDURE:

1. In the shop or library, locate a computer station with Mitchell On-Demand or Alldata.
2. Open the system to locate information for a 1991 Ford Taurus that has an automatic transmission and a 2.5L engine.

Tune-up Specifications

3. Spark plug gap 0. ____ "
4. Spark plug torque ____ ft.-lb
5. Spark plug cable resistance (maximum) ________ ohms per foot
6. Firing order ________________________________
7. Cylinder arrangement. Draw a picture of the engine below that shows the cylinder arrangement.

8. Ignition timing _______ ° BTDC

   When setting the timing on a vehicle with an automatic transmission, the transmission should be in: Neutral ________ Drive ________

General Service Information

9. How often should each of the following be replaced during the first one hundred thousand miles for a vehicle that requires normal maintenance service?

   a. Air filter ______________________ miles

   b. Engine oil ______________________ miles

   c. Fuel filter ______________________ miles

   d. Oil filter ______________________ miles

   e. PCV filter ______________________ miles

   f. Spark plugs ______________________ miles

10. What is the recommended tension for the V belt? ______________

11. What is the cooling system capacity? ______________ qt.

12. What is the engine oil capacity? ______________ qt.

    Does the engine oil capacity include the capacity of the oil filter? Yes ____ No ____

13. What is the capacity of the AXOD automatic transmission? ______________

    What type of transmission fluid should be used? ______________

Tightening specifications

   a. Intake manifold bolt torque ______________________ ft.-lb

   b. Main bearing cap torque ______________________ ft.-lb

   c. Connecting rod bolt torque ______________________ ft.-lb

   d. Cylinder head bolt torque ______________________ ft.-lb

15. Draw a sketch below that shows the cylinder head bolt tightening sequence.

16. Before completing the paperwork, exit the computer program and clean your work area.

STOP

Instructor OK ______________ Score __________

ASE Lab Preparation Worksheet #5-7

FLAT-RATE MANUAL WORKSHEET

Name__________________________ Class ________________

OBJECTIVE:

Upon completion of this assignment, you should be able to use a flat-rate manual to determine the cost of vehicle repairs.

DIRECTIONS:

Before beginning this lab assignment, review the worksheet completely. Fill in the information in the spaces provided as you complete each task.

TOOLS AND EQUIPMENT REQUIRED:

Flat-rate manual

PROCEDURE:

Locate a *Parts and Time Guide* (flat-rate manual) or use an electronic information library. Determine the estimated time for the following repairs. Then multiply the time by a shop rate of $60 per hour to determine the labor estimate for the customer.

1995 Ford F150 2-wheel drive pickup, 4-speed transmission, air conditioning, 5.0L engine.

Example:

1. List the time required to remove and replace (R&R) the engine.

   Labor time 6.2 hours

   Shop rate × $60.00

   Total labor estimate $372.00

   Estimated cost of parts $2178.38

   Total estimate $2550.38

2. List the time required to remove and replace (R&R) a clutch master cylinder.

   Labor time __________ hours

   Shop rate × 60

   Total labor estimate ____________

   Estimated cost of parts ____________

   Total estimate ____________

3. List the time required to replace (R&R) an air-conditioning compressor and the cost of the compressor.

   Labor time __________ hours

   Shop rate × 60

   Total labor estimate __________

   Estimated cost of parts __________

   Total estimate __________

4. List the time and materials required to remove and replace (R&R) the front crankshaft seal.

   Labor time __________ hours

   Shop rate × 60

   Total labor estimate __________

   Estimated cost of parts __________

   Total estimate __________

5. List the time and materials required to remove and replace (R&R) the cylinder head gaskets.

   Labor time __________ hours

   Shop rate × 60

   Total labor estimate __________

   Estimated cost of parts __________

   Total estimate __________

6. After completing this assignment, clean your work area, put the manual in its proper place, or close the computer program.

STOP

Part II

ASE Lab Preparation Worksheets

Service Area 6

Belts, Hoses, and Cooling System Service

Instructor OK ______________ Score __________

ASE Lab Preparation Worksheet #6-1

COOLING SYSTEM INSPECTION

Name____________________________ Class ____________________

OBJECTIVE:

Upon completion of this assignment, you should be able to do a visual inspection of the condition of the cooling system. This task will help prepare you to pass the ASE certification examination in engine repair.

DIRECTIONS:

Before beginning this lab assignment, review the worksheet completely. Fill in the information in the spaces provided as you complete each task.

TOOLS AND EQUIPMENT REQUIRED:

Safety glasses, fender covers, shop towel

PROCEDURE:

Vehicle year ___________ Make _________________ Model ___________________

Repair Order # ________ Engine size_____________ # of Cylinders ___________

1. Open the hood and place fender covers on the fenders and over the front body area.

   How many fender covers were installed? ____

2. Carefully inspect the water pump.

   Drive belt condition OK ____ Loose ____ Glazed ____ Split ____ Damaged ____

   Belt alignment Correct ____ Incorrect ____

   Belt size and length Correct ____ Incorrect ____

CAUTION Do not remove the radiator cap if the system is hot or pressurized.

Radiator Cap:

Is it the correct cap? Yes ____ No ____

What is the condition of the gasket?

OK ____ Worn/damaged ____

What is the condition of the pressure valve?

OK ____ Damaged ____

What is the condition of the vacuum valve? OK ____ Damaged ____

Radiator Condition:

Coolant level OK ____ Low ____

Coolant strength OK ____ Weak ____

TURN

Leaks in radiator Yes ____ No ____

Fan condition OK ____ Damaged ____

Radiator cap seat OK ____ Damaged ____

Drain valve OK ____ Needs service ____

Overflow tube OK ____ Needs service ____

Overflow tank Clean ____ Needs service ____

Radiator core Clean ____ Leaves ____ Bugs ____ Dirt ____

Automatic transmission cooler line condition? OK ____ Damaged ____

Water Pump Condition:

At the weep hole OK ____ Leaks ____

Leakage around gasket OK ____ Leaks ____

Pump bearing OK ____ Loose ____

Cooling Fan Condition:

Fan blades OK ____ Damaged ____

Fan clutch OK ____ Defective ____ N/A ____

Electric fan OK ____ Defective ____ N/A ____

Fan shroud OK ____ Missing/damaged ____

Heater Core Condition:

OK ____ Leaks ____ Debris in core housing ____

Gaskets and Core Plugs:

Thermostat Housing OK ____ Leaking ____

External head gasket leak Yes ____ No ____

Core plugs OK ____ Leaking ____

Intake manifold OK ____ Leaking ____

Temperature Gauge: OK ____ Inoperative ____ N/A ____

Warning Light: OK ____ Inoperative ____ N/A ____

Hoses: OK ____ Leaks ____ Swollen ____ Pinched ____ Collapsed ____ Incorrect fit ____

Hose Clamps: OK ____ Need replacing ____

3. Before completing the paperwork, clean your work area, put the tools in their proper places, and wash your hands.
4. Record your recommendations for additional service or repairs on the repair order.

   Complete the repair order (R.O.).

STOP

Instructor OK ________________ Score ____________

ASE Lab Preparation Worksheet #6-2

PRESSURE TEST A RADIATOR CAP

Name_____________________________ Class ____________________

OBJECTIVE:

Upon completion of this assignment, you should be able to pressure test a radiator cap. This task will help prepare you to pass the ASE certification examination in engine repair.

DIRECTIONS:

Before beginning this lab assignment, review the worksheet completely. Fill in the information in the spaces provided as you complete each task.

TOOLS AND EQUIPMENT REQUIRED:

Safety glasses, fender covers, radiator pressure tester, shop towel

PROCEDURE:

Vehicle year ____________ Make __________________ Model ____________________

Repair Order # _________ Engine size _____________ # of Cylinders ____________

1. Open the hood and install fender covers over the fenders and the front body parts.
2. Squeeze the top hose. Hard ____ Soft ____

CAUTION When removing the radiator cap, the engine must be off and the pressure released. The cooling system is under pressure when the system is at normal operating temperature. As the radiator cap is removed, the coolant may boil. Squeeze the radiator hose to check if the system is under pressure. If the hose is hard, do not remove the radiator cap. See your instructor.

3. Fold a shop rag to 1/4 size. Use it to turn the radiator cap 1/4 turn until its first stop. This will allow any remaining pressure to escape.
4. Squeeze the top radiator hose to check for system pressure. Hard ____ Soft ____
5. If the hose is soft, press down on the cap and remove it.
6. Visually inspect the radiator cap.

   Radiator cap pressure seal:

   OK ____ Worn ____ Missing ____

   Radiator cap pressure spring:

   OK ____ Rusted ____ Broken ____

   Radiator cap vacuum valve:

   OK ____ Stuck ____ Missing ____

   Hint: Check for signs of corrosion under the seal.

7. Wet the rubber pressure seal and install the radiator cap on the adapter.
8. Attach the tester to the adapter.
9. Pump up the tester until the gauge reaches its highest point and holds.
10. What is the highest pressure that the cap will hold?

    15 lb ____ 13–14 lb ____ 7 lb ____ 0 ____
11. Does the cap hold pressure after the high point is reached?

    Yes ____ No ____
12. Remove pressure from the tester by pushing the tester hose sideways at the cap.

    Did it release air? Yes ____ No ____
13. Remove the radiator cap from the tester and return the tester to its container.
14. Put the radiator cap back on the radiator.
15. Does the radiator cap need to be replaced?

    Yes ____ No ____
16. What happens to the boiling point of the coolant if a faulty radiator cap is not replaced?

    ____
17. Before completing the paperwork, clean your work area, put the tools in their proper places, and wash your hands.
18. Were there any other recommendations for needed service or unusual conditions that you noticed while testing the radiator cap?

    Yes ____ No ____
19. Record your recommendations for additional service or repairs on the repair order.
20. Complete the repair order (R.O.).

Instructor OK ______________ Score __________

ASE Lab Preparation Worksheet #6-3

PRESSURE TEST A COOLING SYSTEM

Name____________________________ Class ____________________

OBJECTIVE:

Upon completion of this assignment, you should be able to pressure test a cooling system. This task will help prepare you to pass the ASE certification examination in engine repair.

DIRECTIONS:

Before beginning this lab assignment, review the worksheet completely. Fill in the information in the spaces provided as you complete each task.

TOOLS AND EQUIPMENT REQUIRED:

Safety glasses, fender covers, radiator pressure tester, shop towel

PROCEDURE:

Vehicle year ___________ Make __________________ Model ____________________

Repair Order # ________ Engine size _____________ # of Cylinders ____________

1. Open the hood and install fender covers on the fenders and front body parts.
2. Before starting to work, set the parking brake firmly and put the transmission in park or neutral.

CAUTION When removing the radiator cap, the engine must be off and the pressure released. The cooling system is under pressure when the system is at normal operating temperature. As the radiator cap is removed, the coolant may boil. Squeeze the radiator hose to check if the system is under pressure. If the hose is hard, do not remove the radiator cap. See your instructor.

3. Check to see if the cooling system is pressurized by squeezing the upper radiator hose.
4. Fold a shop rag to 1/4 size. Use the rag to turn the radiator cap 1/4 turn until its first stop. This will allow any remaining pressure to escape.
5. While still holding the cap, check to see if it is loose and all the pressure has escaped.

   Loose ____ Tight ____
6. If the cap is loose, press it down against spring pressure and turn it to remove it.
7. Pressurize the cold system.

   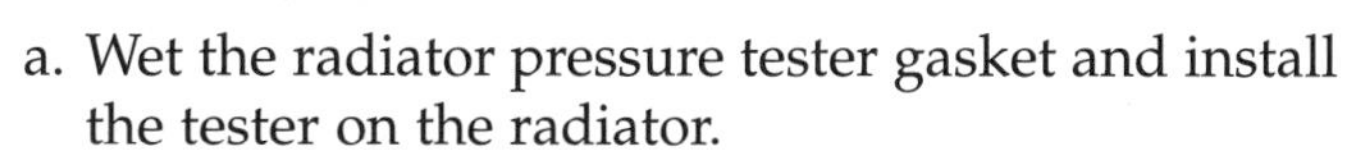

   a. Wet the radiator pressure tester gasket and install the tester on the radiator.

   b. Pressurize the system to the pressure marked on the cap.

   c. Wait 5 minutes. Does the pressure on the gauge remain steady?

   Yes ____ No ____

d. Are there leaks at any of the following?

Heater core ____ Heater hoses ____ Radiator hoses ____

Thermostat housing ____ Core plugs ____ Radiator ____

Water pump ____ Carburetor heater ____ No leaks found ____

Note: If the pressure drops but no leaks are found, do the combustion leak test (Worksheet #6-4) after completing this worksheet.

e. Release the pressure from the cooling system.

8. Start the engine and let it run until it reaches operating temperature, then shut it off.

Note: It is easier to pressure test a cold engine, but some leaks will only show up when the engine is warm.

9. Pressurize the system to the pressure marked on the cap.

   Radiator cap pressure ______ lb ______ bar

10. Wait 5 minutes. Does the pressure on the gauge remain steady? Yes ____ No ____

Note: A very small leak may not be evident in this short a time.

11. If the pressure reading has dropped, look for signs of external leakage.

Note: Some pressure drop may occur as the cooling system temperature drops.

12. Are there leaks at any of the following?

    Heater core ____ Heater hoses ____ Radiator hoses ____

    Thermostat housing ____ Core plugs ____ Radiator ____

    Water pump ____ Carburetor heater ____ No leaks found ____

13. Push the tester hose to the side where it connects to the radiator adapter. This will release the pressure from the system. Remove the tester.
14. Before completing the paperwork, clean your work area, put the tools in their proper places, and wash your hands.
15. Were there any other recommendations for needed service or unusual conditions that you noticed while pressure testing the cooling system?

    Yes ____ No ____

16. Record your recommendations for additional service or repairs on the repair order.
17. Complete the repair order (R.O.).

ASE Lab Preparation Worksheet #6-4

PERFORM A COOLING SYSTEM COMBUSTION LEAK TEST

Name________________________ Class ________________

OBJECTIVE:

Upon completion of this assignment, you should be able to test an engine for internal cooling system combustion leaks. This task will help prepare you to pass the ASE certification examination in engine repair.

DIRECTIONS:

Before beginning this lab assignment, review the worksheet completely. Fill in the information in the spaces provided as you complete each task.

TOOLS AND EQUIPMENT REQUIRED:

Safety glasses, fender covers, drain pan, combustion leak tester, shop towel

PROCEDURE:

Vehicle year __________ Make ________________ Model _________________

Repair Order # _______ Engine size ___________ # of Cylinders __________

1. Install fender covers on the fender and over the front body parts.

CAUTION When removing the radiator cap, the engine must be off and the pressure released. The cooling system is under pressure when the system is at normal operating temperature. As the radiator cap is removed, the coolant may boil. Squeeze the radiator hose to check if the system is under pressure. If the hose is hard, do not remove the radiator cap. See your instructor.

2. Squeeze the top hose. Hard ____ Soft ____
3. Fold a shop rag to 1/4 size. Use it to turn the radiator cap 1/4 turn until its first stop. This will allow any remaining pressure to escape.
4. Check the cap to see that it is loose, indicating that all the pressure has escaped.

   Loose ____ Tight ____
5. If the cap is loose, press down against the spring pressure and turn the cap to remove it.
6. Open the radiator drain valve and release some coolant into a drain pan to lower the water level in the radiator about 2″. Close the drain valve.
7. Start the engine and run it until the engine is warm.

   Did the thermostat open? Yes ____ No ____

   Is the upper radiator hose hot? Yes ____ No ____
8. Pour the testing liquid into the tester until it reaches the "fill" line.
9. Place the tester on the radiator filler neck and pump the bulb several times to suck *air* from above the coolant.

Note: As the coolant gets hotter, its level will rise. Allowing coolant into the test fluid will ruin the fluid and void the test.

10. Did the liquid change color? Yes ____ No ____

    If the liquid did change color, what color is it?

    Blue ____ Green ____ Yellow ____

    What does each color indicate?

    Blue ______________________________

    Green ______________________________

    Yellow ______________________________

11. Refill the radiator and put the radiator cap in place.
12. Drain the used test fluid from the tester and dry the tester.
13. Before completing the paperwork, clean your work area, put the tools in their proper places, and wash your hands.
14. If the test fluid had changed color, what repair would most likely correct the problem?

    ____

15. Were there any other recommendations for needed service or unusual conditions that you noticed while you were testing for combustion leaks?

    Yes ____ No ____

16. Record your recommendations for additional service or repairs on the repair order.
17. Complete the repair order (R.O.).

STOP

Instructor OK ______________ Score __________

ASE Lab Preparation Worksheet #6-5

CHECK COOLANT STRENGTH—HYDROMETER OR REFRACTOMETER

Name______________________________ Class ____________________

OBJECTIVE:

Upon completion of this assignment, you should be able to check the coolant for the proper strength. This task will help prepare you to pass the ASE certification examination in engine repair.

DIRECTIONS:

Before beginning this lab assignment, review the worksheet completely. Fill in the information in the spaces provided as you complete each task.

TOOLS AND EQUIPMENT REQUIRED:

Safety glasses, fender covers, coolant tester, shop towel

PROCEDURE:

Vehicle year ___________ Make _________________ Model __________________

Repair Order # ________ Engine size_____________ # of Cylinders ___________

1. Open the hood and install fender covers over the fenders and front body parts.

CAUTION When removing the radiator cap, the engine must be off and the pressure released. The cooling system is under pressure when the system is at normal operating temperature. As the radiator cap is removed, the coolant may boil. Squeeze the radiator hose to check if the system is under pressure. If the hose is hard, do not remove the radiator cap. See your instructor.

Wire-type clamp

2. Squeeze the top hose. Hard ____ Soft ____
3. Fold a shop rag to 1/4 size. Use it to turn the radiator cap 1/4 turn until its first stop. This will allow any remaining pressure to escape.
4. Check the cap to see that it is loose, indicating that all the pressure has escaped.

   Loose ____ Tight ____
5. If the cap is loose, press down against the spring pressure and turn the cap to remove it.
6. What is the general appearance of the coolant?

   Clean ____ Dirty ____
7. What color is the coolant? ______________________________
8. Draw some coolant into the hydrometer or place a drop on the refractometer and read the gauge to check coolant strength.

TURN

9. Check instructions on the tester. Does coolant have to be at operating temperature for an accurate reading?

   Yes _____ No _____

Coolant hydrometer

10. Coolant strength: Good _____ Weak _____

    Freezing point _____ °F

    Boiling point _____ °F

    Recommendations:

    a. Coolant that is too concentrated may be diluted with water.

    b. A slightly weak concentration can be strengthened by draining off a quart of coolant and adding straight coolant. After running the engine, the strength may be checked again.

    c. If coolant strength is very weak, a coolant flush and change is recommended.

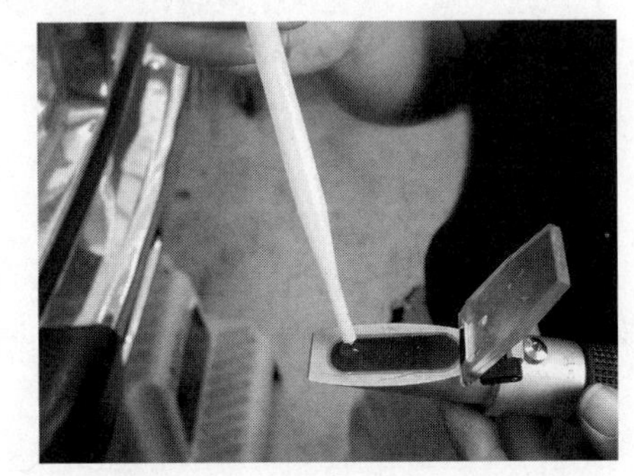

11. Top off the coolant level and reinstall the radiator cap.
12. Clean the coolant tester by flushing it with water.
13. Before completing the paperwork, clean your work area, put the tools in their proper places, and wash your hands.
14. Were there any other recommendations for needed service or unusual conditions that you noticed while checking the coolant?

    Yes _____ No _____

15. Record your recommendations for additional service or repairs on the repair order.
16. Complete the repair order (R.O.).

STOP

Instructor OK ________________ Score ____________
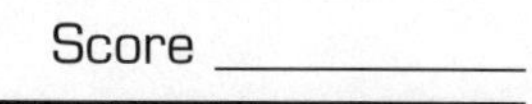

ASE Lab Preparation Worksheet #6-6

CHECK COOLANT STRENGTH—VOLTMETER

Name______________________________ Class ____________________

OBJECTIVE:

Upon completion of this assignment, you should be able to check the coolant strength with a voltmeter. This task will help prepare you to pass the ASE certification examinations in engine repair and engine performance.

DIRECTIONS:

Before beginning this lab assignment, review the worksheet completely. Fill in the information in the spaces provided as you complete each task.

TOOLS AND EQUIPMENT REQUIRED:

Safety glasses, fender covers, coolant tester, shop towel

PROCEDURE:

Vehicle year ___________ Make __________________ Model ___________________

Repair Order # ________ Engine size_____________ # of Cylinders ___________

1. Open the hood and install fender covers over the fenders and front body parts.

CAUTION When removing the radiator cap, the engine must be off and the pressure released. The cooling system is under pressure when the system is at normal operating temperature. As the radiator cap is removed, the coolant may boil. Squeeze the radiator hose to check if the system is under pressure. If the hose is hard, do not remove the radiator cap. See your instructor.

2. Squeeze the top radiator hose. Condition: Hard ____ Soft ____
3. Fold a shop rag to 1/4 size. Use it to rotate the radiator cap 1/4 turn until its first stop. This will allow any remaining pressure to escape.

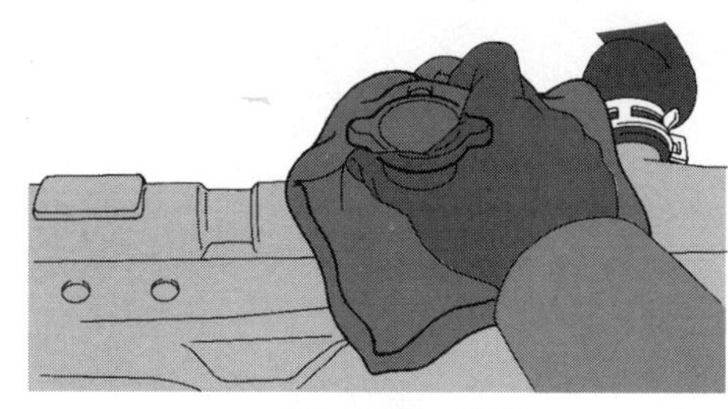

Cover cap with rag

4. Check the cap to see that it is loose, indicating that all the pressure has escaped.

   Loose ____ Tight ____
5. If the cap is loose, press down against the spring pressure and turn the cap to remove it.
6. What is the general appearance of the coolant?

   Clean ____ Dirty ____
7. What color is the coolant? ____________________
8. Connect the ground (black) lead of a voltmeter to the negative (–) terminal of the battery.

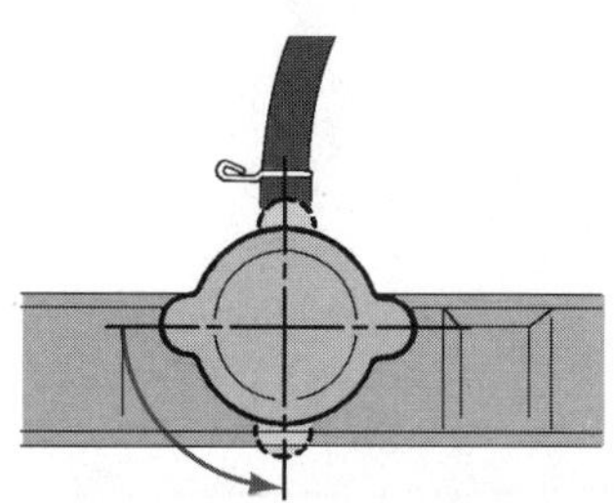

Turn slowly counterclockwise

9. Insert the positive (+) voltmeter lead into the coolant.

   What is the voltage reading? ________

   What is the maximum voltage reading allowable? ________

10. What do you think the results of this test indicate? ____

    ____

11. Before completing the paperwork, clean your work area, put the tools in their proper places, and wash your hands.

12. List any unusual conditions or recommendations for needed service.

    ____

    ____

13. Record your recommendations for needed service or additional repairs on the repair order.

14. Complete the repair order (R.O.).

Instructor OK ____________ Score ____________

ASE Lab Preparation Worksheet #6-7

REPLACE A RADIATOR HOSE

Name____________________ Class ____________

OBJECTIVE:

Upon completion of this assignment, you should be able to properly replace a radiator hose. This task will help prepare you to pass the ASE certification examination in engine repair.

DIRECTIONS:

Before beginning this lab assignment, review the worksheet completely. Fill in the information in the spaces provided as you complete each task.

TOOLS AND EQUIPMENT REQUIRED:

Safety glasses, fender covers, drain pan, slot screwdriver, razor knife, shop towel

PARTS AND SUPPLIES:

Radiator hose, hose clamps

PROCEDURE:

Vehicle year __________ Make ______________ Model ______________

Repair Order # ________ Engine size __________ # of Cylinders __________

1. Obtain the new replacement hose and hose clamps.
2. Open the hood and install fender covers over the fenders and front body parts.
3. Before starting to remove the radiator hose, compare the new hose with the one to be replaced. Do they appear to be the same size and shape?

   Yes ____ No ____

CAUTION When removing the radiator cap, the engine must be off and the pressure released. The cooling system is under pressure when the system is at normal operating temperature. As the radiator cap is removed, the coolant may boil. Squeeze the radiator hose to check if the system is under pressure. If the hose is hard, do not remove the radiator cap. See your instructor.

4. Squeeze the top hose. Hard ____ Soft ____
5. Fold a shop rag to 1/4 size. Use it to turn the radiator cap 1/4 turn until its first stop. This will allow any remaining pressure to escape.
6. The radiator cap should be loose, indicating that all the pressure has escaped.

   Loose ____ Tight ____
7. If the cap is loose, press down against the spring pressure and turn it to remove it.
8. Open the radiator drain valve located in the lower radiator tank and drain some coolant into a clean container. Drain the coolant until its level is below the level of the radiator hose.
9. Loosen the hose clamps on the radiator hose to be removed.

TURN

10. Twist the hose to loosen it. It may be necessary to cut the hose with a sharp knife if it is not easily removed.

Note: Use caution when twisting the hose. The hose fitting on the radiator is easily damaged.

Was the hose easily removed? Yes ____ No ____

Was it necessary to cut the hose? Yes ____ No ____

Was the radiator hose fitting damaged? Yes ____ No ____

11. Clean the radiator hose fittings.
12. Install the new hose. If the hose is difficult to install, apply some rubber lube to the connection. Be sure any bends are properly located. Check to see that the hose will not be damaged by movement of the engine or its accessories and that the clamps do not interfere with the:

Fan belts ____

Fuel lines ____

Fuel pump ____

Fan ____

13. Replace rusted or damaged hose clamps with new clamps and install them on the hose. Position the screw side of the clamp for easy access.

The hose clamps were: Reused ____ Replaced ____

What size are the hose clamps? ________________________

14. Position the hose clamps so they tighten just behind the bead on the connection.

Clamps properly installed? Yes ____ No ____

15. Position and tighten the hose clamps.
16. Refill the system.

ENVIRONMENTAL NOTE Before reusing coolant, check its concentration, appearance, and age. Remember, used coolant must be disposed of properly. Know your local regulations.

17. Run the engine until it is warm. Feel the upper radiator hose. When it is hot, the thermostat has opened. Bleed air from the system when necessary. Then check the coolant level.

OK ____ Low ____

18. Refill the system and check for leaks.
19. Replace the radiator cap.

CONTINUED

Note: After the cooling system has fully warmed up and then cooled again, the hose clamps should be retightened. Hoses may shrink after their first use.

20. Retighten the hose clamps.
21. Before completing the paperwork, clean your work area, put the tools in their proper places, and wash your hands.
22. List any other recommendations for needed service or unusual conditions that you noticed while changing the radiator hose. ____________________

    ____
23. Record your recommendations for additional service or repairs on the repair order.
24. Complete the repair order (R.O.).

STOP

Instructor OK ______________ Score __________

ASE Lab Preparation Worksheet #6-8

ADJUST ALTERNATOR V-BELT TENSION

Name____________________________ Class ____________________

OBJECTIVE:

Upon completion of this assignment, you should be able to adjust an alternator V-belt to the proper tension. This task will help prepare you to pass the ASE certification examination in engine performance.

DIRECTIONS:

Before beginning this lab assignment, review the worksheet completely. Fill in the information in the spaces provided as you complete each task.

TOOLS AND EQUIPMENT REQUIRED:

Safety glasses, fender covers, shop towel, belt tension gauge, hand tools

PROCEDURE:

Vehicle year ___________ Make _________________ Model __________________

Repair Order # ________ Engine size ____________ # of Cylinders ___________

1. Before starting to work, check the service manual for belt tension and adjustment procedures. There are several methods for adjustment. Always check the service manual or computer program for the proper procedure.

   Service manual or computer program used ________________________________ Page # ____

2. Open the vehicle's hood and place fender covers on the fenders and front body parts.
3. Inspect alternator belt condition.

   Good ____

   Glazed ____

   Frayed ____

   Oil-soaked ____

   Cuts or cracks ____

4. Check belt tension.

   Which method of testing was used?

   Belt tension gauge ____

   Belt deflection ____

5. Does the alternator belt need to be adjusted?

   Yes ____ No ____

Cracked Oil-soaked Glazed Torn or split

Note: Consult the service manual before making an adjustment. There are several methods used to adjust accessory V-belts.

6. What is the adjustment method recommended in the service manual?

   Prybar ____

   Jackscrew ____

   Square drive ____
7. Loosen the alternator mounting bolts *slightly.*
8. Adjust the belt using the correct method.

   Note: Do *not* pry anywhere where damage might occur to the alternator.
9. Tighten the mounting bolts/nuts after proper belt tension is reached.
10. The same procedure is used to check and adjust other belt-driven accessories.

    Are there any other belt-driven accessories?

    Yes ____ No ____
11. Before completing the paperwork, clean your work area, put the tools in their proper places, and wash your hands.
12. Were there any other recommendations for needed service or unusual conditions that you noticed while adjusting the alternator belt?

    Yes ____ No ____
13. Record your recommendations for additional service or repairs on the repair order.
14. Complete the repair order (R.O.).

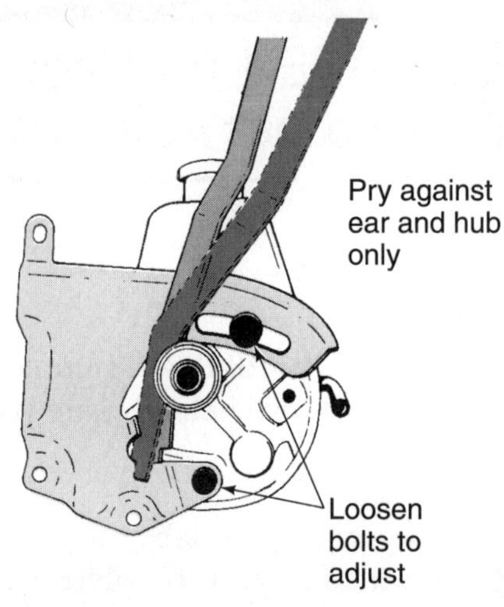

Pulley removed to show bolts

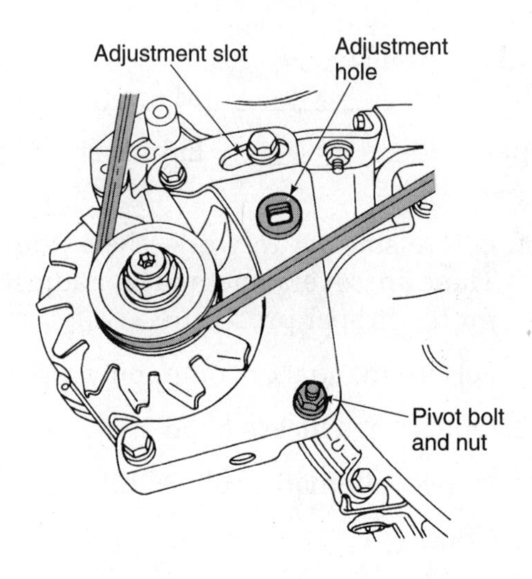

STOP

Instructor OK ______________ Score __________

ASE Lab Preparation Worksheet #6-9

ADJUST ALTERNATOR V-RIBBED BELT TENSION

Name____________________________ Class ____________________

OBJECTIVE:

Upon completion of this assignment, you should be able to adjust an alternator V-ribbed belt to the proper tension. This task will help prepare you to pass the ASE certification examination in engine performance.

DIRECTIONS:

Before beginning this lab assignment, review the worksheet completely. Fill in the information in the spaces provided as you complete each task.

TOOLS AND EQUIPMENT REQUIRED:

Safety glasses, fender covers, shop towel, belt tension gauge, hand tools

PROCEDURE:

Vehicle year __________ Make ________________ Model ________________

Repair Order # ________ Engine size ____________ # of Cylinders __________

1. Before starting to work, check the service manual for belt tension and adjustment procedures. There are several methods for adjustment. Always check the service manual or computer program for the proper procedure.

   Service manual or computer program used ______________________________ Page # ____

2. Open the vehicle's hood and place fender covers on the fenders and front body parts.
3. Inspect alternator belt condition.

   Good ____

   Glazed ____

   Frayed ____

   Oil-soaked ____

   Cuts or cracks ____

4. Check belt tension.

   Which method of testing was used?

   Belt tension gauge ____

   Belt deflection ____

5. Does the alternator belt need to be adjusted?

   Yes ____ No ____

Note: Consult the service manual before making an adjustment. There are several methods used to adjust accessory V-belts.

TURN

6. What is the adjustment method recommended in the service manual?

   Prybar ____

   Jackscrew ____

   Square drive ____

   Automatic adjuster ____

7. Loosen the alternator mounting bolts *slightly.*
8. Adjust the belt using the correct method.

   Note: Do *not* pry anywhere where damage might occur to the alternator.

9. Tighten the mounting bolts/nuts after proper belt tension is reached.
10. The same procedure is used to check and adjust other belt-driven accessories.

    Are there any other belt-driven accessories? Yes ____ No ____

11. Before completing the paperwork, clean your work area, put the tools in their proper places, and wash your hands.
12. Were there any other recommendations for needed service or unusual conditions that you noticed while adjusting the alternator belt?

    Yes ____ No ____

13. Record your recommendations for additional service or repairs on the repair order.
14. Complete the repair order (R.O.).

STOP

Instructor OK ______________ Score __________

ASE Lab Preparation Worksheet #6-10

REPLACE AN ACCESSORY V-BELT

Name______________________________ Class ____________________

OBJECTIVE:

Upon completion of this assignment, you should be able to adjust a V-belt to the proper tension. This task will help prepare you to pass the ASE certification examinations in engine performance, electrical systems, and suspensions.

DIRECTIONS:

Before beginning this lab assignment, review the worksheet completely. Fill in the information in the spaces provided as you complete each task.

TOOLS AND EQUIPMENT REQUIRED:

Safety glasses, fender covers, shop towel, belt tension gauge, hand tools

PARTS AND SUPPLIES:

V-belt

PROCEDURE:

Vehicle year ___________ Make _________________ Model __________________

Repair Order # ________ Engine size ____________ # of Cylinders ___________

1. Which accessories are driven by the belt that is being replaced?

   Air pump ______

   Air-conditioning compressor ______

   Power steering pump ______

2. Before starting to work, check the service manual for belt tension and adjustment procedures. There are several methods for adjustment. Always check the service manual or computer program for the proper procedure.

   Service manual or computer program used ________________________________ Page # ____

3. Obtain the new belt(s) before starting to work.
4. Open the vehicle's hood and place fender covers on the fenders and front body parts.
5. Remove the belt by loosening the adjuster and pivot bolts. Push the component inward and remove the belt. It may be necessary to weave the belt around the fan assembly.

   Was the belt removed? Yes ____ No ____

6. Compare the new belt to the old belt. The width, as well as the length, must be the same. Was the correct belt obtained?

   Yes ____ No ____

Note: If the old belt is not available, use a piece of string in the pulley groove to estimate the replacement belt size. The parts supplier should be able to determine the correct width for the application.

7. Install the belt by weaving it over the fan assembly and onto the pulleys.
8. Adjust belt tension.
9. Check the belt pulley alignment. The pulleys must be in alignment within 1/16″ per foot.

   Are the pulleys in alignment? Yes ____ No ____
10. Inspect the new belt to be sure that it does not rub on a radiator hose, fuel hose, or another belt.

    Rubs ____ OK ____

    Note: If the belt is rubbing, correct the problem.
11. Adjust the belt using the correct method.
12. Tighten the mounting bolts/nuts after proper belt tension is reached.
13. Before completing the paperwork, clean your work area, put the tools in their proper places, and wash your hands.
14. Record your recommendations for additional service or repairs on the repair order.
15. Complete the repair order (R.O.).

ASE Lab Preparation Worksheet #6-11

REPLACE AN ALTERNATOR V-BELT

Name______________________________ Class ____________________

OBJECTIVE:

Upon completion of this assignment, you should be able to replace an alternator V-belt and adjust it to the proper tension. This task will help prepare you to pass the ASE certification examinations in engine performance and electrical systems.

DIRECTIONS:

Before beginning this lab assignment, review the worksheet completely. Fill in the information in the spaces provided as you complete each task.

TOOLS AND EQUIPMENT REQUIRED:

Safety glasses, fender covers, shop towel, belt tension gauge, hand tools

PARTS AND SUPPLIES:

Alternator V-belt

PROCEDURE:

Vehicle year ___________ Make _________________ Model __________________

Repair Order # ________ Engine size ____________ # of Cylinders ___________

1. Before starting to work, check the service manual for belt tension and adjustment procedures. There are several methods for adjustment. Always check the service manual or computer program for the proper procedure.

   Service manual or computer program used ______________________________ Page # _______

2. Open the vehicle's hood and place fender covers on the fenders and front body parts.
3. Inspect alternator belt condition.

   Good ____

   Glazed ____

   Frayed ____

   Oil-soaked ____

   Cuts or cracks ____

4. Does the belt need to be replaced?

   Yes ____ No ____

   Is the belt being removed and replaced for practice?

   Yes ____ No ____

5. If the belt needs to be replaced, obtain a replacement belt before proceeding.
6. Loosen the alternator mounting bolts *slightly.*

7. Move the alternator to loosen the belt. Slip the belt from the pulleys. It may be necessary to feed the belt around the fan to remove the belt from the engine.
8. To install the belt, feed it around the fan and over the pulleys.
9. Pull the alternator back to remove the slack from the belt.
10. Adjust the belt tension. What is the adjustment method recommended in the service manual?

    Prybar ____

    Jackscrew ____

    Square drive ____

    Note: Do *not* pry anywhere where damage might occur to the alternator.
11. Tighten the mounting bolts/nuts after proper belt tension is reached.
12. Check the belt tension.

    a. Which method of testing was used? Belt tension gauge ____ Belt deflection ____

    b. Is the tension correct? Yes ____ No ____
13. The same procedure is used to replace and adjust other belt-driven accessories. Are there any other belt-driven accessories?

    Yes ____ No ____
14. Before completing the paperwork, clean your work area, put the tools in their proper places, and wash your hands.
15. Were there any other recommendations for needed service or unusual conditions that you noticed while replacing the alternator belt?

    Yes ____ No ____
16. Record your recommendations for additional service or repairs on the repair order.
17. Complete the repair order (R.O.).

STOP

Instructor OK ______________ Score __________

ASE Lab Preparation Worksheet #6-12

REPLACE A V-RIBBED BELT (NON-SERPENTINE)

Name______________________________ Class __________________

OBJECTIVE:

Upon completion of this assignment, you should be able to replace a V-ribbed belt and adjust it to the proper tension. This task will help prepare you to pass the ASE certification examinations in engine performance, electrical systems, and suspensions.

DIRECTIONS:

Before beginning this lab assignment, review the worksheet completely. Fill in the information in the spaces provided as you complete each task.

TOOLS AND EQUIPMENT REQUIRED:

Safety glasses, fender covers, shop towel, belt tension gauge, hand tools

PARTS AND SUPPLIES:

V-ribbed belt

PROCEDURE:

Vehicle year __________ Make ________________ Model _________________

Repair Order # _______ Engine size____________ # of Cylinders __________

1. Before starting to work, check the service manual for belt tension and adjustment procedures. There are several methods for adjustment. Always check the service manual or computer program for the proper procedure.

   Service manual or computer program used ______________________________ Page # ____

2. Obtain the new belt(s) before starting to work.
3. Open the vehicle's hood and place fender covers on the fenders and front body parts.
4. Which accessories are driven by the belt that is being replaced?

   Alternator ____ Air-conditioning compressor ____

   Water pump ____ Power steering pump ____

   Air pump ____

Adjustment slot
Adjustment hole (square)
Pivot bolt and nut

5. Remove the belt by loosening the adjuster and pivot bolts. Push the component inward and remove the belt. It may be necessary to weave the belt around the fan assembly.

   Was the belt easily removed? Yes ____ No ____

6. Compare the new belt to the old belt. The width, as well as the length, must be the same. Was the correct belt obtained?

   Yes ____ No ____

TURN

Note: If the old belt is not available, use a piece of string in the pulley groove to estimate the replacement belt size. The parts supplier should be able to determine the correct width for the application.

7. Install the belt by weaving it over the fan assembly and onto the pulleys.

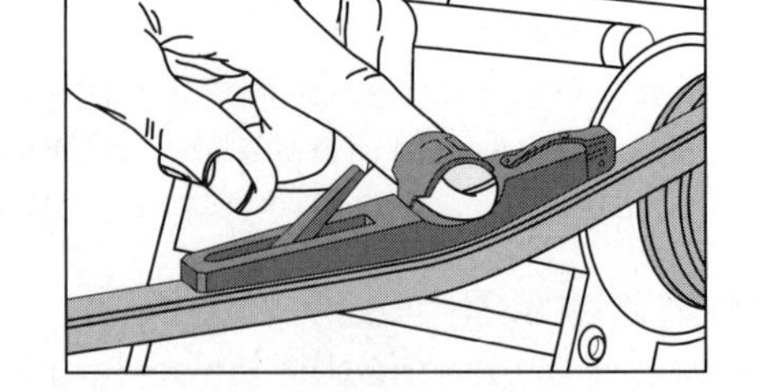

8. Check the belt pulley alignment. The pulleys must be in alignment within 1/16" per foot.

   Are the pulleys in alignment? Yes ____ No ____

9 Inspect the new belt to be sure that it does not rub on a radiator hose, fuel hose, or another belt.

   Rubs ____ OK ____

   Note: If the belt is rubbing, correct the problem.

10. Adjust the belt using the correct method, refer to Worksheet 6-9.
11. Tighten the mounting bolts/nuts after the proper belt tension is reached.

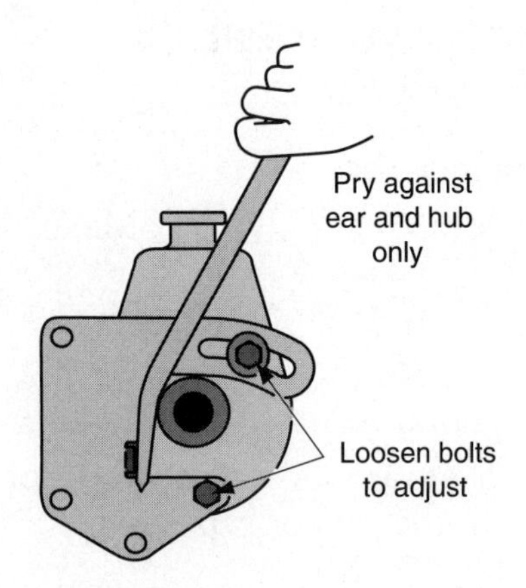

12. Before completing the paperwork, clean your work area, put the tools in their proper places, and wash your hands.
13. Were there any other recommendations for needed service or unusual conditions that you noticed while replacing the belt?

    Yes ____ No ____

14. Record your recommendations for additional service or repairs on the repair order.
15. Complete the repair order (R.O.).

Instructor OK ______________ Score __________

ASE Lab Preparation Worksheet #6-13

REPLACE A SERPENTINE BELT

Name____________________________ Class ________________

OBJECTIVE:

Upon completion of this assignment, you should be able to replace a serpentine belt. This task will help prepare you to pass the ASE certification examinations in engine performance and electrical systems.

DIRECTIONS:

Before beginning this lab assignment, review the worksheet completely. Fill in the information in the spaces provided as you complete each task.

TOOLS AND EQUIPMENT REQUIRED:

Safety glasses, fender covers, shop towel, belt tension gauge, hand tools

PARTS AND SUPPLIES:

Serpentine belt

PROCEDURE:

Vehicle year __________ Make ______________ Model ______________

Repair Order # ________ Engine size __________ # of Cylinders __________

1. Before starting to work, check the service manual for belt tension and adjustment procedures. There are several methods for adjustment. Always check the service manual or computer program for the proper procedure.

   Service manual or computer program used ________

   ____

   Page # ______

Tensioner
Air-conditioning compressor
Alternator
Air pump
Water pump
Belt rotation
Power steering pump
Crankshaft

2. Obtain the new belt before starting to work.

   Brand ______________________ Part #________

3. Open the hood and place fender covers on the fenders and front body parts.
4. Locate the belt routing diagram in the engine compartment. Did you find a diagram?

   Yes ____ No ____

5. In the box draw a sketch of the belt routing before removing the belt.
6. Loosen the adjuster and remove the belt.
7. Compare the new belt to the old belt. The width, as well as the length, must be the same.

   Correct belt obtained? Yes ____ No ____

TURN

Note: If the old belt is not available, use a piece of string in the pulley grooves to estimate the replacement belt size. The parts supplier should be able to determine the correct width for the application.

8. Install the belt by weaving it around the pulleys as pictured in the belt diagram.
9. Adjust belt tension.
10. Check the belt pulley alignment. The pulleys must be in alignment within 1/16" per foot.

    Are the pulleys in alignment? Yes ____ No ____
11. Check the new belt to be sure that it does not rub on a radiator hose, fuel hose, or another belt.

    Rubs ____ OK ____

    If the belt is rubbing, correct the problem.
12. Before completing the paperwork, clean your work area, put the tools in their proper places, and wash your hands.
13. Were there any other recommendations for needed service or unusual conditions that you noticed while replacing the belt?

    Yes ____ No ____
14. Record your recommendations for additional service or repairs on the repair order.
15. Complete the repair order (R.O.).

Instructor OK ______________ Score __________

ASE Lab Preparation Worksheet #6-14

FLUSH A COOLING SYSTEM

Name__________________________ Class __________________

OBJECTIVE:

Upon completion of this assignment, you should be able to flush a cooling system and then add the proper amount of coolant. This task will help prepare you to pass the ASE certification examination in engine repair.

DIRECTIONS:

Before beginning this lab assignment, review the worksheet completely. Fill in the information in the spaces provided as you complete each task.

TOOLS AND EQUIPMENT REQUIRED:

Safety glasses, service publications, fender covers, drain pan, shop towel, jack stands or vehicle lift

PARTS AND SUPPLIES:

Cooling system flush, coolant, flushing-T

PROCEDURE:

Vehicle year ___________ Make _________________ Model __________________

Repair Order # ________ Engine size ____________ # of Cylinders ___________

1. Look up the cooling system capacity in a service manual or computer program.

   Cooling system capacity: ____ qt.

   Which manual or computer program was used? ________________________________

   What page was the information found on? ____

2. Purchase the flushing cleaner and required coolant (50% of cooling system capacity).

   Note: Do not remove the radiator cap if the system is hot or pressurized.

3. Open the hood and install fender covers over the fenders and front body parts.

4. Squeeze the upper radiator hose to check for system pressure.

   Hard ____ Soft ____

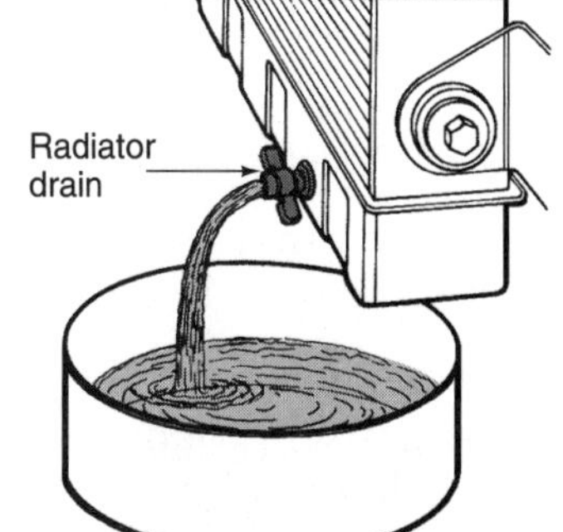

5. Open the radiator cap. Loosen the radiator drain valve and drain the cooling system.

   Note: Sometimes jacking up the rear of the car can allow for a more thorough drain of the block.

6. Lower the car, close the drain valve, and refill the cooling system with water until it is about 2″to 3″ below the filler neck.

7. Read the directions on the cleaning chemical.

8. Add the cleaning chemical to the radiator as directed.

Always wear eye protection (goggles) when working with chemicals.

9. Turn the heater control on to allow coolant to circulate in the heater.
10. With the emergency brake on and the transmission in Park or Neutral, start the engine and run it until operating temperature is reached.
11. Double-check to be sure the water level is correct. OK ____ Low ____
12. Run the engine for the specified period of time.
13. Turn off the engine and drain the cooling system.
14. Add a "flushing-T" to the heater hose coming from the heater.
15. If the cleaning chemical used requires a neutralizer, add it to the radiator.

    Neutralizer required? Yes ____ No ____
16. After neutralizing, flush the system again thoroughly using the "T" connection.
17. Inspect the hose from the radiator fill neck to the overflow tank.

    OK ____ Needs replacement ____

18. Drain and flush the recovery tank with water.
19. Add ethylene glycol coolant to the radiator (50% of the total system capacity).

    Note: The liquid that remains in the block after flushing is 100% water.
20. Top off the radiator with water.
21. Fill the recovery tank to about 1/2 full.
22. Run the engine with the radiator cap off until it is at normal operating temperature. Double-check the coolant level.

    OK ____ Low ____

    Note: Sometimes, the coolant level will drop after the thermostat opens. Many new cars require bleeding air out of the system after a refill and flush.

23. Replace the radiator cap.
24. Before completing the paperwork, clean your work area, put the tools in their proper places, and wash your hands.
25. Were there any other recommendations for needed service or unusual conditions that you noticed while flushing the cooling system?

    Yes ____ No ____

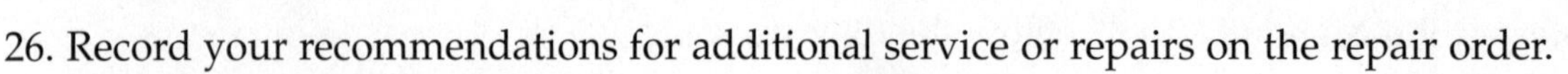

26. Record your recommendations for additional service or repairs on the repair order.
27. Complete the repair order (R.O.).

Instructor OK ______________ Score __________

ASE Lab Preparation Worksheet #6-15

REPLACE A THERMOSTAT

Name___________________________ Class ___________________

OBJECTIVE:

Upon completion of this assignment, you should be able to remove and replace a thermostat. This task will help prepare you to pass the ASE certification examination in engine repair.

DIRECTIONS:

Before beginning this lab assignment, review the worksheet completely. Fill in the information in the spaces provided as you complete each task.

TOOLS AND EQUIPMENT REQUIRED:

Safety glasses, fender covers, drain pan, ratchet, sockets, wrenches, shop towel

PARTS AND SUPPLIES:

Thermostat, gasket, gasket sealer

PROCEDURE:

Vehicle year ___________ Make _________________ Model __________________

Repair Order # ________ Engine size _____________ # of Cylinders ___________

1. Obtain the replacement thermostat and gasket before starting to work.

   Thermostat part number: ____ Brand: ______________________

   Temperature rating: 180° ____ 190° ____ 200° ____ Other ____

2. Open the hood and install fender covers on the fenders and front body parts.

CAUTION When removing the radiator cap, the engine must be off and the pressure released. The cooling system is under pressure when the system is at normal operating temperature. As the radiator cap is removed, the coolant may boil. Squeeze the radiator hose to check if the system is under pressure. If the hose is hard, do not remove the radiator cap. See your instructor.

3. Check to see if the cooling system is pressurized by squeezing the upper radiator hose.

   Is it under pressure? Yes ____ No ____

4. Fold a shop rag to 1/4 size. Use the rag to turn the radiator cap 1/4 turn until its first stop. This will allow any remaining pressure to escape.

5. Check the cap to see if it is loose, indicating that all the pressure has escaped.

6. If the cap is loose, press down against spring pressure and turn it to remove it.

7. Open the drain valve located in the lower radiator tank to allow some coolant to drain into a clean drain pan. Drain the coolant until its level is below the thermostat housing.

   The coolant level is: Below thermostat housing ____ Above thermostat housing ____

8. Disconnect the upper radiator hose from the water outlet housing.

TURN

9. Remove the bolts holding the water outlet housing.
10. Remove the water outlet housing and thermostat.
11. Thoroughly clean the gasket surfaces and the hose fitting.
12. Carefully inspect the thermostat housing for damage.

    OK ____ Damaged ____ 
13. Compare the new thermostat to the old one.

    Is it the same size? Yes ____ No ____

    Is it the same temperature rating? Yes ____ No ____

    Note: Installing a thermostat of a different temperature rating than that specified by the manufacturer can have an adverse effect on the heater, exhaust emissions, and driveability.

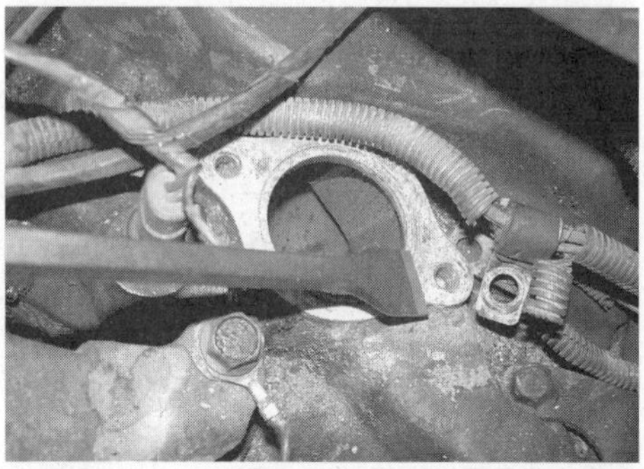

14. Install the thermostat. In which direction is the temperature sensing bulb facing? Toward or away from the engine?

    Toward ____ Away ____
15. Fit the thermostat into the recessed groove.

    The recess is in the:

    Water outlet housing ____ Engine/manifold ____ 
16. Coat a new paper gasket with gasket sealer and install it.
17. Install the water outlet. Before tightening, attempt to rock the water outlet back and forth to be sure it is flush.

    Is the outlet flush? Yes ____ No ____
18. Tighten the bolts evenly and carefully.

    Did the outlet crack? Yes ____ No ____

    Note: The thermostat outlet housing can be broken during this step if the thermostat does not fit in the recess or if the bolts are not tightened evenly.

19. Connect the upper hose and refill the system.
20. Pressurize the system and check for leaks. Refer to Worksheet 6-3 (Pressure Test a Cooling System).

    Are there any leaks? Yes ____ No ____

CONTINUED

21. Run the engine until it is warm to be sure that the thermostat opens and the system is operating properly.

    Did the thermostat open? Yes _____ No _____

    Top off the coolant in the radiator and fill the coolant recovery tank as required.

22. Before completing the paperwork, clean your work area, put the tools in their proper places, and wash your hands.

23. Were there any other recommendations for needed service or unusual conditions that you noticed while replacing the thermostat?

    Yes _____ No _____

24. Record your recommendations for additional service or repairs on the repair order.

25. Complete the repair order (R.O.).

STOP

Instructor OK ______________ Score __________

ASE Lab Preparation Worksheet #6-16

TEST A THERMOSTAT

Name_____________________________ Class ____________________

OBJECTIVE:

Upon completion of this assignment, you should be able to test a thermostat for its proper operating temperature. This task will help prepare you to pass the ASE certification examinations in engine repair and engine performance.

DIRECTIONS:

Before beginning this lab assignment, review the worksheet completely. Fill in the information in the spaces provided as you complete each task.

TOOLS AND EQUIPMENT REQUIRED:

Safety glasses, heat source, container, thermometer, feeler gauge, water

PROCEDURE:

Vehicle year ___________ Make _________________ Model ___________________

Repair Order # ________ Engine size _____________ # of Cylinders ___________

1. Temperature rating of the thermostat being tested: __________
2. Temperature rating of the thermostat required for the vehicle: __________
3. Locate the following items:
   a. Heat source __________
   b. Container __________
   c. Thermometer __________
   d. Feeler gauge __________
   e. Water __________

Check temperature when thermostat opens

Heat

4. Slightly open the thermostat. Slip the feeler gauge in the thermostat and let the thermostat close, holding the feeler gauge in place.
5. Hang the thermostat in the container of water by the feeler gauge and start heating the water.
6. Place a thermometer in the water.
7. Watch the thermostat when it starts to open. The thermostat will fall from the feeler gauge when it starts to open.
8. Note the temperature on the thermometer when the thermostat begins to open.

   ________________________ °Fahrenheit/Celsius.

9. Did the thermostat open at the specified temperature? Yes ____ No ____

10. Before completing the paperwork, clean your work area, put the tools in their proper places, and wash your hands.
11. Did you have any problems completing this worksheet? Yes ____ No ____

    List any problems that you encountered while completing this worksheet.

    ____

    ____

STOP

Instructor OK ______________ Score __________

ASE Lab Preparation Worksheet #6-17

TEST A RADIATOR FAN AND/OR A FAN CLUTCH

Name______________________________ Class __________________

OBJECTIVE:

Upon completion of this assignment, you should be able to test a radiator fan and a fan clutch for the proper operation. This task will help prepare you to pass the ASE certification examinations in engine repair and engine performance.

DIRECTIONS:

Before beginning this lab assignment, review the worksheet completely. Fill in the information in the spaces provided as you complete each task.

TOOLS AND EQUIPMENT REQUIRED:

Safety glasses

PROCEDURE:

Vehicle year ___________ Make _________________ Model __________________

Repair Order # ________ Engine size ____________ # of Cylinders ___________

The vehicle has a: conventional fan _________ , a flex fan _________ , a fan clutch _________ , or an electric fan _________ . If the vehicle you are checking has an electric fan, use Worksheet 6-18.

Conventional and Flex Fan Operation:

1. Check the direction of airflow through the radiator.
   a. Hold a piece of paper in front of the radiator and acclerate the engine to 2000 rpm.
   b. What happened to the paper?
      Nothing _____ Pulled in _____ Blown out _____ Disappeared _____
   c. If the paper was blown out from the radiator, this means
      _____ The fan is working normally.
      _____ The fan is not working.
      _____ The fan is installed backward.

Rigid leading edge
Flexible trailing edge
Water pump pulley
Flex fan assembly
Spacer

Fan Clutch Operation:

2. Before starting the engine, feel the fan by hand.
   a. Do you feel a slight resistance when trying to turn the fan? Yes ____ No ____
   b. Do you feel any roughness when turning the fan by hand? Yes ____ No ____
   c. Rock the fan up and down. Is it loose? Yes ____ No ____
   d. Is there a buildup of greasy dirt in the bearing area of the fan clutch?
      Yes ____ No ____

TURN

3. Run the engine until it reaches normal operating temperature.

   a. Was there an increase in the noise level coming from the fan as the engine temperature increased? Yes ____ No ____

   b. Turn off the engine. Did the fan stop in 4 or 5 revolutions? Yes ____ No ____

4. Before completing the paperwork, clean your work area, put the tools in their proper places, and wash your hands.

5. List any recommendations for needed service or unusual conditions that you noticed while you were checking the fan operation.

   ____

   ____

6. Record your recommendations for needed service or additional repairs on the repair order.

7. Complete the repair order (R.O.).

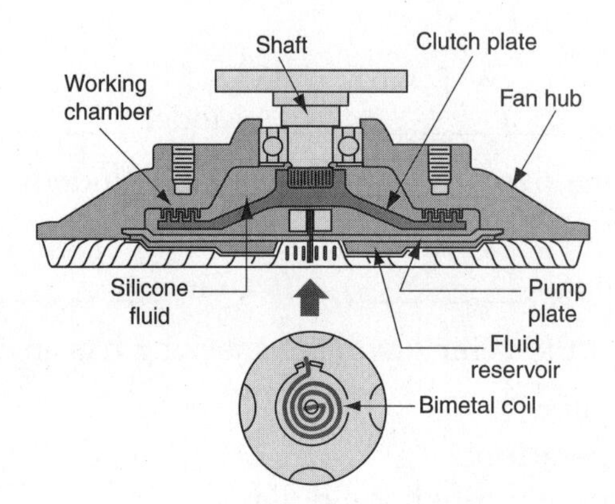

STOP

Instructor OK ______________ Score __________

ASE Lab Preparation Worksheet #6-18

TEST A RADIATOR ELECTRIC FAN

Name__________________________ Class __________________

OBJECTIVE:

Upon completion of this assignment, you should be able to test a radiator electric fan for proper operation. This task will help prepare you to pass the ASE certification examinations in engine repair and engine performance.

DIRECTIONS:

Before beginning this lab assignment, review the worksheet completely. Fill in the information in the spaces provided as you complete each task.

TOOLS AND EQUIPMENT REQUIRED:

Safety glasses

PROCEDURE:

Vehicle year ___________ Make _________________ Model _________________

Repair Order # ________ Engine size ____________ # of Cylinders ___________

Check the operation of a radiator electric fan.

CAUTION Be careful when working around an electric radiator fan. It may come on at any time, whether the engine is running or not.

1. How many electric fans are there on the radiator? __________
2. If there is a problem with the fan, check for obvious problems first:
   a. Electrical connectors: OK _____ Problem _____
   b. Fan fuse in the fuse panel: OK _____ Problem _____
3. If the engine is cold, use an ohmmeter to check the coolant temperature switch:

   Results: Continuity ____ Infinite resistance ____
4. Run the engine until it reaches normal operating temperature. Turn the engine off and recheck the coolant temperature switch.

   Results: Continuity ____ Infinite resistance ____
5. What is the condition of the coolant temperature switch? Good _____ Bad _____
6. Before completing the paperwork, clean your work area, put the tools in their proper places, and wash your hands.

Engine

Cooling fan

A/C condenser fan

7. List any unusual conditions or recommendations for service.

____

____

8. Record your recommendations for needed service or additional repairs on the repair order.
9. Complete the repair order (R.O.).

Instructor OK ______________ Score __________

ASE Lab Preparation Worksheet #6-19

REPLACE A HEATER HOSE

Name____________________________ Class __________________

OBJECTIVE:

Upon completion of this assignment, you should be able to replace a heater hose. This task will help prepare you to pass the ASE certification examination in heating and air conditioning.

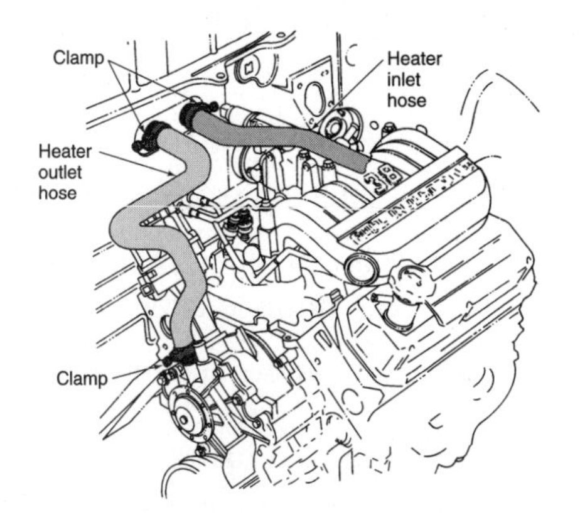

DIRECTIONS:

Before beginning this lab assignment, review the worksheet completely. Fill in the information in the spaces provided as you complete each task.

TOOLS AND EQUIPMENT REQUIRED:

Safety glasses, fender covers, drain pan, shop towel

PARTS AND SUPPLIES:

Heater hose, hose clamps

PROCEDURE:

Vehicle year __________ Make ________________ Model ________________

Repair Order # _______ Engine size ___________ # of Cylinders __________

1. Open the hood and install fender covers over the fenders and front body parts.

CAUTION When removing the radiator cap, the engine must be off and the pressure released. The cooling system is under pressure when the system is at normal operating temperature. As the radiator cap is removed, the coolant may boil. Squeeze the radiator hose to check if the system is under pressure. If the hose is hard, do not remove the radiator cap. See your instructor.

2. Squeeze the top hose. Hard ____ Soft ____
3. Fold a shop rag to 1/4 size. Use it to turn the radiator cap 1/4 turn until its first stop. This will allow any remaining pressure to escape.
4. Check the cap to see that it is loose, indicating that all the pressure has escaped.

   Loose ____ Tight ____
5. If the cap is loose, press down against the spring pressure and turn the cap to remove it.
6. Open the radiator drain valve located in the lower radiator tank and drain some coolant into a clean container. Drain the coolant until the coolant level is below the level of the heater hose.

Before reusing coolant, check its concentration, appearance, and age. Remember, used coolant must be disposed of properly.

7. Remove the hose clamps from the heater hose.

The heater core inlet or outlet is easily damaged or deformed from rough handling of the hoses.

8. Twist the hose to loosen it. It may be necessary to cut the hose with a sharp knife if it does not come off easily.

   Was the hose easily removed? Yes ____ No ____

   Was it necessary to cut the hose? Yes ____ No ____

9. The diameter of the hose is determined by its:

   Outside diameter (O.D.) ____ Inside diameter (I.D.) ____

10. What is the diameter of the heater hose? ________

11. Cut a piece of replacement hose slightly longer than the old one. It can be trimmed later if necessary.

    How long is the new replacement hose? ________

    Note: Many vehicles use molded heater hoses. Match the new hose to the old one before installation.

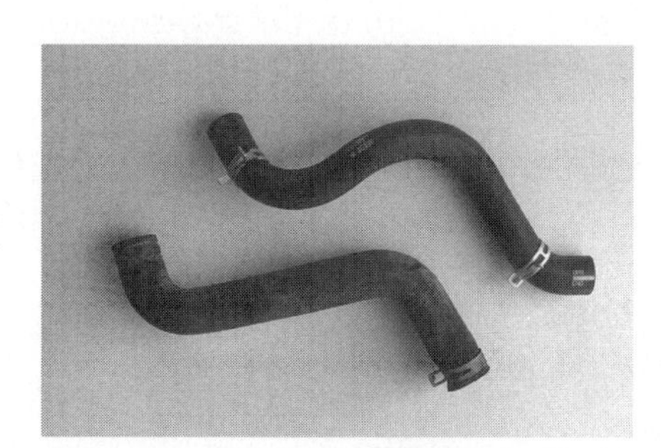

12. Clean the hose connection prior to installing the new hose.

    Note: If the hose is difficult to install, apply some soap to the connection.

Be certain the hose clamps do not interfere with manifolds, belts, or spark plug wiring, and that they will not be damaged by movement of the engine or its accessories.

13. Replace rusted or damaged hose clamps with new clamps and install them on the hose. Position the screw side of the clamp for easy access.

    What size are the replacement hose clamps? ________ "

    Clamps properly installed? Yes ____ No ____

14. Position and tighten the clamps.

15. Refill the system and check for leaks.

    Note: After the cooling system has fully warmed up and then cooled again, the hose clamps should be retightened. Hoses may shrink after their first use.

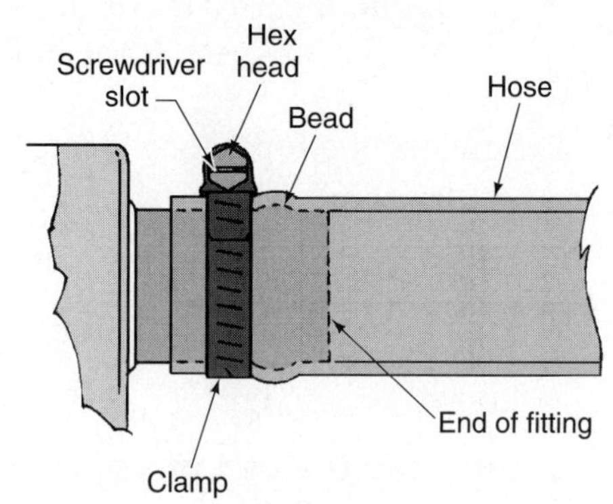

16. Did you retighten the hose clamps? Yes ____ No ____

17. Before completing the paperwork, clean your work area, put the tools in their proper places, and wash your hands.

18. Were there any other recommendations for needed service or unusual conditions that you noticed while replacing the heater hose?

    Yes _____ No _____

19. Record your recommendations for additional service or repairs on the repair order.
20. Complete the repair order (R.O.).

STOP

Instructor OK ______________ Score __________

ASE Lab Preparation Worksheet #6-20

METAL TUBING SERVICE

Name______________________________ Class ____________________

OBJECTIVE:

Upon completion of this assignment, you should be able to fabricate a replacement metal tube with double flared fittings. This task will help prepare you to pass the ASE certification examination in brakes.

DIRECTIONS:

Before beginning this lab assignment, review the worksheet completely. Fill in the information in the spaces provided as you complete each task.

TOOLS AND EQUIPMENT REQUIRED:

Safety glasses, flaring tool set, tubing cutter, tube bender, shop towel

PARTS AND SUPPLIES:

Six inches of copper or steel tubing

PROCEDURE:

Vehicle year ___________ Make _________________ Model __________________

Repair Order # ________ Engine size ____________ # of Cylinders ___________

1. Locate the following:

   Six inches of tubing ____

   Tubing cutter ____

   Flaring tool ____

   Tube bender ____

2. What is the tubing's diameter? ______ "

   Note: The size of tubing or hose is determined by its inside diameter.

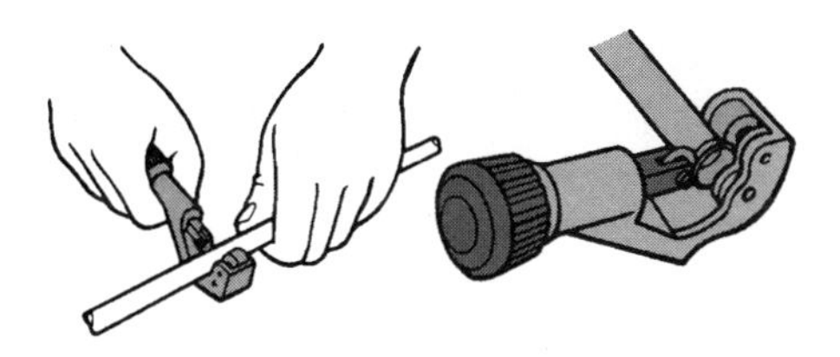

3. Use the tubing cutter to cut off 1" from the tubing and discard it.

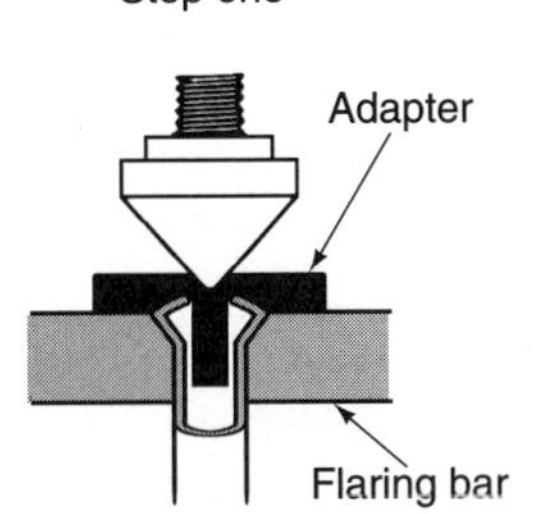

4. Deburr the end of the tubing using the deburring tool.
5. Clamp the line in the flaring tool bar. It should protrude above the bar by the width of the flaring tool adapter.
6. Insert the adapter in the end of the tubing and tighten down on it with the threaded flaring tool until it bottoms out.
7. Remove the adapter and tighten the flaring tool against the tubing again to complete the flare.

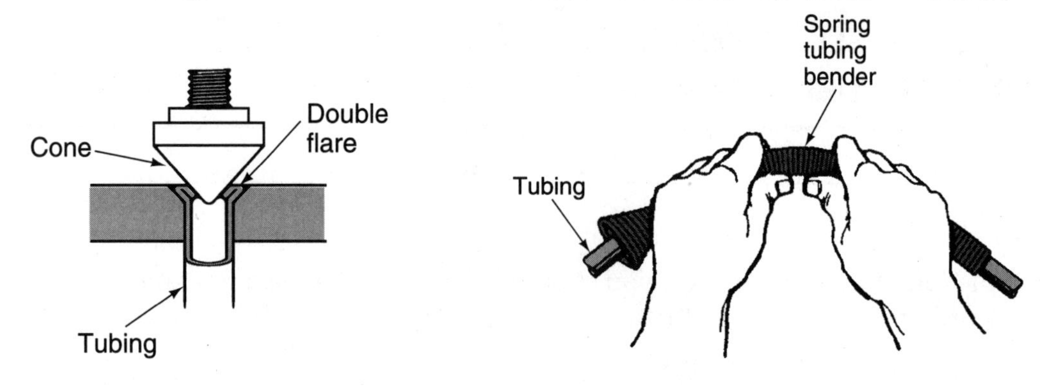

8. Use the tubing bender to bend the tubing in the middle to a 90-degree angle.
9. Before completing the paperwork, clean your work area, put the tools in their proper places, and wash your hands.
10. Did you have any problems completing this worksheet? Yes ____ No ____

    Explain ____

    ____

    ____

STOP

Instructor OK ______________ Score __________

ASE Lab Preparation Worksheet #6-21

CARBURETOR IN-LINE FUEL FILTER SERVICE

Name____________________________ Class ____________________

OBJECTIVE:

Upon completion of this assignment, you should be able to replace an in-line fuel filter on a carbureted vehicle. This task will help prepare you to pass the ASE certification examination in engine performance.

DIRECTIONS:

Before beginning this lab assignment, review the worksheet completely. Fill in the information in the spaces provided as you complete each task.

TOOLS AND EQUIPMENT REQUIRED:

Safety glasses, fender covers, shop towel, screwdriver, drain pan, tape measure, hose cutter

PARTS AND SUPPLIES:

In-line fuel filter, fuel hose, hose clamps

PROCEDURE:

Vehicle year __________ Make ________________ Model ________________

Repair Order # _______ Engine size ____________ # of Cylinders __________

1. Open the hood and place fender covers on the fenders and front body parts.
2. Locate and inspect all synthetic rubber fuel hoses in the fuel system.

   OK ____ Need replacement ____
3. How many synthetic rubber fuel hoses are there on the vehicle? _________

SAFETY NOTE: Any fuel hose that appears to be old or damaged should be replaced immediately, after getting the customer's approval.

4. Fuel hose is sold in bulk. Measure the length of the fuel hoses that need to be replaced.

   How many inches of hose are needed? _________ "
5. Purchase the correct filter, any fuel hose, and clamps that are to be replaced.

   New fuel filter part # _________

   How much hose did you purchase (length)? _________

   What size hose did you purchase (diameter)? _________

   How many hose clamps did you purchase? _________

   What size hose clamps did you purchase? _________

Note: When a new filter is installed, it is a good practice to install new hoses and hose clamps.

TURN

6. Is the filter installed before or after the fuel pump? Before ____ After ____
7. Position something under the filter to catch any leaking fuel (a drain pan, if possible).
8. Apply pinch pliers to the hose on the fuel tank side of the filter.
9. Release fuel pressure by disconnecting the electric wiring to shut down the fuel pump, then run the vehicle until it dies because of lack of fuel.
10. Slowly loosen the clamp and hose on the carburetor end of the filter.

    Was there any residual pressure in the hose?

    Yes ____ No ____

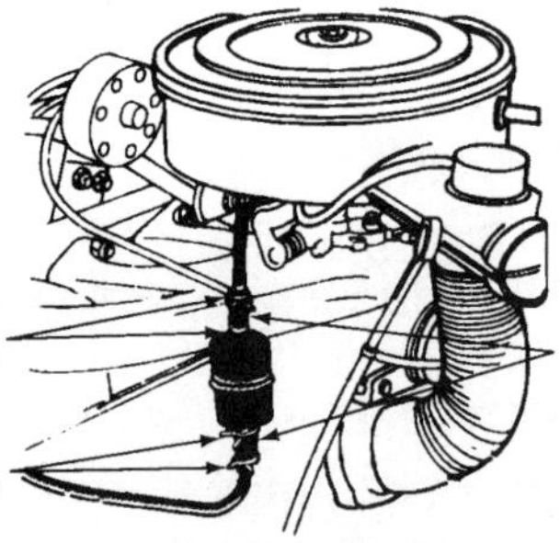

    Spring clamps
    Spring clamps
    Hose

    Note: Sometimes, the gasoline is under *residual pressure* in the fuel line. This will result in a squirt of fuel out of the hose as it is removed.

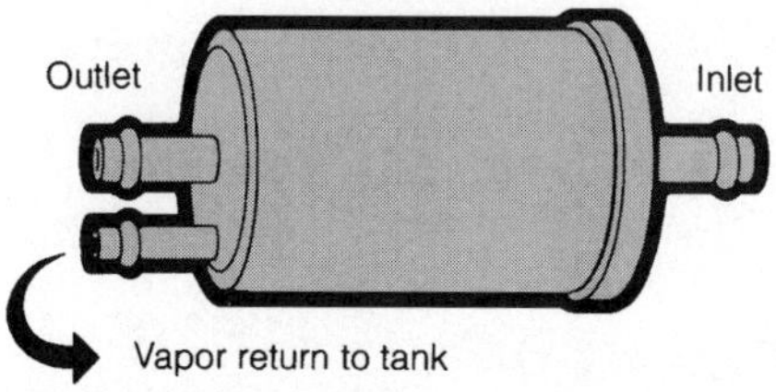

    Outlet
    Inlet
    Vapor return to tank

11. Does the filter have an arrow that indicates the correct direction of installation?

    Arrow ____ No arrow ____

12. Is the filter a "vapor return" type? Yes ____ No ____

    2-line ____ 3-line ____

13. Install the new hoses on the new filter.
14. Install the filter on the carburetor end of the fuel line first.
15. Loosen and remove the old fuel filter and hose. Then *quickly* install the new filter.

    Did you get the new filter in place before any substantial amount of fuel leaked out?

    Yes ____ No ____

16. Tighten the hose clamps and run the engine to check for leaks.

    Are there any leaks? Leaks ____ No leaks ____

SAFETY NOTE Always start the engine and check for leaks after replacing a filter or any fuel lines. Ask your instructor to check your work.

17. Before completing the paperwork, clean your work area, put the tools in their proper places, and wash your hands.
18. Were there any other recommendations for needed service or unusual conditions that you noticed while changing the fuel filter?

    Yes ____ No ____

19. Record your recommendations for additional service or repairs on the repair order.
20. Complete the repair order (R.O.).

STOP

Instructor OK ______________ Score __________

ASE Lab Preparation Worksheet #6-22

FUEL FILTER SERVICE (FUEL INJECTION)

Name______________________________ Class ____________________

OBJECTIVE:

Upon completion of this assignment, you should be able to replace a fuel filter on a fuel-injected vehicle. This task will help prepare you to pass the ASE certification examination in engine performance.

DIRECTIONS:

Before beginning this lab assignment, review the worksheet completely. Fill in the information in the spaces provided as you complete each task.

TOOLS AND EQUIPMENT REQUIRED:

Safety glasses, fender covers, shop towel, screwdriver, drain pan, tape measure, hose cutter, hand tools, spring clock clamp tool, flare nut wrenches

PARTS AND SUPPLIES:

In-line fuel filter, fuel hose, hose clamps

PROCEDURE:

Vehicle year ___________ Make __________________ Model ___________________

Repair Order # ________ Engine size _____________ # of Cylinders ___________

1. Open the hood and place fender covers on the fenders and front body parts.
2. Locate and inspect all synthetic rubber hoses in the fuel system.

   OK ____ Need replacement ____

   Note: Fuel injection systems can use one of several methods to attach the fuel lines to the fuel filter. Hose clamps, special clamps, and banjo fittings are common.

   ❑ Check the service manual for the correct procedure for removing the special clamps.
   ❑ Use the correct wrenches when removing the banjo bolts.

   How are the fuel lines attached to the fuel filter on the vehicle that you are servicing?

   Banjo bolts _______ Hose clamps _______ Spring lock _______

3. How many synthetic rubber fuel hoses are there on the vehicle? ______________

SAFETY NOTE

Any fuel hose that appears to be old or damaged should be replaced immediately, after getting the customer's approval. Always start the engine and check for leaks after replacing any fuel lines.

4. Fuel hose is sold in bulk. Measure the length of the fuel hoses that need to be replaced.

   How many inches of hose are needed? ____ "

TURN

5. Purchase the correct filter, any fuel hose, and clamps to be replaced.

   New fuel filter part # _________

   How much hose did you purchase (length)? _________

   What size hose did you purchase (diameter)? _________

   How many hose clamps did you purchase? _________

   What size hose clamps did you purchase? _________

   Note: When a new filter is installed, it is a good practice to install new hoses and hose clamps.

6. Is the fuel filter installed before or after the fuel pump? Before ____ After ____

7. Relieve any pressure from the fuel tank by removing the fuel filter cap.

   Fuel filter cap removed? Yes ____ No ____

8. Position something under the filter to catch any leaking fuel (a drain pan, if possible).

   Note: Most fuel injection systems maintain some residual pressure in the fuel system. It is recommended that the pressure be relieved before the fuel hoses are removed.

9. There are several ways that the fuel pressure can be bled from the system. Check the service manual for the recommended procedure. Indicate which method below is recommended for the vehicle you are servicing:

   ____ Remove the fuel pump fuse and crank the engine for a few seconds.

   ____ Use a jumper wire to bypass the fuel pump relay.

   ____ Some fuel injection systems have a *Schrader valve* that can be used to bleed off pressure from the system before disassembly. (A Schrader valve is the kind of valve that is found on tire valve stems.)

   ____ Use jumper wires to energize an injector.

   ____ Other (describe) ____

   ____

10. If there is a synthetic rubber hose on the fuel tank side of the filter, apply pinch pliers to the hose. This will minimize leakage of fuel.

    Is there a synthetic rubber hose? Yes ____ No ____

    Were pinch pliers applied to the hose? Yes ____ No ____

FUEL FEED PIPE

FRT

FUEL FILTER

11. Slowly loosen the clamp or banjo bolt and hose on the injector end of the filter.

    Was there any residual pressure in the hose?

    Yes ____ No ____

    Note: Use caution when removing the hose. The gasoline may still be under *residual pressure* in the fuel line. This will result in a squirt of fuel out of the hose as it is removed. Use caution when removing the hose.

CONTINUED

12. Does the filter have an arrow that indicates the correct direction of installation?

    Arrow ____ No arrow ____

13. Install the new hoses on the new filter.
14. Install the filter on the injector end of the fuel line first.
15. Loosen and remove the old fuel filter and hose. Then *quickly* install the new filter.

    Did you get the new filter in place before any substantial amount of fuel leaked out?

    Yes ____ No ____

16. Tighten the hose clamps or banjo bolts and clean up any fuel that leaked while you were changing the fuel filter.
17. Replace the fuel filler cap and run the engine to check for leaks. Leaks ____ No leaks ____
18. Before completing the paperwork, clean your work area, put the tools in their proper places, and wash your hands.
19. Record your recommendations for additional service or repairs on the repair order.
20. Complete the repair order (R.O.).

STOP

ASE Lab Preparation Worksheet #6-23

CHECK FUEL PRESSURE—FUEL INJECTION

Name______________________________ Class ____________________

OBJECTIVE:

Upon completion of this assignment, you should be able to check the fuel pressure in a fuel injection system. This task will help prepare you to pass the ASE certification examination in engine performance.

DIRECTIONS:

Before beginning this lab assignment, review the worksheet completely. Fill in the information in the spaces provided as you complete each task.

TOOLS AND EQUIPMENT REQUIRED:

Safety glasses, fender covers, shop towel

PROCEDURE:

Vehicle year ____________ Make __________________ Model ___________________

Repair Order # ________ Engine size _____________ # of Cylinders ____________

1. Open the hood and place fender covers on the fenders and front body parts.
2. Locate the appropriate place to attach the fuel pressure tester. Indicate which you choose.

   Fuel line pressure port (Schrader valve) ____________

   Fuel line (in series) ____________

   Cold start injector ____________

   Fuel pressure gauge

   Fuel rail test port

   Note: Most fuel injection systems maintain some residual pressure in the fuel system. It is recommended that the pressure be relieved before the fuel system is opened to attach the fuel pressure gauge.

3. There are several ways that the fuel pressure can be bled from the system. Check the service manual for the recommended procedure. Indicate which method below is recommended for the vehicle that you are servicing.

   ________ Remove the fuel pump fuse and crank the engine for a few seconds.

   ________ Use a jumper wire to bypass the fuel pump relay.

   ________ Bleed the Schrader valve.

   Note: Cover the valve with a clean rag to prevent fuel from spraying on you or the engine.

   ________ Use jumper wires to energize an injector.

   ________ Other (describe) ____

   ____

TURN

4. Connect the fuel pressure gauge to the fuel injection system.

   Note: The pressure gauge must be able to read at least 100 psi.

5. Start the engine and let it run at idle. If the engine will not run, check the service manual for the proper way to energize the fuel pump.
6. Read the pressure on the fuel gauge.
   a. What did it read? ____________
   b. What is the specification for the fuel pressure? ____________
7. Was the pressure: High _________ Low _________ OK _________
8. What could cause high fuel pressure? ________________________________
9. What could cause the fuel pressure to be too low? ________________________________

   ____
10. Remove the fuel gauge and replace the fuel line or valve cap.
11. Start the engine and check for fuel leaks. Is there any fuel leaking? Yes ____ No ____
12. Before completing the paperwork, clean your work area, put the tools in their proper places, and wash your hands.
13. Were there any recommendations for needed service or unusual conditions that you noticed while you were checking the fuel pressure? Yes ____ No ____
14. Record your recommendations for needed service or additional repairs on the repair order.
15. Complete the repair order (R.O.).

STOP

ASE Lab Preparation Worksheet #6-24

PCV VALVE INSPECTION

Name______________________________ Class ____________________

OBJECTIVE:

Upon completion of this assignment, you should be able to inspect and service the positive crankcase ventilation (PCV) system. This task will help prepare you to pass the ASE certification examination in engine performance.

DIRECTIONS:

Before beginning this lab assignment, review the worksheet completely. Fill in the information in the spaces provided as you complete each task.

TOOLS AND EQUIPMENT REQUIRED:

Safety glasses, fender covers, shop towel

PROCEDURE:

Vehicle year ___________ Make _________________ Model __________________

Repair Order # ________ Engine size _____________ # of Cylinders ____________

1. Open the hood and place fender covers on the fenders and front body parts.
2. Locate the PCV system air filter. Where is it located?

   Air filter housing ____ Oil filler cap ____ Other ____

3. Condition of crankcase vent filter:

   OK ____ Needs service ____

   Note: Oil in the air cleaner can indicate a plugged PCV system or excessive blowby. Further testing of the PCV system or engine will be required.

Vapor separator
Air management valve
EGR valve
Evaporative canister
Purge valve
PCV valve
Catalytic converter
Fuel tank

4. Locate the PCV valve. Where is it located? ______________________________________
5. Which of the following engine gaskets show signs of leakage?

   Valve cover ____ Rear crankshaft seal ____

   Oil pan ____ Camshaft seal ____

   Timing cover ____ Other ____

   Front crankshaft seal ____

6. Inspect the condition of the PCV hoses.

   Good ____ Cracked ____

   Deteriorated ____ Loose connections ____

7. Disconnect the PCV valve from the engine.
8. Shake the valve. Does it rattle? Yes ____ No ____

Rpm Drop Test

9. Be certain the car is in Park (automatic transmission)/Neutral (manual transmission) with the parking brake firmly set.

   Park ____ Neutral ____ Parking brake set ____

10. Start the engine and let it idle.
11. Cover the end of the PCV valve with your thumb. Does the engine idle change (engine speed should drop)?

    Yes ____ No ____

12. Did you feel a strong vacuum when the valve was restricted? Yes ____ No ____
13. Reinstall the PCV valve. How is it attached to the engine?

    Rubber grommet ____ Threaded connection ____ Other ____

Vacuum Test

14. Remove the oil filler cap.
15. Position a piece of paper or a dollar bill over the oil filler opening. If the system is operating properly, engine vacuum will pull the paper against the opening. If the system fails to operate properly, replace the valve.

    OK ____ Needs valve replaced ____

16. Before completing the paperwork, clean your work area, put the tools in their proper places, and wash your hands.
17. Record your recommendations for additional service or repairs on the repair order.
18. Complete the repair order (R.O.).

STOP

Instructor OK ______________ Score __________

ASE Lab Preparation Worksheet #6-25

OXYGEN SENSOR TEST (ZIRCONIUM-TYPE SENSOR)

Name_____________________________ Class ____________________

OBJECTIVE:

Upon completion of this assignment, you should be able to test the oxygen sensor(s). This task will help prepare you to pass the ASE certification examination in engine performance.

DIRECTIONS:

Before beginning this lab assignment, review the worksheet completely. Fill in the information in the spaces provided as you complete each task.

TOOLS AND EQUIPMENT REQUIRED:

Safety glasses, digital multimeter, fender covers, shop towel

Note: A *high impedance voltmeter* must be used to perform the following tests. Most *digital* meters are high impedance.

PROCEDURE:

Vehicle year ___________ Make _________________ Model __________________

Repair Order # ________ Engine size ____________ # of Cylinders ___________

1. Open the hood and install fender covers over the fenders.
2. Locate the oxygen sensor(s).

   How many oxygen sensors are used on the vehicle? ____
3. How many wires does each of the sensors have?

   1 wire ____ 2 wires ____

   3 wires ____ 4 wires ____

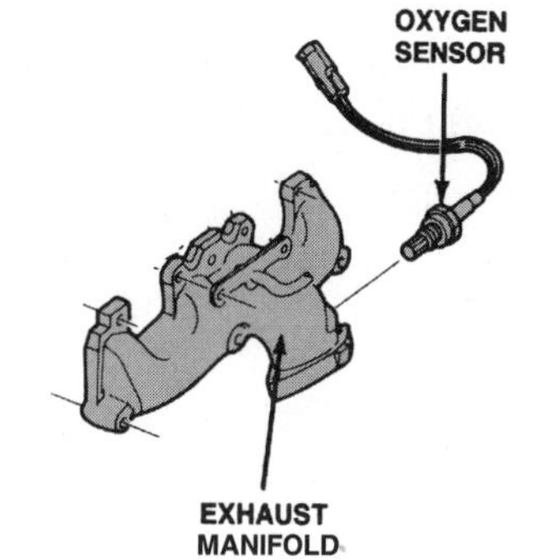

Note: Oxygen sensors have many variations in the number of wires they use. Single-wire systems use the wire for signal voltage. In multiwire systems, one of the wires provides power to heat the sensor, one or two provide the ground paths, and the other wire provides the computer signal.

4. Connect the voltmeter ground lead to ground and the positive lead to the signal wire coming from the sensor.

Note: If there is more than one wire, checking with a voltmeter will determine the signal lead. The engine must be running at normal operating temperature to make this test.

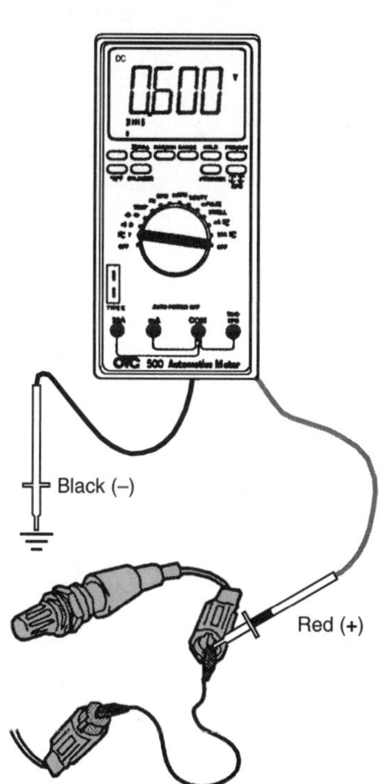

5. Start the engine and run it at fast idle (2000 rpm) for 2 minutes to heat the exhaust system.

   Hint: Feel the top radiator hose to see if the engine is getting hot. Also, cars with smog pumps often have switching for the pump that changes when the fuel system goes into closed loop. This can be heard or felt through the hose from the smog pump.

6. When the oxygen sensor is hot enough, it will begin to operate.

   Note the voltage readings.

   Does voltage move constantly between 0.2 volt and 0.8 volt? Yes ____ No ____

   Does the voltage hold steady? Yes ____ No ____

   Note: A good oxygen sensor will vary between low and high voltage at least 10 times a second at 2000 rpm. This is too fast to measure with a voltmeter, but a scan tool or oscilloscope can measure this "crosscount" speed.

Test Results:

Good oxygen sensor, operating correctly ____

Requires more testing, the system is not operating correctly ____

7. Before completing the paperwork, clean your work area, put the tools in their proper places, and wash your hands.
8. Were there any other recommendations for needed service or unusual conditions that you noticed while testing the oxygen sensor?

   Yes ____ No ____

9. Record your recommendations for additional service or repairs on the repair order.
10. Complete the repair order (R.O.).

STOP

ASE Lab Preparation Worksheet #6-26

IDENTIFY AND INSPECT EMISSION CONTROL SYSTEMS

Name______________________________ Class ____________________

OBJECTIVE:

Upon completion of this assignment, you should be able to identify emission control systems and inspect for their presence on a vehicle. This task will help prepare you to pass the ASE certification examination in engine performance.

DIRECTIONS:

Before beginning this lab assignment, review the worksheet completely. Fill in the information in the spaces provided as you complete each task.

TOOLS AND EQUIPMENT REQUIRED:

Safety glasses, fender covers

PROCEDURE:

Vehicle year ___________ Make __________________ Model ___________________

1. Open the hood and place fender covers on the fenders and front body parts.
2. Locate the vehicle's underhood emission label.
3. Record the information from the underhood emission label below:

   Note: Not all vehicles require all of the adjustments requested. If requested information is not on the underhood label, write N/A (not applicable) in the answer space.

Air management valve
EGR valve
Vapor separator
Evaporative canister
Purge valve
PCV valve
Catalytic converter
Fuel tank

   Engine size ______________________

   Timing specifications ______________________

   Special ignition timing instructions ____

   ____

   Fast idle speed ______________________

   Curb idle speed ______________________

   Valve lash

   Intake _________ Exhaust _________

   Spark plug gap _________

   Other _________

4. Read the emission control label and check below the emission control systems that are required on the vehicle:

Note: Not all emission control systems are identified on the underhood label. Some emission systems are required on all vehicles and these are not identified on the label. It may be necessary to use a service manual to identify which systems are required on the vehicle you are inspecting.

Positive Crankcase Ventilation (PCV) Yes ____ No ____

Thermostatic Air Cleaner (TAC) Yes ____ No ____

Fuel Evaporation System (EVAP) Yes ____ No ____

Catalytic Converter (CAT) Yes ____ No ____

Exhaust Gas Recirculation System (EGR) Yes ____ No ____

Air Injection (AIR) Yes ____ No ____

Early Fuel Evaporation (EFE) Yes ____ No ____

5. Does the vehicle have an underhood vacuum routing label? Yes ____ No ____
6. Without removing the air cleaner, check the routing of the vacuum hoses. Do they match the routing on the underhood label?

   Yes ____ No ____
7. Locate and inspect each of the emission control systems. Indicate below the condition of the system. Use the following terms to describe the condition of the systems: pass, missing, disconnected, defective, or not used.

   Positive Crankcase Ventilation (PCV) ____________________

   Thermostatic Air Cleaner (TAC) ____________________

   Fuel Evaporation System (EVAP) ____________________

   Catalytic Converter (CAT) ____________________

   Exhaust Gas Recirculation System (EGR) ____________________

   Air Injection (AIR) ____________________

   Early Fuel Evaporation (EFE) ____________________
8. List any other recommendations for needed service or unusual conditions that you noticed while inspecting the emission systems.

Recommendations: ____

____

STOP

Part II

ASE Lab Preparation Worksheets

Service Area 7

Electrical Services

Instructor OK ______________ Score ___________

ASE Lab Preparation Worksheet #7-1

GLASS AND CERAMIC FUSE TESTING AND SERVICE

Name_____________________________ Class _____________________

OBJECTIVE:

Upon completion of this assignment, you should be able to determine if a fuse is faulty and be able to replace it. This task will help prepare you to pass the ASE certification examination in electrical systems.

DIRECTIONS:

Before beginning this lab assignment, review the worksheet completely. Fill in the information in the spaces provided as you complete each task.

TOOLS AND EQUIPMENT REQUIRED:

Safety glasses, fender covers, circuit tester (test light), shop towel

Note: A voltmeter can also be used for this test.

PARTS AND SUPPLIES:

Glass or ceramic fuses

PROCEDURE:

Vehicle year ___________ Make _________________ Model ___________________

1. Open the hood and place fender covers on the fenders and front body parts.
2. Locate the fuse block (usually under the instrument panel). Consult a service manual if the fuse block is not located quickly.

   Was a service manual used?

   Yes ____ No ____
3. Does the vehicle have glass cartridge, ceramic, or blade fuses?

   Glass ____

   Ceramic ____

   Blade ____

   Note: If the fuse is the glass or ceramic type, complete this worksheet. If the fuse is a blade fuse, complete Worksheet 7-2.
4. Use a test light. Check tester operation by connecting it across the terminals of a battery.

   Does the tester light up? Yes ____ No ____
5. Locate the windshield wiper fuse. Fuse identification and rating is usually labeled inside of the fuse box cover.

TURN

6. What fuse rating is required for the wiper circuit? _____ amps
7. Visually inspect the fuse. Does the fuse appear to be good? Yes _____ No _____
8. With the test light clipped to ground, does the tester light up when both sides of the fuse are probed?

   Yes _____ No _____

   Note: If the test light does not light when either end of the fuse is probed, turn the ignition to the *ON* position and try the test again.

9. Does the tester light up when the ends of the fuse are probed with the key on?

   Yes _____ No _____
10. Remove the fuse with a fuse removal tool.
11. How many amps is the wiper fuse? The fuse rating is marked on the metal ends of the fuse.

    _____amps
12. What is the letter designation on the fuse? Is it the correct fuse? Yes _____ No _____
13. Probe both sides of the fuse socket with the test light.

    Did the tester light when both ends of the fuse socket were probed? Yes _____ No _____
14. Which side of the fuse socket is the ground side?

    Top _____Bottom _____Right _____Left _____
15. Which side of the fuse socket is the battery (B+) side?

    Top _____Bottom _____Right _____Left _____
16. Replace the windshield wiper fuse.
17. Turn the wipers on. Do they work? Yes _____ No _____

Interpretation of Fuse Test Results

- ❑ If the tester glows when touched to either side of an installed fuse, the fuse is good and the current is flowing.
- ❑ If the tester glows only when touched to one end of the fuse, the fuse is defective. (The side that does not light is the ground side of the circuit.)
- ❑ If the tester does not glow on either side of the circuit, the circuit is shut off, the circuit is open, or the tester does not have a good ground connection. There might be nothing wrong.

18. Before completing the paperwork, clean your work area, put the tools in their proper places, and wash your hands.
19. Record any other recommendations below for needed service or unusual conditions that you noticed while checking the fuses.

Recommendations: ____

____

____

____

STOP

Instructor OK ______________ Score __________

ASE Lab Preparation Worksheet #7-2

BLADE FUSE TESTING AND SERVICE

Name____________________________ Class ____________________

OBJECTIVE:

Upon completion of this assignment, you should be able to determine if a blade fuse is faulty and be able to replace it. This task will help prepare you to pass the ASE certification examination in electrical systems.

DIRECTIONS:

Before beginning this lab assignment, review the worksheet completely. Fill in the information in the spaces provided as you complete each task.

TOOLS AND EQUIPMENT REQUIRED:

Safety glasses, fender covers, circuit tester (test light), shop towel

PARTS AND SUPPLIES:

Blade fuses

PROCEDURE:

Vehicle year ___________ Make _________________ Model ___________________

1. Open the hood and place fender covers on the fenders and front body parts.
2. Locate the fuse block (usually under the instrument panel). Consult a service manual if the fuse block is not located quickly.

   Was a service manual used? Yes ____ No ____
3. Does the vehicle have glass cartridge, ceramic, or blade fuses?

   Glass ____ Ceramic ____ Blade ____

   Note: If the fuses are of the blade type, complete this worksheet. If the fuses are glass or ceramic, complete Worksheet 7-1.
4. Use a test light. Check tester operation by connecting it across the terminals of a battery.

   Does the tester light up? Yes ____ No ____
5. Locate the windshield wiper fuse. Fuse identification and rating is usually labeled inside of the fuse box cover.
6. What fuse rating is required for the wiper circuit? ____amps
7. What is the rating of the fuse in the windshield wiper socket? ____amps

   Is it the correct fuse? Yes ____ No ____
8. What color is the fuse? __________

9. Blade fuses are color coded for easy identification. Use lines to connect the fuse color to the fuse rating.

5 amps	Yellow
7.5 amps	Brown
10 amps	Green
15 amps	Yellowish brown
20 amps	Blue
25 amps	Colorless (clear)
30 amps	Red

10. Connect the test light (–) to ground. Does the tester light up when both sides of the fuse are probed?

    Yes ____ No ____

    Note: If the test light does not light when either side of the fuse is probed, turn the ignition to the *ON* position and try the test again.

12-volt test light

11. Does the tester light up when the ends of the fuse are probed with the key on?

    Yes ____ No ____

12. Remove the fuse with a fuse removal tool.
13. Visually inspect the fuse. Does the fuse appear to be good? Yes ____ No ____
14. Probe both sides of the fuse socket with the test light. Did the tester light up when both sides of the fuse socket were probed?

    Yes ____ No ____

15. Which side of the fuse socket is the ground side?

    Top ____Bottom ____Right ____Left ____

16. Which side of the fuse socket is the battery (B+) side?

    Top ____Bottom ____Right ____Left ____

17. Replace the windshield wiper fuse.
18. Turn the wipers on. Do they work? Yes ____ No ____

Interpretation of Fuse Test Results

- ❑ If the tester glows when touched to either side of an installed fuse, the fuse is good and the current is flowing.
- ❑ If the tester glows only when touched to one end of the fuse, the fuse is defective. (The side that does not light is the ground side of the circuit.)
- ❑ If the tester does not glow on either side of the circuit, the circuit is shut off, the circuit is faulty, or the tester does not have a good ground connection. There could be nothing wrong.

19. Before completing the paperwork, clean your work area, put the tools in their proper places, and wash your hands.
20. Record any other recommendations below for needed service or unusual conditions that you noticed while checking the fuses.

Recommendations: ____

____

____

____

STOP

Instructor OK _______________ Score ___________

ASE Lab Preparation Worksheet #7-3

SPLICE A WIRE WITH A CRIMP CONNECTOR

Name_____________________________ Class ____________________

OBJECTIVE:

Upon completion of this assignment, you should be able to repair a damaged wire by splicing it together with a crimp connector. This task will help prepare you to pass the ASE certification examination in electrical systems.

DIRECTIONS:

Before beginning this lab assignment, review the worksheet completely. Fill in the information in the spaces provided as you complete each task.

TOOLS AND EQUIPMENT REQUIRED:

Safety glasses, fender covers, crimping-stripper tool, shop towel

PARTS AND SUPPLIES:

Crimp-type wire terminals

PROCEDURE:

Note: If you are doing this worksheet for practice, write N/A in the spaces for vehicle identification.

Vehicle year ___________ Make __________________ Model ___________________

Repair Order # ________ Engine size _____________ # of Cylinders ____________

1. Locate the two wires to be spliced together.

   Are the wires part of a repair or are the wires being spliced together for practice?

   Repair _____ Practice _____

2. Use a crimping tool to strip the insulation (about 1/4") from the ends of the two wires that are to be spliced. If the wire is not clean and shiny, cut the wire back until the wires are clean and shiny.

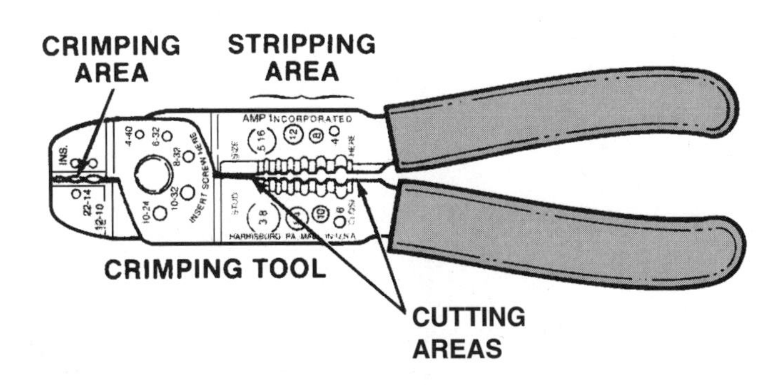

3. Select the proper crimp connector.

   What type of crimp connector was chosen?

   ____

4. Locate the proper crimp area on the crimp tool that corresponds to the gauge of the wire being spliced.
5. Crimp the wire into the connector. The dimple from the crimp tool should be opposite the seam in the connector.
6. Insert the other wire end into the crimp.
7. Test the crimp by lightly pulling on the wires.

   Did the crimp repair hold? Yes ____ No ____
8. Before completing the paperwork, clean your work area, put the tools in their proper places, and wash your hands.
9. If you completed this task as a repair, complete the repair order (R.O.).

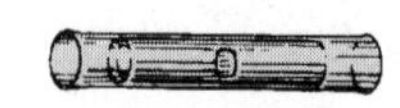

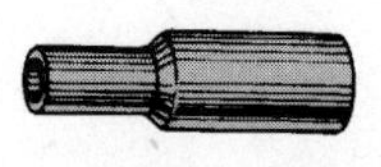

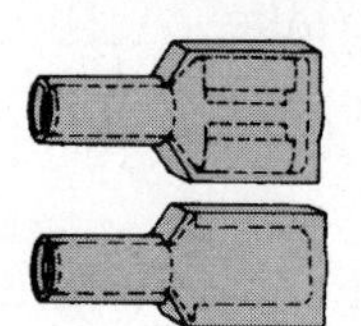

Instructor OK ____________ Score ____________

ASE Lab Preparation Worksheet #7-4

SOLDER A WIRE CONNECTION

Name______________________ Class ______________

OBJECTIVE:

Upon completion of this assignment, you should be able to solder a wire to make an electrical repair. This task will help prepare you to pass the ASE certification examination in electrical systems.

DIRECTIONS:

Before beginning this lab assignment, review the worksheet completely. Fill in the information in the spaces provided as you complete each task.

TOOLS AND EQUIPMENT REQUIRED:

Safety glasses, shop towel, soldering gun, wire strippers, heat gun

PARTS AND SUPPLIES:

Shrink tube or electrical tape, rosin core solder, six-inch (6″) piece of wire

PROCEDURE:

If a repair is being made to a vehicle, complete the following. Otherwise list "N/A" in the blanks.

Vehicle year __________ Make ______________ Model ______________

Repair Order # ________ Engine size __________ # of Cylinders __________

1. Obtain a six-inch (6″) length of wire.
2. Strip the ends of the wire using wire strippers.
3. Bring the ends of the wire together to form a loop and twist the wires together like a pigtail.

   Note: Shrink tubing must be installed on the wire before splicing and soldering the connection.

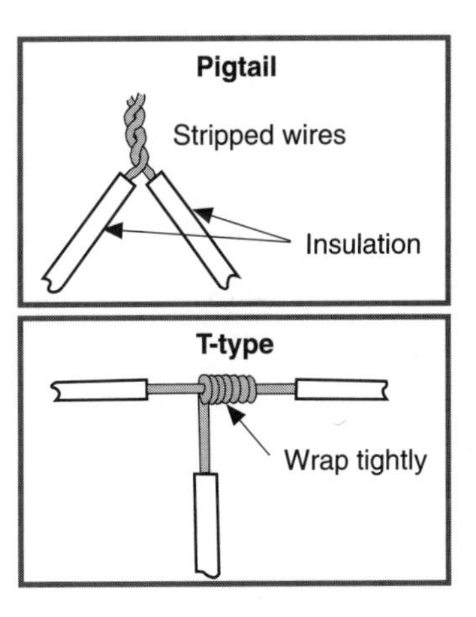

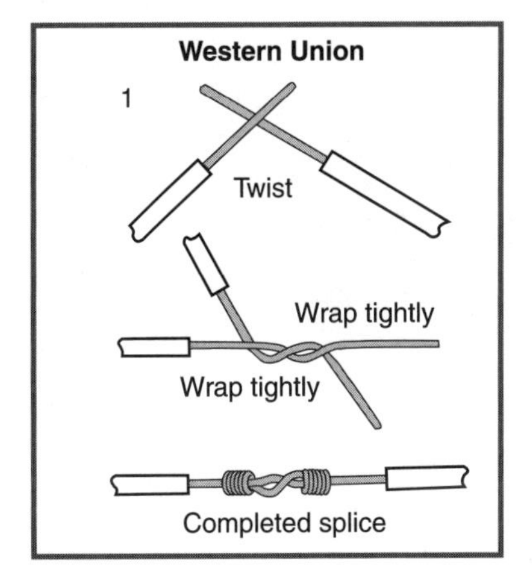

4. When the soldering gun is hot, apply solder to the soldering tip. Clean the hot tip with a damp towel. Reheat the tip and place a little solder (tinning) on the tip.

   Note: Use rosin core or solid core solder for electrical connections.

5. Hold the tip of the soldering gun against the wire. Depress the trigger to heat the wire.
6. Hold the solder against the opposite side of the connection from the soldering gun.

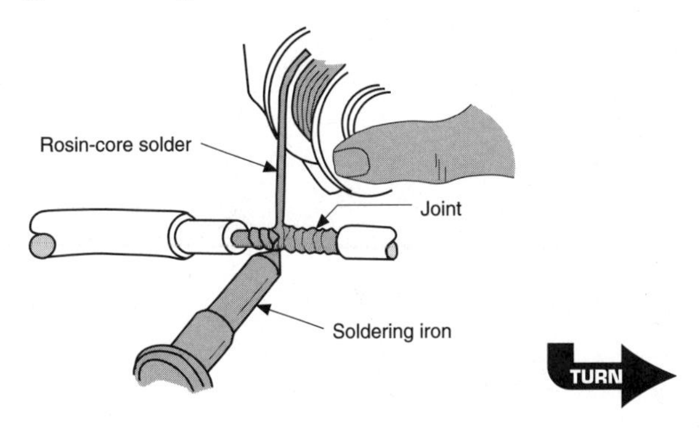

TURN

7. When the wires are heated, the solder will melt, saturating the connection.

   Note: Crimp connections can be soldered to wire in a similar manner.

8. After the soldered wire cools, insulate the connection with electrical tape or shrink tube.

   What insulation was used?

   Electrical tape ___________

   Shrink tubing ___________

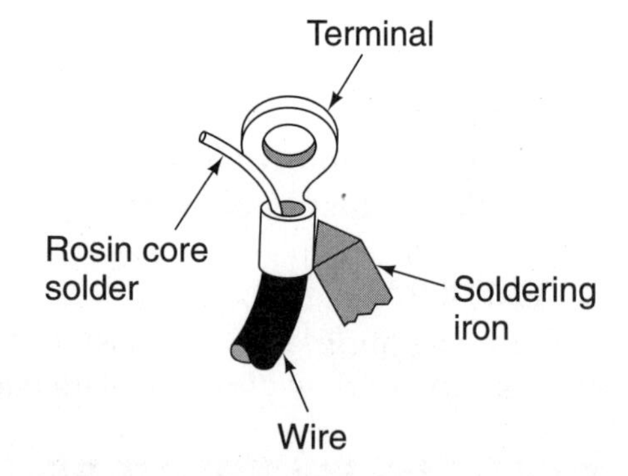

9. Before completing the paperwork, clean your work area, put the tools in their proper places, and wash your hands.

   Note: If the repair was done on a vehicle, complete the following questions.

10. Were there any other recommendations for needed service or unusual conditions that you noticed while soldering the wires?

    Yes ____ No ____

11. Record your recommendations for additional service or repairs on the repair order.
12. Complete the repair order (R.O.).

STOP

Instructor OK ______________ Score __________

ASE Lab Preparation Worksheet #7-5

BATTERY SERVICE

Name____________________________ Class ________________

OBJECTIVE:

Upon completion of this assignment, you should be able to service a battery. This task will help prepare you to pass the ASE certification examinations in electrical and engine performance.

DIRECTIONS:

Before beginning this lab assignment, review the worksheet completely. Fill in the information in the spaces provided as you complete each task.

TOOLS AND EQUIPMENT REQUIRED:

Safety glasses, fender covers, shop towel, computer memory retaining tool, battery pliers, battery terminal cleaner, battery terminal puller, battery carrier or strap

PARTS AND SUPPLIES:

Baking soda, paint

PROCEDURE:

Vehicle year __________ Make ________________ Model ________________

Repair Order # _______ Engine size ___________ # of Cylinders __________

Note: Parking the vehicle outdoors near a water drain for battery tray washing will make cleanup easier.

1. Open the hood and place fender covers on the fenders and front body parts.
2. Check that all electrical circuits are off.
3. If there are any electronic memory circuits (radio stations, radio security, seats, computers, etc.), install a computer memory retaining tool into the cigarette lighter socket. Otherwise the memory circuits will need to be reset after the battery service.
4. Loosen the ground cable first. Which is the ground cable?

   (+) _____ (–) _____

5. Inspect the condition of the terminal clamp nut.

   Worn _____ OK _____

6. If the nut is worn, use battery pliers to loosen it.
7. Remove the terminal clamp.

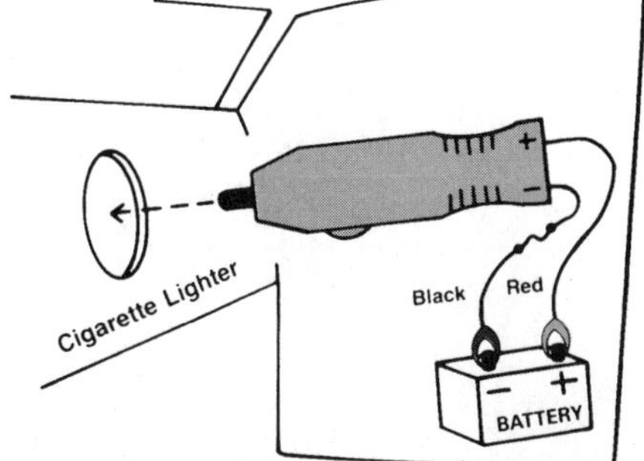

Note: If clamp does not come off easily, use a battery terminal puller.

8. Inspect the condition of the terminal clamp.

   Oxidized _____ Clean _____ Worn _____

TURN

Note: If the clamp is not serviceable, the cable should be replaced. Refer to Worksheet 7-6, Replace a Battery Cable.

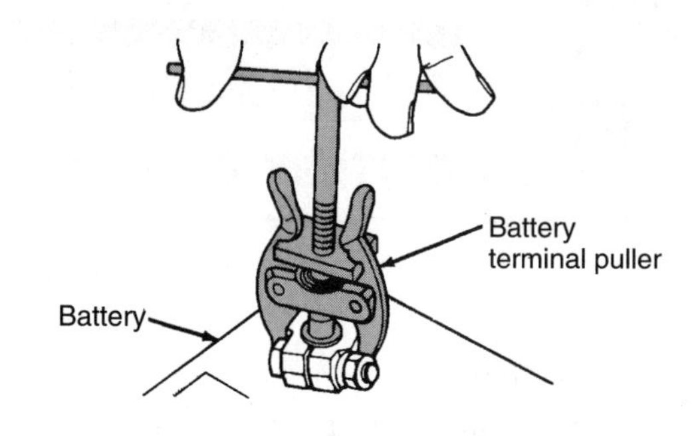

9. Remove the remaining terminal clamp.
10. Inspect the condition of the terminal post.

    Oxidized _____ Clean _____ Worn _____

11. Use a battery terminal cleaning tool to carefully clean both posts and the insides of the terminal clamps.
12. Lubricate the threads on the battery holddown clamp.
13. Remove the battery holddown clamp.
14. Lift the battery from the vehicle. Use a battery carrier to avoid dropping the battery.
15. Clean the battery, its tray, and the battery holddown with baking soda and water. Hose them off thoroughly. Do not allow baking soda solution to enter the battery cells.
16. Repaint the tray and holddown after they are completely dry.
17. Check the electrolyte level and refill the battery as needed.
18. Reinstall the battery and holddown.
19. Tighten the holddown fasteners until they are snug. *Do not overtighten.*

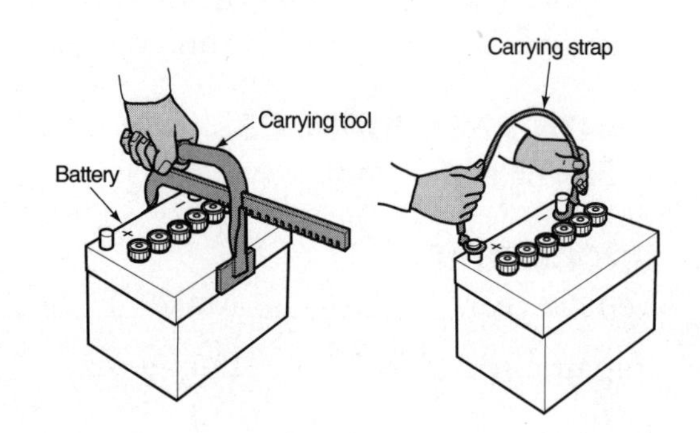

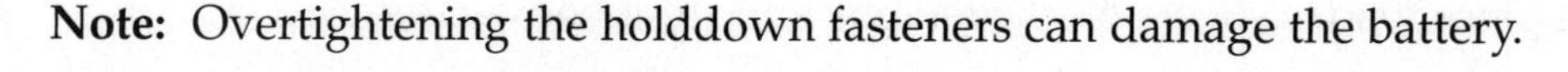

Note: Overtightening the holddown fasteners can damage the battery.

20. Reinstall the positive (+) battery cable.
21. Reinstall the ground (-) battery cable.

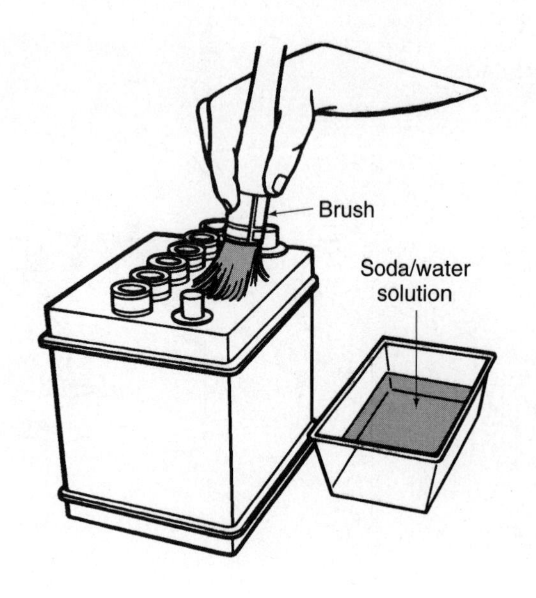

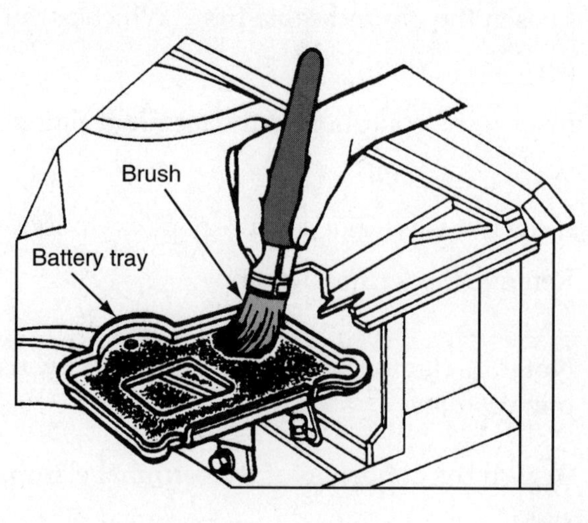

22. Use battery spray, treated felt washers, Silicone RTV, or apply some grease around the base of the terminal to prevent oxidation. Method used:

    Battery spray _____ Felt washers _____ Grease _____ Silicone RTV _____ Other _____

23. Remove the computer memory retaining tool from the cigarette lighter.
24. Check that the vehicle starts.
25. Before completing the paperwork, clean your work area, put the tools in their proper places, and wash your hands.
26. Were there any other recommendations for needed service or unusual conditions that you noticed while servicing the battery? Yes ____ No ____
27. Record your recommendations for additional service or repairs on the repair order.
28. Complete the repair order (R.O.).

Instructor OK ______________ Score __________

ASE Lab Preparation Worksheet #7-6

REPLACE A BATTERY CABLE

Name________________________ Class ________________

OBJECTIVE:

Upon completion of this assignment, you should be able to replace a battery cable. This task will help prepare you to pass the ASE certification examinations in electrical and engine performance.

DIRECTIONS:

Before beginning this lab assignment, review the worksheet completely. Fill in the information in the spaces provided as you complete each task.

TOOLS AND EQUIPMENT REQUIRED:

Safety glasses, fender covers, shop towel, computer memory retaining tool, battery pliers, battery terminal cleaner, battery terminal puller

PROCEDURE:

Vehicle year __________ Make ________________ Model ________________

Repair Order # ________ Engine size____________ # of Cylinders ___________

1. Which cable is being replaced?

   Positive ____ Negative ____

2. How did you determine that the battery cable needed to be replaced?

   ____

3. Open the hood and place fender covers on the fenders and front body parts.
4. Remove the terminal clamp from the battery post. Refer to Worksheet 7-5, Battery Service.

   Note: Always remove the ground cable before working with the positive battery cable. Also, remember to protect the memory circuits.

5. Remove the other end of the cable from the:

   Starter solenoid ____

   Cylinder block ____

   Other ____

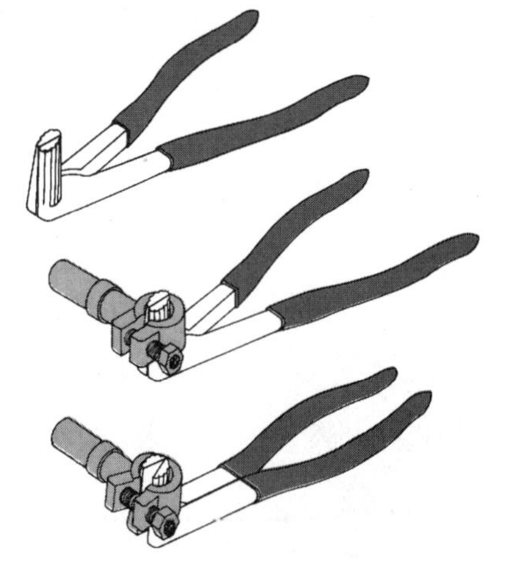

6. Clean the battery post with a terminal cleaner.
7. Expand the eye of the terminal using the expanding tool.
8. Install the battery terminal clamp over the battery post.
9. Install the other end of the cable to the:

   Cylinder block ____

   Starter solenoid ____

   Other ____

10. Which cable do you always install last?

    Positive ____ Negative ____ Ground ____

11. Check that the cables are properly routed so that they will not be damaged.

    Are the cables properly routed?

    Yes ____ No ____

12. Treat the terminal connection to prevent corrosion. Which method was used?

    Grease ________

    Felt washers ____

    Spray _________

13. Does the vehicle start?

    Yes ____ No ____

14. Before completing the paperwork, clean your work area, put the tools in their proper places, and wash your hands.

15. Were there any other recommendations for needed service or unusual conditions that you noticed while replacing the battery cable?

    Yes ____ No ____

16. Record your recommendations for additional service or repairs on the repair order.

17. Complete the repair order (R.O.).

Instructor OK ________________ Score ____________

ASE Lab Preparation Worksheet #7-7

REPLACE A BATTERY TERMINAL CLAMP

Name_______________________________ Class ____________________

OBJECTIVE:

Upon completion of this assignment, you should be able to repair a starting system problem by replacing a battery terminal clamp. This task will help prepare you to pass the ASE certification examinations in electrical and engine performance.

DIRECTIONS:

Before beginning this lab assignment, review the worksheet completely. Fill in the information in the spaces provided as you complete each task.

TOOLS AND EQUIPMENT REQUIRED:

Safety glasses, fender covers, shop towel, diagonal cutting pliers, sharp knife, computer memory retaining tool, battery pliers, battery terminal cleaner, battery terminal puller

A propane torch will be needed if a permanent terminal is being installed.

PARTS AND SUPPLIES:

Battery terminal clamp

PROCEDURE:

Vehicle year ___________ Make __________________ Model ____________________

Repair Order # ________ Engine size_____________ # of Cylinders ____________

1. Obtain a new battery terminal clamp.
2. Which type of terminal clamp is being used as the replacement?

   Temporary bolt-on clamp ____

   Solder-type clamp ___________

3. Open the hood and place fender covers on the fenders and front body parts.

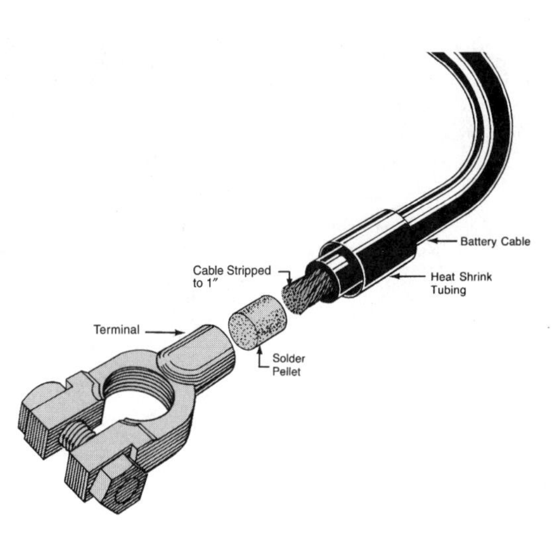

   Note: If there are any electronic memory circuits (radio stations, radio security, seats, computers, etc.), install a computer memory retaining tool into the cigarette lighter socket. Refer to Worksheet 7-5, Battery Service.

4. Remove the battery ground cable.

   Note: Always remove the ground cable first when working on the battery.

5. Remove the terminal clamp that will be replaced.
6. Use diagonal cutting pliers to cut the old clamp from the cable.

7. Use a sharp knife to strip back about 1/2" of insulation from the end of the cable.

   Note: Before either type of terminal clamp is used, the end of the cable must be bright and clean.

8. Are the connecting surfaces clean and bright? Yes ____ No ____
9. Install the new clamp on the cable following the terminal clamp manufacturer's installation instructions.
10. Install the clamp onto the battery post.
11. Which cable is always connected last? Positive _____ Negative _____ Ground _____
12. After both cables have been connected to the battery, start the vehicle.

    Does it start? Yes ____ No ____
13. Before completing the paperwork, clean your work area, put the tools in their proper places, and wash your hands.
14. Were there any other recommendations for needed service or unusual conditions that you noticed while replacing the battery terminal clamp?

    Yes ____ No ____
15. Record your recommendations for additional service or repairs on the repair order.
16. Complete the repair order (R.O.).

STOP

Instructor OK ______________ Score __________

ASE Lab Preparation Worksheet #7-8

BATTERY SPECIFIC GRAVITY TEST

Name________________________________ Class ____________________

OBJECTIVE:

Upon completion of this assignment, you should be able to test the battery's state of charge using a hydrometer. This task will help prepare you to pass the ASE certification examinations in electrical and engine performance.

DIRECTIONS:

Before beginning this lab assignment, review the worksheet completely. Fill in the information in the spaces provided as you complete each task.

TOOLS AND EQUIPMENT REQUIRED:

Safety glasses, fender covers, shop towel, hydrometer

RELATED INFORMATION:

Not all batteries have vent caps. There are two ways to test a battery's state of charge. One method is to measure the specific gravity of the electrolyte. The other method is to measure the open-circuit voltage of the battery. Either method will give a good indication of the battery's state of charge.

If the battery being tested has vent caps, perform both tests. When a battery has no vent caps, do only the open-circuit voltage test.

Be sure to wear eye protection when performing this test. Splashed electrolyte can cause serious eye injuries.

PROCEDURE:

Vehicle year ___________ Make __________________ Model ____________________

Repair Order # ________ Engine size_____________ # of Cylinders ____________

Hydrometer Test (specific gravity of electrolyte)

1. Install a fender cover on the fender nearest the battery.
2. Remove the vent caps and set them on top of the battery.
3. Check the level of the electrolyte.

Note: If the electrolyte level is too low to perform this test, add water to the cells and recharge the battery. Refer to Worksheet 7-10, Battery Slow Charge.

Electrolyte level: OK ____ Add water ____ Recharge ____

4. Draw electrolyte into the hydrometer to the line on the tester bulb (if so equipped). The gauge must float freely.
5. Hold the hydrometer vertically so the float can rise to its proper level.
6. Read the specific gravity on the float and record it below. Make a temperature correction, if necessary. The compensation factor is more important at temperature extremes.

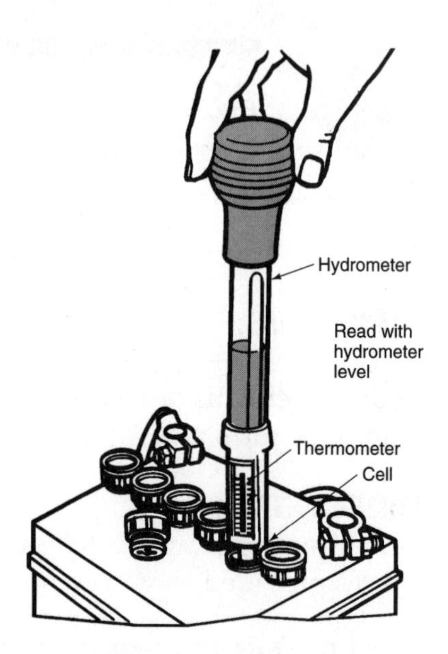

Note: Most hydrometers are temperature compensated. If not, a correction of +0.004 is made for each 10°F change above 80°F. Subtract 0.004 for each 10°F drop in temperature below 80°F.

7. Put the hydrometer tube back into the cell and squeeze it gently to return the electrolyte to the cell. Repeat steps 4 through 6 for each of the cells.

Record the readings below:

1 ________ 2 ________ 3 ________ 4 ________ 5 ________ 6 ________

Indicate the state of charge in the space provided.

(1.260-1.280) 100% ____

(1.240-1.260) 75% ____

(1.220-1.240) 50% ____

(1.200-1.220) 25% ____

(1.180-1.200) Dead ____

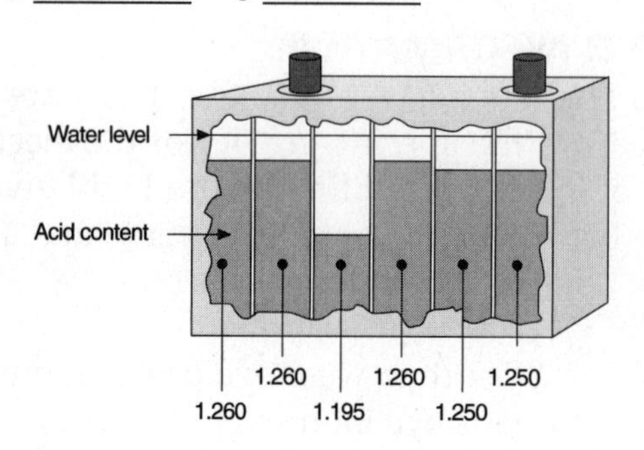

8. Are differences between the cell readings less than 0.050? Yes ____ No ____

Note: A faulty battery can be determined by the specific gravity readings. If the differences among any of the cells are greater than 0.050, a cell is damaged and the battery must be replaced.

9. What service does the battery need?

None. The battery is fully charged. ____

Recharging. The specific gravity is below 75%. ____

Replacement. The battery has a damaged cell. ____

10. Before completing the paperwork, clean your work area, put the tools in their proper places, and wash your hands.
11. Were there any other recommendations for needed service or unusual conditions that you noticed while checking the battery's state of charge?

Yes ____ No ____

12. Record your recommendations for additional service or repairs on the repair order.
13. Complete the repair order (R.O.).

ASE Lab Preparation Worksheet #7-9

BATTERY OPEN-CIRCUIT VOLTAGE TEST

Name__________________________ Class __________________

OBJECTIVE:

Upon completion of this assignment, you should be able to test the battery using a digital voltmeter and determine the state of charge. This task will help prepare you to pass the ASE certification examinations in electrical and engine performance.

DIRECTIONS:

Before beginning this lab assignment, review the worksheet completely. Fill in the information in the spaces provided as you complete each task.

TOOLS AND EQUIPMENT REQUIRED:

Safety glasses, fender covers, shop towel, digital voltmeter

RELATED INFORMATION:

Not all batteries have vent caps. There are two ways to test a battery's state of charge. One method is to measure the specific gravity of the electrolyte. The other method is to measure the open-circuit voltage of the battery. Either method will give a good indication of the battery's state of charge.

If the battery being tested has vent caps, perform both tests. When a battery has no vent caps, do only the open-circuit voltage test.

PROCEDURE:

Vehicle year ___________ Make ________________ Model _________________

Repair Order # ________ Engine size____________ # of Cylinders ___________

Open Circuit Voltage Test

1. Install a fender cover on the fender nearest the battery.
2. Turn on the headlights for 30 seconds to remove the surface charge from the battery.
3. Set the digital voltmeter to the next voltage scale higher than the rated voltage of the battery.

   Voltage scale selected __________
4. Connect the voltmeter in parallel with the battery. This means to connect the positive meter lead to the positive battery post and the negative meter lead to the negative battery post.
5. Read and record the voltage. _______ volts

6. The battery's state of charge relates to the measured voltage as follows:

   12.6 = 100% charged

   12.4 = 75% charged

   12.2 = 50% charged

   12.0 = discharged

   What is the state of charge of the battery being tested? ________ %

7. What service does the battery need?

   None. The battery is fully charged. _____

   Recharging. The voltage is below 75% charge. _____

   Note: If the battery measured below 75% charge and the reason is not known, more complete testing will be necessary. A problem with the battery, the charging system, or the vehicle's electrical system could be the reason why the battery was not completely charged.

8. Is more testing necessary to determine the reason for the battery's low state of charge?

   Yes _____ No _____

9. Before completing the paperwork, clean your work area, put the tools in their proper places, and wash your hands.

10. Were there any other recommendations for needed service or unusual conditions that you noticed while checking the battery's state of charge?

    Yes _____ No _____

11. Record your recommendations for additional service or repairs on the repair order.

12. Complete the repair order (R.O.).

STOP

Instructor OK ______________ Score __________

ASE Lab Preparation Worksheet #7-10

BATTERY SLOW CHARGE

Name______________________________ Class ____________________

OBJECTIVE:

Upon completion of this assignment, you should be able to charge a battery. This task will help prepare you to pass the ASE certification examination in electrical systems.

DIRECTIONS:

Before beginning this lab assignment, review the worksheet completely. Fill in the information in the spaces provided as you complete each task.

TOOLS AND EQUIPMENT REQUIRED:

Safety glasses, fender covers, shop towel, battery charger

SAFETY NOTE

A battery being charged gives off hydrogen gas. A spark can cause a dangerous explosion!

PROCEDURE:

Vehicle year ___________ Make __________________ Model ___________________

Repair Order # ________ Engine size_____________ # of Cylinders ____________

1. Open the hood and place fender covers on the fenders and front body parts.
2. Check the electrolyte level and fill the battery only enough to cover the plates.

   Low ____ Plates are covered ____

   If the battery is the sealed type and its electrolyte level is low, do *not* attempt to recharge it.

   Is it a sealed battery? Yes ____ No ____
3. Disconnect the ground cable from the battery.

   Which is the ground cable? (+) ____ (–) ____
4. Connect the charger to the battery (before plugging it into wall current).
5. What color charger clamp goes to the positive battery terminal?

   Red ____ Black ____
6. Plug the charger into wall current.
7. Turn the charger on.
8. Set the charging current to approximately 1% of the battery's cold cranking amps. Example, a 650 cold cranking amp battery should be charged at 6.5 amps.

   What is the CCA of the battery? ____ amps

   How many amps should your battery be charged at? ____ amps

9. Check the voltage at regular intervals. When the voltage does not change for one hour, the battery is fully charged.

   Did the voltage stabilize for one hour? Yes ____ No ____

   Note: If the battery will not take a charge, double-check to see that the polarity is correct. Hold down the jump start button for one minute and then release it. The battery should begin to take a charge.

10. When the desired charge is completed, shut off the charger.

    How long was the battery on the charger? ____________________

11. Unplug the charger from the wall current.
12. Disconnect the charger cables and reattach the battery ground cable.
13. Adjust the battery electrolyte level as needed. Low ____ OK ____
14. Before completing the paperwork, clean your work area, put the tools in their proper places, and wash your hands.
15. Were there any other recommendations for needed service or unusual conditions that you noticed while charging the battery?

    Yes ____ No ____

16. Record your recommendations for additional service or repairs on the repair order.
17. Complete the repair order (R.O.).

STOP

Instructor OK ______________ Score __________

ASE Lab Preparation Worksheet #7-11

BATTERY CHARGING (FAST CHARGE)

Name______________________________ Class ____________________

OBJECTIVE:

Upon completion of this assignment, you should be able to return a battery to service quickly by fast charging. This task will help prepare you to pass the ASE certification examination in electrical systems.

DIRECTIONS:

Before beginning this lab assignment, review the worksheet completely. Fill in the information in the spaces provided as you complete each task.

TOOLS AND EQUIPMENT REQUIRED:

Safety glasses, fender covers, battery charger, shop towel

SAFETY NOTE

A battery being charged gives off hydrogen gas. A spark can cause a dangerous explosion!

PROCEDURE:

Vehicle year ___________ Make _________________ Model _________________

Repair Order # ________ Engine size_____________ # of Cylinders ___________

1. Open the hood and place fender covers on the fenders and front body parts.
2. Check the electrolyte level and fill the battery only enough to cover the plates.

   Low ____ Plates are covered ____

   If the battery is the sealed type and its electrolyte level is low, do *not* attempt to recharge it.

   Is it a sealed battery? Yes ____ No ____
3. Disconnect the ground cable from the battery.

   Which is the ground cable? (+) ____ (–) ____
4. Connect the charger to the battery (before plugging it into wall current).
5. What color charger clamp goes to the positive battery terminal?

   Red ____ Black ____
6. Plug the charger into wall current.
7. Turn the charger on and set the charger control to maximum.

   How many amps does the gauge indicate? 20-30 amps ____ Under 20 amps ____
8. If the amp gauge reads over 30 amps, adjust charger output to a lower setting.

   Is the charger reading over 30 amps? Yes ____ No ____

Control panel
(-) Cable
(+) Cable
Meter

Note: If the battery will not take a charge, double-check to see that the polarity is correct. Hold down the jump start button for one minute and then release it. The battery should begin to take a charge.

9. When the desired charge is completed, shut off the charger.

   Note: Do not fast charge a battery for an extended time. Heat is generated in the battery during charging. Overheating a battery can severely shorten its useful life.

10. Unplug the charger from the wall current.
11. Disconnect the charger cables and reattach the battery ground cable.
12. Adjust the battery electrolyte level as needed. Low ____ OK ____
13. Before completing the paperwork, clean your work area, put the tools in their proper places, and wash your hands.
14. Were there any other recommendations for needed service or unusual conditions that you noticed while charging the battery?

    Yes ____ No ____
15. Record your recommendations for additional service or repairs on the repair order.
16. Complete the repair order (R.O.).

Instructor OK ______________ Score __________

ASE Lab Preparation Worksheet #7-12

BATTERY JUMP STARTING (LOW-MAINTENANCE BATTERY)

Name__________________________ Class ________________

OBJECTIVE:

Upon completion of this assignment, you should be able to jump a vehicle that has a low-maintenance battery. This task will help prepare you to pass the ASE certification examinations in engine performance and electrical systems.

DIRECTIONS:

Before beginning this lab assignment, review the worksheet completely. Fill in the information in the spaces provided as you complete each task.

TOOLS AND EQUIPMENT REQUIRED:

Safety glasses, fender covers, jumper cables, shop towel

PROCEDURE:

Vehicle year __________ Make ________________ Model ________________

Repair Order # ________ Engine size____________ # of Cylinders __________

CAUTION Use extra care when jump starting a vehicle with electronic components. Misconnected cables or electrical surges can damage electronic components.

1. Wear eye protection. Batteries produce explosive gases, which may accidentally explode.

   Type of eye protection worn:

   Safety glasses ____

   Goggles ____

   Face shield ____

2. Be sure the transmission is in neutral or park and the emergency brake is set firmly.

   Transmission: Park ____ Neutral ____

   Parking brake: On ____ Off ____

3. Are all electrical loads off? Yes ____ No ____

4. Check the battery's electrolyte level. If the level is low, do not attempt to jump start the car. Refill and recharge the battery as needed.

   Electrolyte level: OK ____ Needs water ____

5. Keep the vent caps in place on the battery. The vent caps act as spark arrestors.

6. Attach one end of the jumper cable to the positive terminal on the booster battery and the other end to the positive terminal on the discharged battery.

NEG − + POS
Smaller Larger
Top-terminal batteries

TURN

The vehicles should not be touching each other. This could provide an unwanted ground path and a spark could result.

7. Attach one end of the negative cable to the negative terminal on the booster (fully charged) battery.
8. Connect the other end of the negative cable to a ground on the engine. A metal bracket or the end of the negative battery cable that is attached to the block are good ground points.

A spark can occur as the negative cable is attached to the dead battery as it tries to equalize its voltage with the booster battery. Making a connection at a point away from the battery avoids the possibility of a dangerous spark near the battery.

Note: Low-maintenance batteries have higher internal resistance than conventional lead-antimony batteries. The jumper cables may need to remain in place for a minute or so before attempting to start the disabled vehicle. This will allow the dead battery to take on a charge.

9. When the dead vehicle starts, immediately disconnect the jumper cable from the block.

   Note: If the vehicle does not start quickly, do not crank the engine for longer than 30 seconds. The starter motor can overheat, resulting in starter failure.

10. Run the host vehicle at 2000 rpm to allow its charging system to recharge the battery.
11. Before completing the paperwork, clean your work area, put the tools in their proper places, and wash your hands.

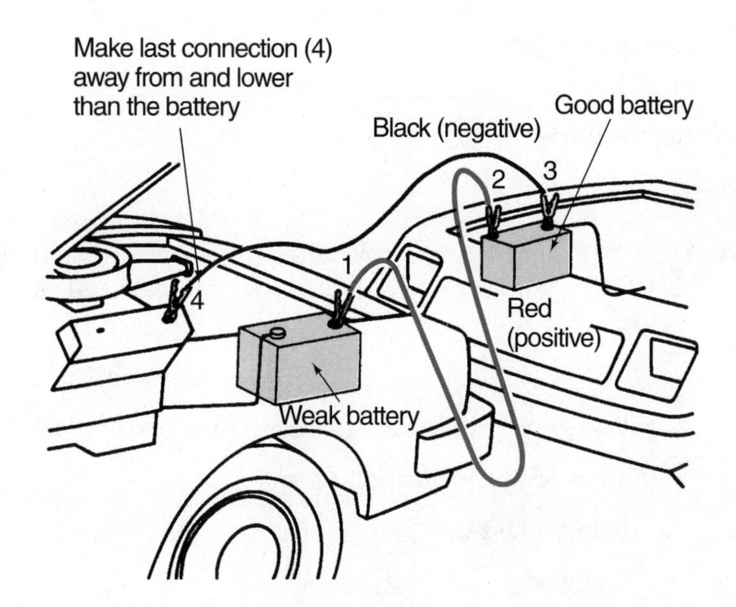

Instructor OK ____________ Score ________

ASE Lab Preparation Worksheet #7-13

BATTERY CAPACITY/LOAD TESTING (WITH VAT)

Name________________________ Class ________________

OBJECTIVE:

Upon completion of this assignment, you should be able to load test a battery. This task will help prepare you to pass the ASE certification examination in electrical systems.

DIRECTIONS:

Before beginning this lab assignment, review the worksheet completely. Fill in the information in the spaces provided as you complete each task.

TOOLS AND EQUIPMENT REQUIRED:

Safety glasses, fender covers, shop towel, battery load tester

PROCEDURE:

Vehicle year __________ Make ________________ Model ________________

Repair Order # _______ Engine size____________ # of Cylinders __________

Note: Before doing a load/capacity test on a battery, the battery must be in good physical condition and be more than 75% charged.

1. Open the hood and place fender covers on the fender nearest the battery.
2. Inspect the battery for the following:

Cracks in the case	OK	____	Needs attention	____
Leaking electrolyte	OK	____	Needs attention	____
Battery tray condition	OK	____	Needs attention	____
Battery cables	OK	____	Need attention	____
Posts and terminals	OK	____	Need attention	____
Electrolyte level	OK	____	Needs attention	____
Caps and/or covers	In place	____	Missing	____
Holddown	Secure	____	Loose	____
Corrosion	Clean	____	Built up	____

3. Cracks or leaks mean the battery must be replaced.

   Cracks ____ Leaks ____ OK ____

4. What is the battery's state of charge? Refer to Worksheets 7-8 and 7-9.

   100% ____

   75% ____

   Less than 75% ____

Note: The battery must be 75% or more charged to continue load testing.

How was the state of charge measured?

Specific gravity ____ Open circuit voltage ____

5. Remove the ignition key.
6. Disconnect the battery ground cable.

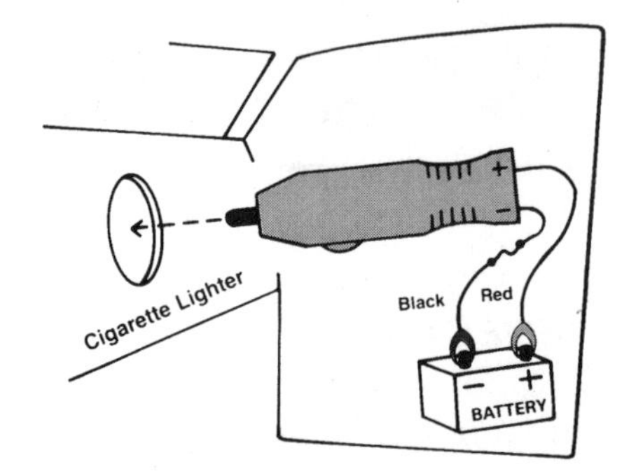

Note: If the vehicle has computer controls, connect a memory retaining tool to the vehicle.

Battery Capacity Test Using a Carbon Pile (VAT 40)

(This test is only accurate if the battery is more than 75% charged.)

7. Check that the control knob on the carbon pile is off (counterclockwise).
8. Turn test selector switch to **battery test** or **starting system.**
9. Turn voltage selector switch to a voltage higher than battery voltage.

   What range was selected? ____________

10. Connect the large tester leads to the battery. Connect the red (positive) tester lead to the positive post of the battery and connect the black (negative) tester lead to the negative post of the battery.
11. Clamp the inductive ammeter pickup around the negative tester lead.

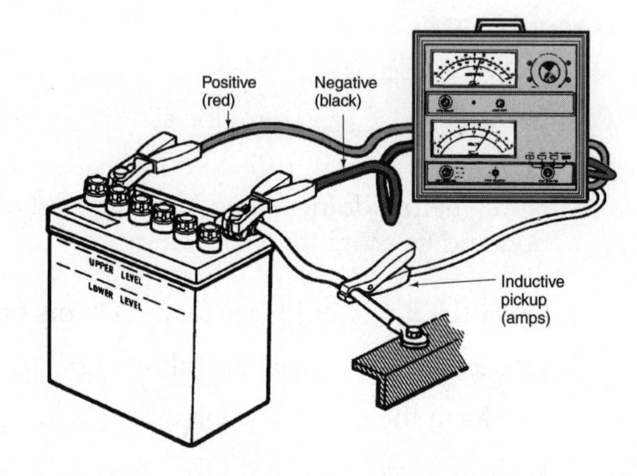

Note: Side terminal batteries require a special adapter when the battery cables are disconnected.

What kind of terminals does the battery being tested have?

Side terminals ____ Top terminals ____

12. Determine the load to use during the battery capacity test in one of the following ways:
    a. Amperage needed to crank the engine.
    b. The CCA of the battery divided by 2.

Note: Use the vehicle CCA specifications. Do not use the CCA listed on the battery.

c. Load required for capacity test: ____ amps

Note: The large knob on the tester will cause a rapid discharge of the battery when turned clockwise. This device is known as a "carbon-pile." As the layers of carbon under the knob are compressed, more current can pass through it to ground. Watch the ammeter to determine the amount of current that is flowing.

13. To test the battery, turn the large control knob clockwise to load the battery. The ammeter needle should move. If not, determine the cause before proceeding.

    Ammeter moves? Yes ____ No ____

    Maintain the desired amperage for 15 seconds, while watching the *voltmeter.* Then turn off the amp draw control knob (counterclockwise).

14. What is the voltmeter reading at the end of the test (before turning off the carbon-pile)?

    ____ volts

Test results:

The battery passed the capacity test if the voltage remained above 9.6 volts during the test.

Note: For every 10 degrees below 70 degrees, voltage may be 0.1 less than 9.6.

Passed ____ Failed ____

15. If the battery *passed* the capacity test, disconnect the tester and reconnect the battery negative cable.

    If the battery *failed* the capacity test, the next step is to do a 3-minute charge test.

16. Before completing the paperwork, clean your work area, put the tools in their proper places, and wash your hands.

17. Were there any other recommendations for needed service or unusual conditions that you noticed while testing the battery?

    Yes ____ No ____

18. Record your recommendations for additional service or repairs on the repair order.

19. Complete the repair order (R.O.).

STOP

Instructor OK ______________ Score __________

ASE Lab Preparation Worksheet #7-14

BATTERY CAPACITY/LOAD TESTING (WITHOUT VAT)

Name___________________________ Class ___________________

OBJECTIVE:

Upon completion of this assignment, you should be able to load test a battery. This task will help prepare you to pass the ASE certification examination in electrical systems.

DIRECTIONS:

Before beginning this lab assignment, review the worksheet completely. Fill in the information in the spaces provided as you complete each task.

TOOLS AND EQUIPMENT REQUIRED:

Safety glasses, fender covers, shop towel

PROCEDURE:

Vehicle year ___________ Make _________________ Model __________________

Repair Order # ________ Engine size_____________ # of Cylinders ___________

Note: Before doing a load/capacity test on a battery, the battery must be in good physical condition and be more than 75% charged.

1. Open the hood and place a fender cover on the fender nearest the battery.
2. Inspect the battery for the following:

Cracks in the case	OK	____	Needs attention	____
Leaking electrolyte	OK	____	Needs attention	____
Battery tray condition	OK	____	Needs attention	____
Battery cables	OK	____	Need attention	____
Posts and terminals	OK	____	Need attention	____
Electrolyte level	OK	____	Needs attention	____
Caps and/or covers	In place	____	Missing	____
Holddown	Secure	____	Loose	____
Corrosion	Clean	____	Built up	____

3. Cracks or leaks mean the battery must be replaced.

   Cracks ____ Leaks ____ OK ____

4. What is the battery's state of charge? Refer to Worksheets 7-8 and 7-9.

   100% ____

   75% ____

   Less than 75% ____

TURN

Note: The battery must be 75% or more charged to continue load testing.

How was the state of charge measured?

Specific gravity ____ Open circuit voltage ____

Battery capacity test without a carbon-pile

Note: If the battery is less than 75% charged, charge the battery before testing or do a 3-minute charge (sulfation) test to determine the battery's condition. Refer to Worksheet 7-15.

The battery requires:

A charge ____________________

A 3-minute charge test ____

Nothing ____________________

5. If the battery is more than 75% charged, continue by disabling the engine's ignition system. See an instructor for specific instructions.

   Ignition disabled? Yes ____ No ____

6. Connect a voltmeter to the battery.
7. Use the starter motor to crank the engine for 15 seconds.

   Record the voltmeter reading. ____ volts

   a. If the battery voltage is 9.6 volts or higher during the test, the battery is good.
   b. If battery voltage falls below 9.6 volts, recharge the battery and repeat the test. If the result is still below 9.6 volts, the starter motor draw should be checked to see that its current draw is not excessive. If the starter is good, replace the battery.
8. Before completing the paperwork, clean your work area, put the tools in their proper places, and wash your hands.
9. Were there any other recommendations for needed service or unusual conditions that you noticed while testing the battery?

   Yes ____ No ____

10. Record your recommendations for additional service or repairs on the repair order.
11. Complete the repair order (R.O.).

Instructor OK ______________ Score ____________

ASE Lab Preparation Worksheet #7-15

BATTERY SULFATION TEST (THREE-MINUTE CHARGE TEST)

Name____________________________ Class ____________________

OBJECTIVE:

Upon completion of this assignment, you should be able to test a battery to ensure that it is serviceable. This task will help prepare you to pass the ASE certification examination in electrical systems.

DIRECTIONS:

Before beginning this lab assignment, review the worksheet completely. Fill in the information in the spaces provided as you complete each task.

TOOLS AND EQUIPMENT REQUIRED:

Safety glasses, fender covers, shop towel, battery charger

PROCEDURE:

Vehicle year ____________ Make __________________ Model ___________________

Repair Order # ________ Engine size_____________ # of Cylinders ____________

Note: Before doing a 3-minute charge test on a battery, the battery must be in good physical condition and be less than 75% charged or have failed the capacity test.

1. Put a fender cover on the fender nearest the battery.

Do this test only if the battery fails the capacity test or is less than 75% charged.

2. Connect the battery charger to the battery. Connect the battery charger positive cable clamp to the battery positive terminal and battery charger negative cable clamp to the battery negative terminal.
3. Connect a voltmeter to the battery terminals (positive to positive and negative to negative).

   What scale is the meter set on? ______________________
4. Plug the charger into the wall outlet.
5. Adjust the charger to its highest charging rate.

   What is the charger meter reading? ____ amps
6. Charge the battery for 3 minutes. At the end of 3 minutes, what is the voltage?

   ____ volts

   The battery passes the sulfation (3-minute charge) test if the voltage is below 15.5 at the end of 3 minutes.

   Passed ____ go to step 7

   Failed ____ go to step 8

TURN

7. If the battery passes the sulfation (3-minute charge) test, recharge it and do a battery capacity test. Discard the battery if it fails the capacity test.
8. If the battery fails the sulfation (3-minute charge) test (above 15.5 volts at the end of 3 minutes), replace the battery.
9. Disconnect the charger. Turn it off, and unplug it before disconnecting it from the battery.
10. Disconnect the voltmeter and turn it off.
11. Before completing the paperwork, clean your work area, put the tools in their proper places, and wash your hands.
12. Were there any other recommendations for needed service or unusual conditions that you noticed while testing the battery?

    Yes ____ No ____
13. Record your recommendations for additional service or repairs on the repair order.
14. Complete the repair order (R.O.).

Instructor OK ______________ Score __________

ASE Lab Preparation Worksheet #7-16

BATTERY DRAIN TEST

Name____________________________ Class ____________________

OBJECTIVE:

Upon completion of this assignment, you should be able to test a vehicle's electrical system for excessive battery drain. This task will help prepare you to pass the ASE certification examination in electrical systems.

DIRECTIONS:

Before beginning this lab assignment, review the worksheet completely. Fill in the information in the spaces provided as you complete each task.

TOOLS AND EQUIPMENT REQUIRED:

Safety glasses, fender covers, shop towel, test light, digital multimeter

PROCEDURE:

Vehicle year ___________ Make _________________ Model __________________

Repair Order # ________ Engine size_____________ # of Cylinders ___________

1. Open the hood and place fender covers on the fenders and front body parts.
2. With the key off, be sure all lights and accessories are off.

   Off ____ On ____
3. Check all courtesy lights (those that come on when a car door is open). They should not be on.

   Off ____ On ____
4. Disconnect the battery ground cable.
5. Connect a test light in series between the battery ground cable end and the battery post.

   No light ____ Lights up ____

   Note: If the bulb lights, additional testing is necessary to locate the problem.

Another method to more accurately check for a drain would be to use an ammeter in place of the test light.

6. In place of the test light, connect an ammeter between the negative cable and the negative post of the battery. The meter should read zero.

   Ammeter reading ____

   Note: Excessive amps can damage the meter or blow the meter fuse.

Do not try to start the vehicle or turn on any of the accessories while the ammeter is connected.

Note: A drain of less than 50 milliamps (0.050 amp) is acceptable on **some** computer-controlled vehicles.

7. Is the drain excessive? Yes ____ No ____

   Is further testing required to locate the problem? Yes ____ No ____

8. Before completing the paperwork, clean your work area, put the tools in their proper places, and wash your hands.

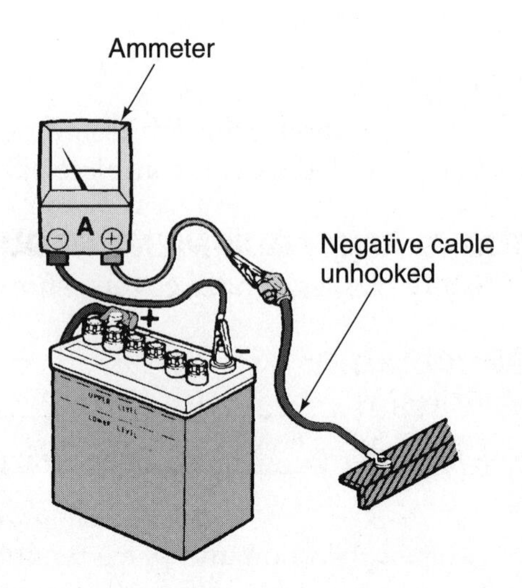

9. Were there any other recommendations for needed service or unusual conditions that you noticed while you were testing for excessive battery drain?

   Yes ____ No ____

10. Record your recommendations for additional service or repairs on the repair order.
11. Complete the repair order (R.O.).

STOP

Instructor OK ______________ Score ____________

ASE Lab Preparation Worksheet #7-17

STARTER CIRCUIT VOLTAGE DROP TEST

Name______________________________ Class ____________________

OBJECTIVE:

Upon completion of this assignment, you should be able to test a starter circuit for high resistance by measuring the voltage drop. This task will help prepare you to pass the ASE certification examinations in electrical systems and engine performance.

DIRECTIONS:

Before beginning this lab assignment, review the worksheet completely. Fill in the information in the spaces provided as you complete each task.

TOOLS AND EQUIPMENT REQUIRED:

Safety glasses, fender covers, voltmeter, shop towel

PROCEDURE:

Vehicle year ____________ Make __________________ Model ____________________

Repair Order # ________ Engine size______________ # of Cylinders ____________

1. Open the hood and place fender covers on the fenders and front body parts.

Note: The starter circuit voltage drop test is a valuable test for locating hard starting problems.

Positive Side Voltage Drop Test:

Note: Do not crank the starter for periods longer than 30 seconds or the starter can overheat.

2. Set the parking brake. Position the transmission selector in Park or Neutral.
3. Disable the ignition system so that the engine can be cranked but will not start. See your instructor if you need help.
4. Set the voltmeter on the low scale (2 volts).
5. Connect the positive lead of the voltmeter to the positive post of the battery. Be sure it is connected to the post, not the clamp.
6. Connect the negative voltmeter lead to the point where the positive lead enters the starter. (This will be past the solenoid or relay.) The circuit is open at the solenoid, so the meter should read battery voltage.

   Voltmeter reading _____ volts

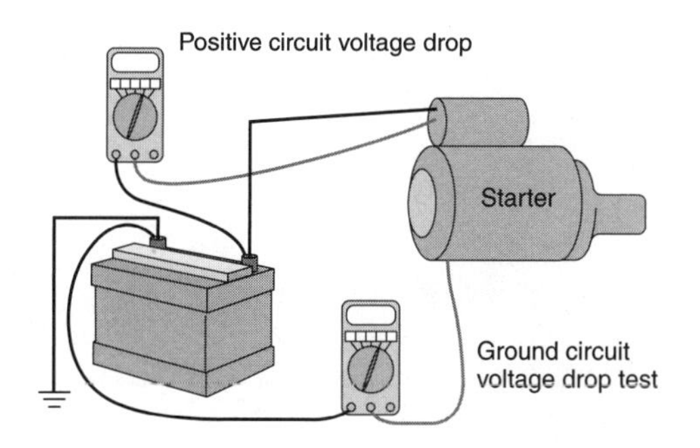

7. Crank the engine. The voltage drop will appear on the meter while cranking. An acceptable reading is less than 0.5 volt.

   Voltmeter reading _____ volts

Negative Side Voltage Drop Test:

8. Connect the negative lead of the voltmeter to the negative post of the battery. Be sure it is connected to the post, not the clamp.
9. Connect the positive voltmeter lead to the case of the starter. The meter will read "0" volt because there is not any electrical potential difference between the connections.

   Voltmeter reading ________ volts
10. Crank the engine. The voltage drop will appear on the meter while cranking. An acceptable reading is less than 0.2 volt.

    Voltmeter reading ________ volts
11. Reconnect the ignition system and start the vehicle.

    Did the engine start? Yes ____ No ____
12. Before completing the paperwork, clean your work area, put the tools in their proper places, and wash your hands.
13. Were there any other recommendations for needed service or unusual conditions that you noticed while checking the starter circuit voltage drop?

    Yes ____ No ____
14. Record your recommendations for additional service or repairs on the repair order.
15. Complete the repair order (R.O.).

Instructor OK ______________ Score __________

ASE Lab Preparation Worksheet #7-18

REPLACE A TAIL/BRAKE LIGHT BULB

Name__________________________ Class ___________________

OBJECTIVE:

Upon completion of this assignment, you should be able to replace a signal, tail, or brake light bulb. This task will help prepare you to pass the ASE certification examination in electrical systems.

DIRECTIONS:

Before beginning this lab assignment, review the worksheet completely. Fill in the information in the spaces provided as you complete each task.

TOOLS AND EQUIPMENT REQUIRED:

Safety glasses, shop towel, screwdriver, ohmmeter

PARTS AND SUPPLIES:

Light bulb(s)

PROCEDURE:

Vehicle year ___________ Make _________________ Model __________________

Repair Order # ________ Engine size____________ # of Cylinders ___________

Signal lights (lights used to signal a driver's intentions) must be checked on a regular basis.

1. Check all exterior lights. Does each of the bulbs light when it is switched on?

   Yes ____ No ____

   Remember that the key must be on for some of the lights to work.

2. List all of the exterior lights found on the vehicle being inspected.

   a. ______________________ e. ______________________

   b. ______________________ f. ______________________

   c. ______________________ g. ______________________

   d. ______________________ h. ______________________

   List any lights that are not operating properly below.

   a. ____

   b. ____

   c. ____

   d. ____

Note: To replace a bulb, it may be necessary to remove the lens. Some bulbs can be replaced from the back of the light assembly, either under the vehicle or in the fender well. Many rear lights are replaced by accessing them from inside the trunk.

Note: Usually if the lens has visible screws, the bulb is replaced by removing the lens. Before removing the lens, check the back of the assembly. If there is an easily removed cover or if the bulb socket is exposed, the bulb is probably removed from the back. Check carefully before proceeding.

How does the bulb that you are changing appear to be removed?

____

Note: Many bulbs are replaced by pushing them in and turning counterclockwise. Some smaller bulbs are removed by pulling them straight out of their sockets.

3. To replace a bulb that is removed from the back of the assembly, twist the bulb socket counterclockwise. The socket and bulb assembly will come out of the light housing. Hold the socket and push in on the bulb as you turn it counterclockwise. Reverse the procedure to install the new bulb.

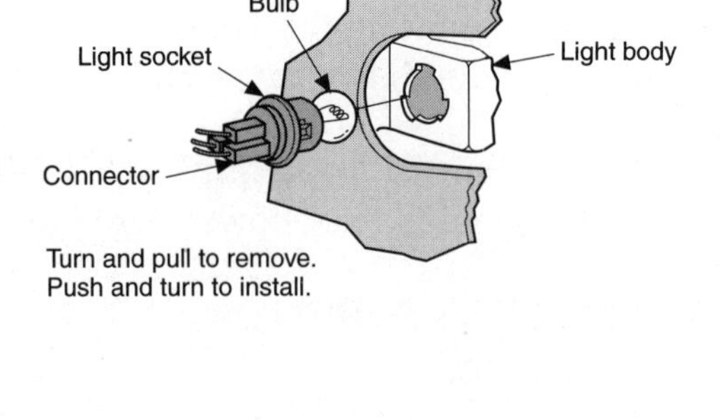

Turn and pull to remove.
Push and turn to install.

4. To replace a bulb that requires removal of the lens, remove the mounting screws and carefully remove the lens. Be careful not to damage the gasket. Now, push in on the bulb and turn it counterclockwise. Reverse the procedure to install the new bulb.

Note: When installing the lens, be careful not to overtighten the lens screws. If the lens cracks, moisture will be able to enter the light assembly, which can damage the light and socket.

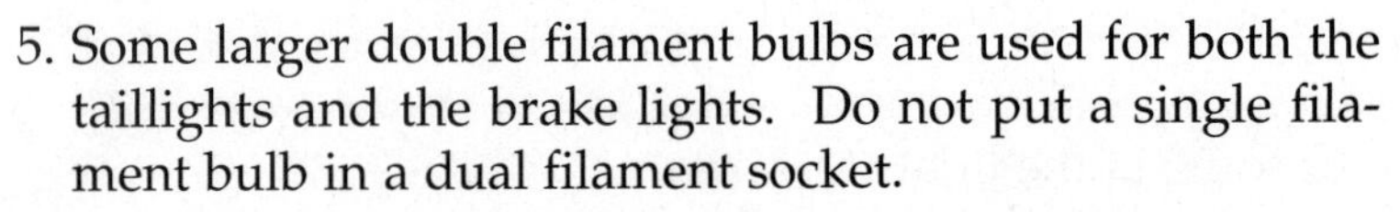

Bulb
Bulb socket
Seal
Lens

5. Some larger double filament bulbs are used for both the taillights and the brake lights. Do not put a single filament bulb in a dual filament socket.

The bulb that was changed was a: Single filament ____ Double filament ____

6. Check the operation of all of the lights.
7. Use an ohmmeter to check the bulb that was replaced. Set the meter to the highest ohm scale. Connect the red lead to the center electrode of the bulb and the black lead to the side of the bulb.

   Is the resistance: High (infinity) _____ ? Low _____ ?

   Was the bulb:

   Good _____ ? Bad _____ ?

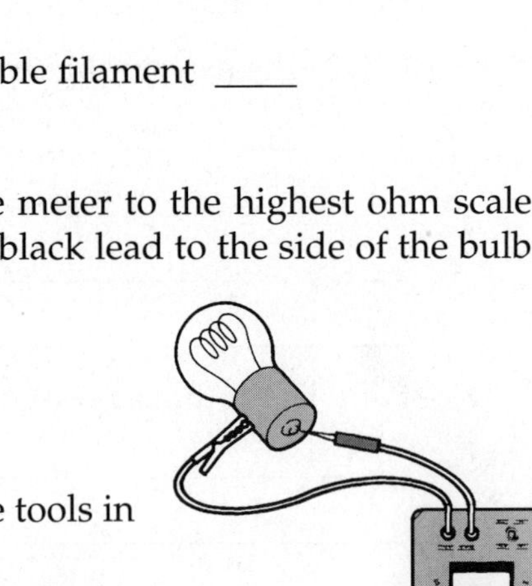

8. Before completing the paperwork, clean your work area, put the tools in their proper places, and wash your hands.
9. Were there any other recommendations for needed service or unusual conditions that you noticed while checking and replacing the light bulbs?

   Yes ____ No ____

10. Record your recommendations for additional service or repairs on the repair order.
11. Complete the repair order (R.O.).

STOP

Instructor OK ______________ Score __________

ASE Lab Preparation Worksheet #7-19

REPLACE A TURN SIGNAL FLASHER

Name__________________________ Class __________________

OBJECTIVE:

Upon completion of this assignment, you should be able to replace a turn signal flasher. This task will help prepare you to pass the ASE certification examination in electrical systems.

DIRECTIONS:

Before beginning this lab assignment, review the worksheet completely. Fill in the information in the spaces provided as you complete each task.

TOOLS AND EQUIPMENT REQUIRED:

Safety glasses, shop towel

PROCEDURE:

Vehicle year __________ Make ________________ Model __________________

Repair Order # ________ Engine size____________ # of Cylinders ___________

1. Use a service manual or computer program to locate the turn signal flasher and the hazard flasher.
   a. What service manual or computer program did you use to find the flasher locations?

      Service manual ____

      Page ________

   b. Where is the turn signal flasher located?

      ____

   c. Where is the hazard flasher located?

      ____

2. Is a flasher located in the fuse panel? Yes ____ No ____
3. Turn on the ignition switch and operate the turn signal.

   Can you hear the flasher? Yes ____ No ____

4. Remove the turn signal flasher from its mount. How many electrical connector prongs does it have?

   Two ____ Three ____

5. Reinstall the turn signal flasher.
6. Check the operation of the turn signals.

Right front	OK ____	Bad ____
Right rear	OK ____	Bad ____
Left front	OK ____	Bad ____
Left rear	OK ____	Bad ____

Note: Turn signal flashers that do not flash or that flash too rapidly can indicate a defective turn signal bulb.

7. Before completing the paperwork, clean your work area, put the tools in their proper places, and wash your hands.
8. Were there any other recommendations for needed service or unusual conditions that you noticed while replacing the turn signal flasher?

   Yes _____ No _____
9. Record your recommendations for additional service or repairs on the repair order.
10. Complete the repair order (R.O.).

Instructor OK ______________ Score __________

ASE Lab Preparation Worksheet #7-20

REPLACE A SEALED BEAM HEADLAMP

Name____________________________ Class ____________________

OBJECTIVE:

Upon completion of this assignment, you should be able to replace a sealed beam headlamp. This task will help prepare you to pass the ASE certification examination in electrical systems.

DIRECTIONS:

Before beginning this lab assignment, review the worksheet completely. Fill in the information in the spaces provided as you complete each task.

TOOLS AND EQUIPMENT REQUIRED:

Safety glasses, fender covers, shop towel, screwdrivers

PARTS AND SUPPLIES:

Sealed beam headlamp

PROCEDURE:

Vehicle year ___________ Make _________________ Model ___________________

Repair Order # ________ Engine size_____________ # of Cylinders ___________

Note: Vehicles use either two or four sealed beam headlamps. A two-headlamp system has both high and low beams in one lamp.

1. Verify that the headlights operate on low and high beam.

Right high beam	OK ____	Bad ____
Right low beam	OK ____	Bad ____
Left high beam	OK ____	Bad ____
Left low beam	OK ____	Bad ____

Note: If both high beams or both low beams fail to operate, or if one beam is bright and the other is dim, perform basic electrical testing. This is done to determine whether the bulbs are faulty or there is a problem with an electrical circuit before replacing the bulb(s).

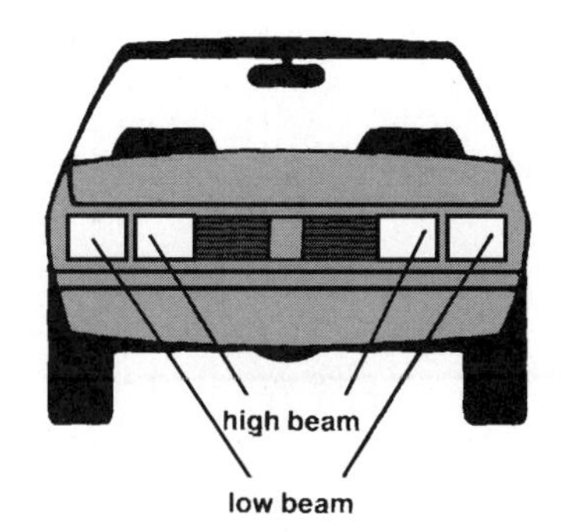

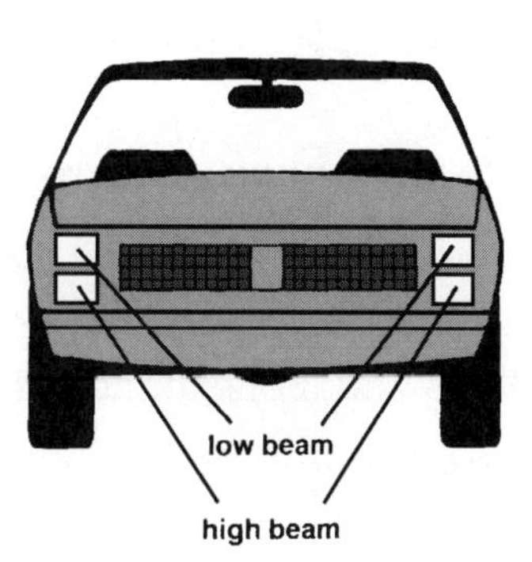

2. Turn off the lights.
3. Open the hood and place fender covers on the fenders.
4. Are the headlight bulbs of the sealed beam design or the composite design?

   Sealed beam ____ Composite ____

   Note: Use Worksheet 7-21 to replace a composite-type headlight bulb.

5. An identification number is embossed on the lens of the sealed beam bulb. List the number on each of the bulbs.

   Left __________ Right __________
6. Obtain the replacement bulb(s) before removal of the originals. Check that they are the correct replacements. Are they?

   Yes ____ No ____
7. Remove any trim (bezel) or grill pieces that will interfere with the removal of the headlight.
8. Locate the headlight adjusting screws. Do *not* turn these screws.

   Did you locate the adjusting screws? Yes ____ No ____

   Note: Turning the adjustment screws will make it necessary to adjust the headlights after the bulb(s) are replaced.

9. Locate and remove the small screws that hold the headlight retaining ring.
10. If necessary, unhook the headlight retaining ring from its spring and remove it.
11. Remove the headlight and disconnect its electrical connection.
12. Locate the alignment tabs on the rear of the new bulb.
13. Install the bulb in the bracket.

    Note: Align the bulb carefully. It only fits properly one way.

14. Reinstall the headlight retaining ring and reconnect the bulb to the electrical connector.
15. Reinstall any trim or grill pieces that were removed.
16. Check the operation of the headlight. Good ____ Bad ____
17. Before completing the paperwork, clean your work area, put the tools in their proper places, and wash your hands.
18. Were there any other recommendations for needed service or unusual conditions that you noticed while replacing the headlight?

    Yes ____ No ____
19. Record your recommendations for additional service or repairs on the repair order.
20. Complete the repair order (R.O.).

STOP

Instructor OK _______________ Score ___________

ASE Lab Preparation Worksheet #7-21

REPLACE A COMPOSITE HEADLAMP BULB

Name___________________________ Class ____________________

OBJECTIVE:

Upon completion of this assignment, you should be able to replace a composite headlight bulb. This task will help prepare you to pass the ASE certification examination in electrical systems.

DIRECTIONS:

Before beginning this lab assignment, review the worksheet completely. Fill in the information in the spaces provided as you complete each task.

TOOLS AND EQUIPMENT REQUIRED:

Safety glasses, fender covers, shop towel

PARTS AND SUPPLIES:

Composite headlight bulb(s)

PROCEDURE:

Vehicle year ___________ Make __________________ Model ___________________

Repair Order # ________ Engine size_____________ # of Cylinders ___________

1. Verify that the headlights operate on low and high beam.

   Right high beam OK ____ Bad ____

   Right low beam OK ____ Bad ____

   Left high beam OK ____ Bad ____

   Left low beam OK ____ Bad ____

   Note: If both high beams or both low beams fail to operate, or if one beam is bright and the other is dim, perform basic electrical testing. This is done to determine whether the bulbs are faulty or there is a problem with an electrical circuit before replacing the bulb(s).

   high/low beam

   high beam

   low beam

2. Turn off the lights.
3. Open the hood and place fender covers on the fenders.
4. Are the headlight bulbs of the sealed beam design or the composite design?

   Sealed beam ____ Composite ____

Note: Use Worksheet 7-20 to replace a sealed beam headlight bulb.

5. Obtain the replacement bulb(s) before removal of the originals. Check that they are the correct replacements. Are they?

   Yes _____ No _____

6. Disconnect the electrical connection to the bulb.
7. Unscrew the bulb retaining ring and remove the bulb.

   Note: Handle the new bulb carefully. Do not touch the glass part of the bulb. The oil from your hands may cause it to explode as it heats up.

Hermetically sealed
Retaining ring
Replaceable dual filament bulb
Metal reflector
Glass balloon

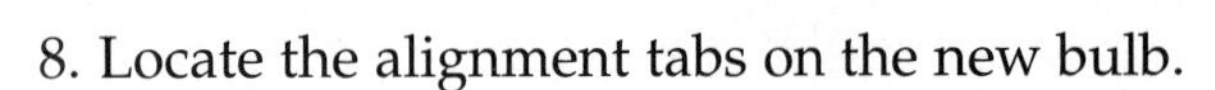

8. Locate the alignment tabs on the new bulb.

9. Install the bulb in the bracket.

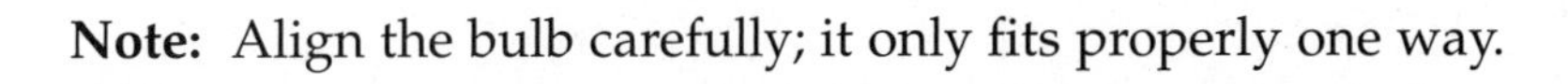

   Note: Align the bulb carefully; it only fits properly one way.

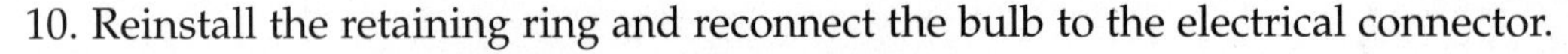

10. Reinstall the retaining ring and reconnect the bulb to the electrical connector.
11. Check the operation of the headlight. Good _____ Bad _____
12. Before completing the paperwork, clean your work area, put the tools in their proper places, and wash your hands.
13. Were there any other recommendations for needed service or unusual conditions that you noticed while replacing the headlight bulb?

    Yes _____ No _____

14. Record your recommendations for additional service or repairs on the repair order.
15. Complete the repair order (R.O.).

Instructor OK ________________ Score ____________

ASE Lab Preparation Worksheet #7-22

HEADLIGHT ADJUSTING WITH AIMING TOOLS

Name________________________________ Class ____________________

OBJECTIVE:

Upon completion of this assignment, you should be able to use headlight aiming tools to adjust a vehicle's headlights. This task will help prepare you to pass the ASE certification examination in electrical systems.

DIRECTIONS:

Before beginning this lab assignment, review the worksheet completely. Fill in the information in the spaces provided as you complete each task.

TOOLS AND EQUIPMENT REQUIRED:

Safety glasses, fender covers, headlight aimer, Phillips screwdriver, shop towel

PROCEDURE:

Vehicle year ____________ Make ___________________ Model ____________________

Repair Order # _________ Engine size______________ # of Cylinders ____________

Number of headlights: Two _____ Four _____

Headlight shape: Round _____ Rectangular _____ Curved _____

Bulb type: Sealed beam _____ Sealed halogen _____ Halogen _____

1. Select a level work area and park the vehicle there.
2. Check that there are no unusual loads in the vehicle.

   Is the trunk empty? Yes ____ No ____
3. What is the fuel level?

   Full ____ 3/4 full ____

   1/2 full ____ 1/4 full ____ Empty ____

   Note: Ideally the vehicle should have a half tank of fuel.
4. Jounce the vehicle to settle the suspension.
5. Clean the headlight lenses.
6. Gently rock the headlight to see if there is any slack between the adjusting screw and the headlight mounting bracket. Is there any slack or looseness?

   Yes ____ No ____

Note: If the headlight is loose, the problem must be corrected before adjusting it.

7. Select the correct adapter for the size and shape of the headlight. Mount it on the aimer.
8. Clean the headlamp lenses.

   Hold the suction cup against the headlight while pushing forward on the handle on the bottom of the aimer.

   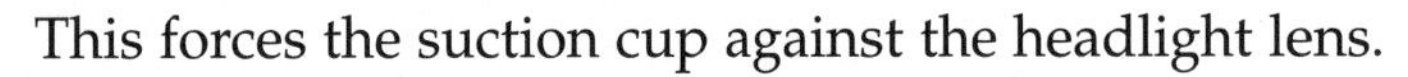

   This forces the suction cup against the headlight lens.

   Pull the handle back until it locks in place.

   Rock the aimer back and forth gently to check that it is attached to the headlight securely .

   Repeat the procedure for the other side.

   Note: When making adjustments, the final adjustment should be made by turning the adjusting screw clockwise.

9. If the aimers have not already been calibrated to the slope of the floor, follow the manufacturer's instructions to calibrate the aimers.

   Are they calibrated? Yes ____ No ____

10. Horizontal adjustment:

    Observe the split image lines through the viewing port on the top of the aimer.

    Not aligned

    Aligned

    Note: If the lines are not showing, the bulb may be improperly installed or the headlights are far out of adjustment.

    Use a screwdriver to turn the horizontal adjustment screw until the split image aligns.

    Are the images aligned? Yes ____ No ____

    Repeat the adjustment for the other headlight.

    Are the images aligned? Yes ____ No ____

11. Vertical Adjustment:

    Turn the vertical adjusting screw until the vertical level bubble is centered.

    Is the bubble centered? Yes ____ No ____

12. Repeat the procedure on the other headlight.

    Is the bubble centered? Yes ____ No ____

LEVEL BUBBLE
VERTICAL DIAL
HORIZONTAL DIAL
VIEWING PORT
SIGHT OPENINGS (MUST FACE EACH OTHER)

13. Double-check to see that the split images are still aligned. If they are not aligned, readjust the horizontal adjustment.

    Image aligned? Yes ____ No ____

    Additional adjustment required? Yes ____ No ____

14. When a vehicle has four headlights, repeat the adjusting procedure for the remaining two headlights.
15. To remove the aimers, press the vacuum handle.
16. Carefully put the aimers in their box.
17. Before completing the paperwork, clean your work area, put the tools in their proper places, and wash your hands.
18. Were there any other recommendations for needed service or unusual conditions that you noticed while adjusting the headlights?

    Yes _____ No _____
19. Record your recommendations for additional service or repairs on the repair order.
20. Complete the repair order (R.O.).

STOP

ASE Lab Preparation Worksheet #7-23

HEADLAMP ADJUSTING WITHOUT AIMING TOOLS

Name_____________________________ Class ____________________

OBJECTIVE:

Upon completion of this assignment, you should be able to adjust a vehicle's headlights without the use of aiming tools. This task will help prepare you to pass the ASE certification examination in electrical systems.

DIRECTIONS:

Before beginning this lab assignment, review the worksheet completely. Fill in the information in the spaces provided as you complete each task.

TOOLS AND EQUIPMENT REQUIRED:

Safety glasses, fender covers, Phillips screwdriver, shop towel, chalk or masking tape

PROCEDURE:

Vehicle year ___________ Make __________________ Model ___________________

Repair Order # ________ Engine size_____________ # of Cylinders ___________

Number of Headlights:	Two	_____	Four	_____		
Headlight shape:	Round	_____	Rectangular	_____	Curved	_____
Bulb type:	Sealed beam	_____	Sealed halogen	_____	Halogen	_____

1. Select a level work area in front of a wall.
2. Position the vehicle about 3 feet from the wall.
3. Mark the wall with large crosses (+) directly opposite the center of each headlight bulb. There should be one cross for each headlight.
4. Move the vehicle back 25 feet from the wall.
5. Clean the headlight lenses.
6. Gently rock the headlight to see if there is any slack between the adjusting screw and the headlight mounting bracket.

   Is there any slack or looseness? Yes ____ No ____

   Note: If the headlight is loose, the problem must be corrected before adjusting the headlights.

7. Turn on the headlights to low beam. The low beam lights should be shining within 6 inches below and to the right of the crosses on the wall.
8. To adjust the low beams, start with the light on the right side of the vehicle.
9. Turn the horizontal adjusting screw in or out to move the beam on the wall.

Note: When making adjustments, the final adjustment should always be made by turning the screw clockwise.

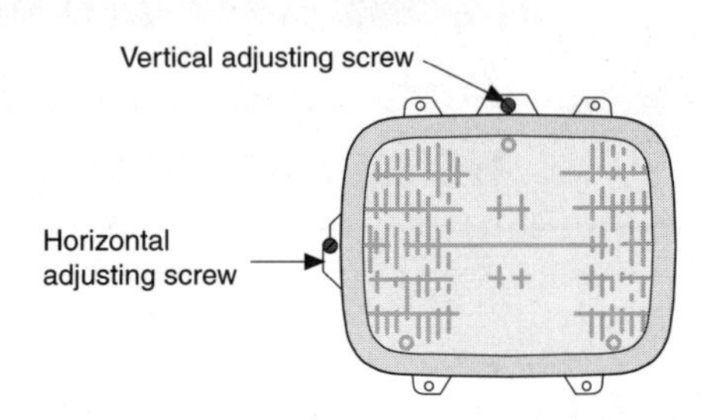

10. Next, turn the vertical adjusting screw in or out to move the beam on the wall.
11. Repeat the procedure on the left side headlight.
12. Check the high beam adjustment. The centers of the light from the high beams should be located near the center of the crosses.

Note: On two-bulb systems, the high beam will be adjusted when the low beam is adjusted.

13. On four-bulb systems, the high beams will need to be adjusted. Repeat the adjustment process for each of the high beams.

Note: Since the low beams will be on while the high beams are being adjusted, it may be necessary to cover the low beam lights. The high beams will be easier to see on the wall with the low beams covered.

14. Before completing the paperwork, clean your work area, put the tools in their proper places, and wash your hands.
15. Were there any other recommendations for needed service or unusual conditions that you noticed while adjusting the headlights?

    Yes ____ No ____

16. Record your recommendations for additional service or repairs on the repair order.
17. Complete the repair order (R.O.).

Part II

ASE Lab Preparation Worksheets

Service Area 8

Tune-Up Services

Instructor OK ______________ Score __________

ASE Lab Preparation Worksheet #8-1

REPLACE SPARK PLUGS

Name_____________________________ Class ___________________

OBJECTIVE:

Upon completion of this assignment, you should be able to replace an engine's spark plugs. This task will help prepare you to pass the ASE certification examination in engine performance.

DIRECTIONS:

Before beginning this lab assignment, review the worksheet completely. Fill in the information in the spaces provided as you complete each task.

TOOLS AND EQUIPMENT REQUIRED:

Safety glasses, fender covers, spark plug wrench, ratchet, extension, vacuum hose, anti-seize, shop towel

PARTS AND SUPPLIES:

Spark plugs

PROCEDURE:

Vehicle year ___________ Make _________________ Model __________________

Repair Order # ________ Engine size_____________ # of Cylinders ___________

1. Open the hood and place fender covers on the fenders and front body parts.
2. List the part number and brand name of one of the spark plugs installed in the engine.

   Number _________ Brand ______________________________
3. Purchase the specified spark plugs for the vehicle.
4. Be certain that the plugs used previously are of the same heat range. The parts person can use a cross reference if a different brand is used.

   Is the heat range the same? Yes ____ No ____

   Note: Changing heat ranges can result in serious engine damage or poor cold engine operation.

R46TS

R = resistor
4 = 14mm thread
6 = heat range
T = taper seat
S = extended tip

5. Open each of the spark plug packages and check the gaps. See that they are set at the manufacturer's specification. Adjust as needed.

   Gap specification 0. ____ "

   Note: Removing and replacing spark plugs is best done on a cold engine.
6. Use eye protection and blow around all of the plugs with compressed air.

Note: This prevents dirt from entering the cylinder when the plugs are removed. This is especially important on transverse mounted engines.

7. Remove *one* spark plug wire at a time. Grasp the rubber boot on the wire and twist it loose before pulling it off the spark plug.

   Note: Remove and replace only one spark plug at a time to prevent accidental switching of the spark plug wires.

8. Use a spark plug socket to remove the plug.

   Socket size: ____ 5/8″ ____ 13/16″ ____ Other ____

9. Inspect each spark plug and note its condition.

INCORRECT WAY

CORRECT WAY

	#1	#2	#3	#4	#5	#6	#7	#8
Normal	____	____	____	____	____	____	____	____
Blistered	____	____	____	____	____	____	____	____
Gray/white	____	____	____	____	____	____	____	____
Carbon fouled	____	____	____	____	____	____	____	____
Oil fouled	____	____	____	____	____	____	____	____
Damaged	____	____	____	____	____	____	____	____
Other	____	____	____	____	____	____	____	____

10. Install a new plug. Do *not* use a ratchet to turn the spark plug until the plug is hand-tightened all of the way.

SHOP TIP A short piece of vacuum hose installed on the top of the plug makes a handy installation tool. Install the spark plug carefully. Cross-threading a spark plug can result in a very expensive repair.

11. If the cylinder head is aluminum, use "anti-seize" on the spark plug threads.

    Aluminum ____ Cast Iron ____

    Anti-seize installed on the spark plugs? Yes ____ No ____

    Note: If in doubt about whether or not the head is aluminum, check it with a magnet.

12. Tighten the spark plug to the recommended torque. Torque specification ____

    In the absence of the correct torque specification, the following recommendation can be used:

    ❑ Tighten gasketed plugs 1/4 turn (90° or 0-3 o'clock) past finger-tight.

    ❑ Tighten tapered seat plugs 1/16 turn (23° or 12-12:45 o'clock) past finger-tight.

    The spark plugs are: Gasketed ____ Tapered ____

CONTINUED

13. Install the spark plug wire, making sure the metal clip inside the boot firmly grasps the metal end of the spark plug.

    Note: Some manufacturers recommend the use of a dielectric compound in spark plug boots.

14. Repeat the spark plug replacement procedure for the remaining plugs.
15. Before completing the paperwork, clean your work area, put the tools in their proper places, and wash your hands.
16. Were there any other recommendations for needed service or unusual conditions that you noticed while changing the spark plugs?

    Yes ____ No ____

17. Record your recommendations for additional service or repairs on the repair order.
18. Complete the repair order (R.O.).

STOP

Instructor OK ______________ Score ____________

ASE Lab Preparation Worksheet #8-2

INSPECT SPARK PLUG CABLES

Name____________________________ Class ____________________

OBJECTIVE:

Upon completion of this assignment, you should be able to test and replace spark plug cables. This task will help prepare you to pass the ASE certification examination in engine performance.

DIRECTIONS:

Before beginning this lab assignment, review the worksheet completely. Fill in the information in the spaces provided as you complete each task.

TOOLS AND EQUIPMENT REQUIRED:

Safety glasses, digital multimeter, shop towel

PROCEDURE:

Vehicle year ___________ Make _________________ Model __________________

Repair Order # ________ Engine size_____________ # of Cylinders ___________

Before starting to test the spark plug cables, identify the type of ignition system used on the vehicle. There are many different types of ignition systems. The basic test procedures are the same for most of them, but knowledge of the systems and the location of their components is important for successful completion of the task. On some vehicles, it may be easier to test the distributor end of the wires from the inside of the distributor cap. Some vehicles do not even have a distributor.

1. Open the hood and place fender covers on the fenders and front body parts.
2. Identify the type of ignition system on the vehicle.

   Breaker points ____ Electronic ignition ____

   Distributorless ignition ____ Coil on Plug (COP) ____
3. Remove and test the spark plug wires, one cylinder at a time. This prevents the cables from being accidentally installed on the wrong spark plugs.

CAUTION When removing a spark plug cable from a spark plug, remember to hold onto the spark plug boot and give it a twist before trying to pull the wire off the spark plug.

CAUTION Some spark plug cables are permanently attached to the distributor cap. Do not try to remove them.

Correct Wrong

4. Visually inspect the condition of each of the spark plug cables. List any defects.

   Dirty ____ Insulation burned ____ Boot damaged ____

   Terminals corroded ____ Other ____

TURN

Note: Check for cracks in the boots. Boots are often cracked. This can result in a miss under hard acceleration or heavy load.

5. Select the ohms × 1000 scale on the ohmmeter.

Note: Always choose the scale that is to be used before connecting a meter to a component or part to be tested.

6. If an analog meter is used, zero the meter before testing the cables. Connect the two leads together and zero the scale. Digital meters are self-zeroing. It will not be necessary to complete this step if a DMM is used.

   What kind of meter is being used? Digital ____ Analog ____

   Was it necessary to zero the meter? Yes ____ No ____

7. Connect the ohmmeter leads to both ends of the cable.

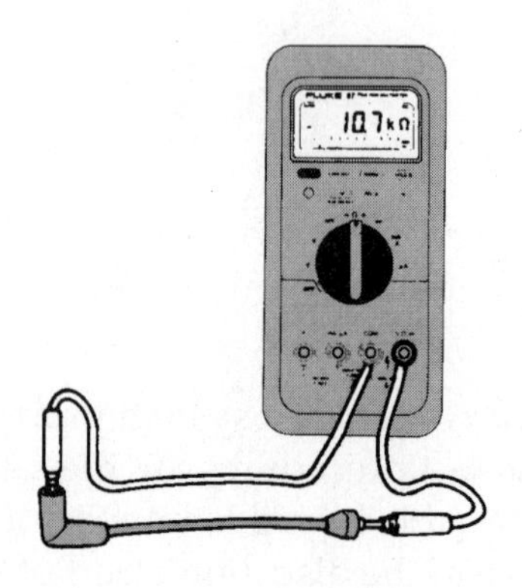

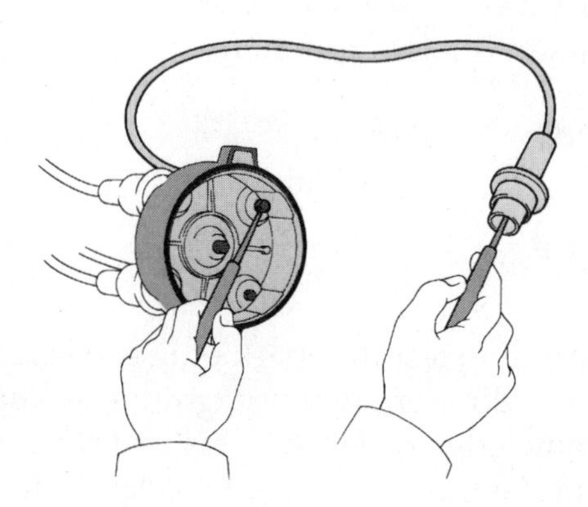

Note: Sometimes it is necessary to insert a paper clip or jumper wire into the plug boot end of the cable to complete the connection.

Record resistances measured below:

Coil Wire

Length ________ Specs. ________ Reading ________ Results ________

Spark Plug Cables

#1 Length ________ Specs. ________ Reading ________ Results ________

#2 Length ________ Specs. ________ Reading ________ Results ________

#3 Length ________ Specs. ________ Reading ________ Results ________

#4 Length ________ Specs. ________ Reading ________ Results ________

#5 Length ________ Specs. ________ Reading ________ Results ________

#6 Length ________ Specs. ________ Reading ________ Results ________

#7 Length ________ Specs. ________ Reading ________ Results ________

#8 Length ________ Specs. ________ Reading ________ Results ________

8. The resistance measures approximately 1000 (1 k) ohms per inch of cable for each of the cables. The total resistance must be 20,000 ohms or less.

9. Condition of the cables: Good ____ Need replacement ____

   Number of cables to be replaced ____

   Note: If more than one cable is "open" (infinite resistance), replacement of all the cables is recommended.

10. Before completing the paperwork, clean your work area, put the tools in their proper places, and wash your hands.

11. Were there any other recommendations for needed service or unusual conditions that you noticed while testing the spark plug cables?

    Yes ____ No ____

12. Record your recommendations for additional service or repairs on the repair order.

13. Complete the repair order (R.O.).

STOP

Instructor OK ________________ Score ____________

ASE Lab Preparation Worksheet #8-3

REPLACE SPARK PLUG CABLES

Name________________________________ Class ____________________

OBJECTIVE:

Upon completion of this assignment, you should be able to replace spark plug cables. This task will help prepare you to pass the ASE certification examinations in engine performance and engine repair.

DIRECTIONS:

Before beginning this lab assignment, review the worksheet completely. Fill in the information in the spaces provided as you complete each task.

TOOLS AND EQUIPMENT REQUIRED:

Safety glasses, shop towel

PROCEDURE:

Vehicle year ____________ Make ____________________ Model ____________________

Repair Order # _________ Engine size_______________ # of Cylinders _____________

Note: It is easiest to replace the cables one at a time. This will prevent mixing them up or misrouting them. If you choose to remove all of the cables at once, this worksheet will help you properly reinstall the cables.

Common firing orders

1. Check a service manual for the correct firing order, cylinder numbering, and direction of distributor rotation.

   Firing order ________________________

   Distribution rotation:

   Clockwise _________ Counterclockwise _________

   Cylinder numbering: Draw a sketch showing cylinder numbering.

2. Crank the engine until the #1 cylinder is at top dead center on the compression stroke.
3. The distributor rotor should now be pointing at the distributor cap terminal for the #1 cylinder's cable.
4. Install the #1 spark plug cable in the distributor cap and carefully route it to the #1 spark plug.
5. Install the next cable in the firing order beside the #1 cable in the distributor cap. Make sure it follows the correct direction of distributor rotation. Carefully route the other end of the cable to the second spark plug in the firing order.
6. Repeat this procedure until all of the cables are installed.

   Note: Route the cables so they are clear of the drive belt(s) and exhaust manifold(s). Cylinders that fire one after the other should not be positioned next to each other. Keep the plug cables away from the vehicle's electrical wiring.

7. Start the vehicle. Is the idle smooth? Yes ____ No ____
8. Accelerate the engine. Does it accelerate smoothly? Yes ____ No ____
9. Before completing the paperwork, clean your work area, put the tools in their proper places, and wash your hands.
10. List any unusual conditions or recommendations for needed service.

    ____

    ____

11. Record your recommendations for needed service or additional repairs on the repair order.
12. Complete the repair order (R.O.).

Instructor OK ______________ Score ___________

ASE Lab Preparation Worksheet #8-4

REPLACE A DISTRIBUTOR CAP AND ROTOR

Name ______________________________ Class ____________________

OBJECTIVE:

Upon completion of this assignment, you should be able to replace a distributor cap and rotor. This task will help prepare you to pass the ASE certification examination in engine performance.

DIRECTIONS:

Before beginning this lab assignment, review the worksheet completely. Fill in the information in the spaces provided as you complete each task.

TOOLS AND EQUIPMENT REQUIRED:

Safety glasses, fender covers, shop towel

PARTS AND SUPPLIES:

Distributor cap, distributor rotor

PROCEDURE:

Vehicle year ___________ Make _________________ Model ___________________

Repair Order # ________ Engine size_____________ # of Cylinders ____________

1. Open the hood and place fender covers on the fenders and front body parts.
2. Remove the holddown clips or screws from the distributor cap.

   Screws ____ Clips ____

3. Remove the cap and inspect its condition:

   Dirty ____

   Corroded terminals ____

   Cracks (carbon trails) ____

   Carbon button burned ____

4. Remove the rotor. How is the rotor mounted to the distributor shaft?

   Holddown screws ____

   Snug fit ____

   Note: Both rotor styles have an alignment feature that keys them to the proper place on the distributor shaft.

SCREWDRIVER

L-SHAPED LUG HOOK

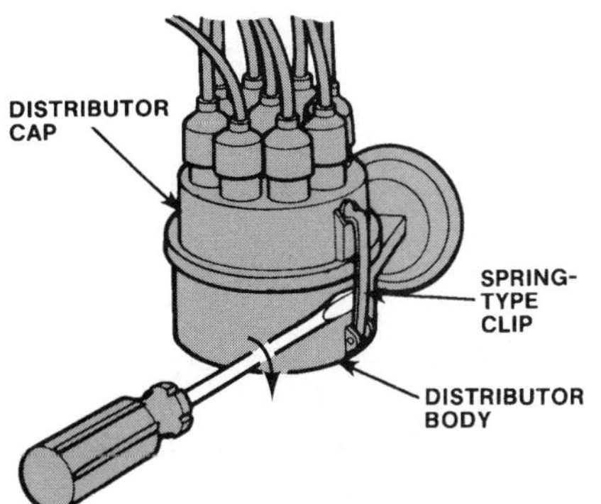

5. Inspect the rotor.

   Tip corroded ____

   Cracked ____

   Physical damage ____

   Burn or puncture marks ____

   Note: Some cars with electronic ignition have a tendency to burn through rotors. This indicates another problem, which should be diagnosed, or the condition will repeat.

6. Install the new rotor.
7. Install the new distributor cap on the distributor. Place the old cap next to it. Remove the #1 cable from the distributor cap and install it in the correct position in the new cap. Install the rest of the spark plug cables into the cap one at a time.

   Why is it necessary to move the cables one at a time?

   ____

8. Start the vehicle. Did it start? Yes ____ No ____
9. Before completing the paperwork, clean your work area, put the tools in their proper places, and wash your hands.
10. Were there any other recommendations for needed service or unusual conditions that you noticed while replacing the distributor cap and rotor?

    Yes ____ No ____

11. Record your recommendations for additional service or repairs on the repair order.
12. Complete the repair order (R.O.).

STOP

Instructor OK ______________ Score ___________

ASE Lab Preparation Worksheet #8-5

CHECK IGNITION TIMING USING A TIMING LIGHT

Name___________________________ Class ___________________

OBJECTIVE:

Upon completion of this assignment, you should be able to use a timing light to check a vehicle's ignition timing. This task will help prepare you to pass the ASE certification examinations in engine performance and engine repair.

DIRECTIONS:

Before beginning this lab assignment, review the worksheet completely. Fill in the information in the spaces provided as you complete each task.

TOOLS AND EQUIPMENT REQUIRED:

Safety glasses, fender covers, timing light, shop towel

PROCEDURE:

Vehicle year ___________ Make _________________ Model __________________

Repair Order # ________ Engine size_____________ # of Cylinders ___________

Check Base Timing

1. Locate the specifications and timing mark diagram in the repair manual.

   Specification _________ ° *before/after* TDC (circle one)

2. Draw a sketch below of the timing marks as they appear in the service manual.

3. Open the hood and place fender covers on the fenders and front body parts.
4. Does the vehicle have an underhood emission label listing the timing specifications?

   Yes ____ No ____

5. Use a flashlight to help locate the timing mark on the crankshaft pulley and timing cover.
6. Use chalk, a crayon, correction fluid, or appliance paint to highlight the specified marks.
7. The timing mark is easier to read from:

   Under the hood ____ Under the car ____

8. Connect the timing light's two power leads to the battery.
9. Connect the inductive pickup to the #1 spark cable.

10. On most older vehicles the vacuum advance must be disconnected to check timing accurately.

Note: This information is available in the manual or on the underhood emissions sticker.

Does the vacuum advance need to be disconnected? Yes ____ No ____ N/A ____

11. With the engine at an idle, aim the light at the mark and note the reading.

Advanced ____ Retarded ____ Ok ____

Note: Be sure the idle speed is correct or the mechanical advance might begin operating before it is supposed to.

Did the timing mark remain steady during the timing check?

Yes ____ No ____

Inductive pickup on number 1 spark plug cable

Gauge

Adjusting knob

Timing mark

Related Information:

- ❏ If the timing mark jumps around as the timing is checked, this can indicate a worn distributor bushing or timing chain. Use Worksheet 8-8, Read an Oscilloscope, and Worksheet 8-13, Check Timing Chain for Wear (Pushrod-Type Engine) to help determine the cause of the problem.
- ❏ Timing too far advanced can cause a high idle or engine damage from abnormal combustion.
- ❏ Timing too far retarded can cause elevated temperatures of engine parts and possible engine damage.
- ❏ Incorrect timing can cause poor performance, poor economy, and increased exhaust emissions.

Mechanical Advance Check

12. Raise the engine speed while observing the timing marks to check the distributor mechanical advance.

Timing advances smoothly ______ Advances in jerks ______

No change ______ N/A ______

Vacuum Advance Check (If Equipped)

13. Allow the engine to idle. Then apply vacuum to the vacuum advance diagram with a vacuum pump to check the vacuum advance. The idle speed should increase as the timing advances.
14. Before completing the paperwork, clean your work area, put the tools in their proper places, and wash your hands.
15. Were there any other recommendations for needed service or unusual conditions that you noticed while checking the timing?

Yes ____ No ____

16. Record your recommendations for additional service or repairs on the repair order.
17. Complete the repair order (R.O.).

STOP

Instructor OK ______________ Score ____________

ASE Lab Preparation Worksheet #8-6

CHECK IGNITION TIMING USING A TIMING LIGHT (COMPUTER-CONTROLLED VEHICLE)

Name__________________________ Class ____________________

OBJECTIVE:

Upon completion of this assignment, you should be able to use a timing light to check ignition timing on a computer-controlled vehicle. This task will help prepare you to pass the ASE certification examinations in engine performance and engine repair.

DIRECTIONS:

Before beginning this lab assignment, review the worksheet completely. Fill in the information in the spaces provided as you complete each task.

TOOLS AND EQUIPMENT REQUIRED:

Safety glasses, fender covers, timing light, shop towel

PROCEDURE:

Vehicle year ___________ Make _________________ Model __________________

Repair Order # ________ Engine size_____________ # of Cylinders ___________

Check Base Timing

1. Locate the specifications and timing mark diagram in the repair manual.

   Specification __________ ° *before/after* TDC (circle one)

2. Sketch the timing marks as they appear in the service manual.

3. Open the hood and place fender covers on the fenders and front body parts.
4. Does the vehicle have an underhood emissions label listing the timing specifications?

   Yes _____ No _____

5. Use a flashlight to help locate the timing mark on the crankshaft pulley and timing cover.
6. Use chalk, a crayon, correction fluid, or appliance paint to highlight the specified marks.
7. The timing mark is easier to read from:

   Under the hood _____ Under the car _____

8. Connect the timing light's two power leads to the battery.
9. Connect the inductive pickup to the #1 spark plug cable.

Timing marks aligned at 3

10. The computer must be relieved of timing control to accurately check the timing.

    Note: The procedure for this is available in the service manual or on the underhood emissions label.

    How is the computer disconnected from the circuit during the timing check?

    ____

    Disconnect the computer from the circuit.

11. With the engine at an idle, aim the light at the timing mark and note the reading.

    Advanced ____ OK __________ Retarded _____

    Did the timing mark remain steady during the timing check? Yes ____ No ____

Related Information:

- ❑ If the timing mark jumps around as the timing is checked, this can indicate a worn distributor bushing or timing chain. Use Worksheet 8-8, Read an Oscilloscope, and Worksheet 8-13, Check Timing Chain for Wear to help determine the cause of the problem.
- ❑ Timing too far advanced can cause a high idle or engine damage from abnormal combustion.
- ❑ Timing too far retarded can cause elevated temperatures of engine parts and possible engine damage.
- ❑ Incorrect timing can cause poor performance, poor economy, and increased exhaust emissions.

12. Before completing the paperwork, clean your work area, put the tools in their proper places, and wash your hands.

13. Were there any other recommendations for needed service or unusual conditions that you noticed while checking the timing?

    Yes ____ No ____

14. Record your recommendations for additional service or repairs on the repair order.

15. Complete the repair order (R.O.).

STOP

Instructor OK ______________ Score __________

ASE Lab Preparation Worksheet #8-7

MEASURE ENGINE VACUUM

Name____________________________ Class __________________

OBJECTIVE:

Upon completion of this assignment, you should be able to measure engine vacuum and analyze the results. This task will help prepare you to pass the ASE certification examinations in engine performance and engine repair.

DIRECTIONS:

Before beginning this lab assignment, review the worksheet completely. Fill in the information in the spaces provided as you complete each task.

TOOLS AND EQUIPMENT REQUIRED:

Safety glasses, shop towel

PROCEDURE:

Vehicle year __________ Make ________________ Model ________________

Repair Order # ________ Engine size____________ # of Cylinders __________

1. Open the hood and place fender covers on the fenders.
2. Locate a vacuum source at the intake manifold.
3. Contact a vacuum gauge to the intake manifold.
4. Start the engine and read the gauge. What is the vacuum reading? _______

   If the reading is zero inch, open the throttle. Did the reading increase? Yes ____ No ____

   If the reading increased when the engine was accelerated, connect the gauge to another vacuum source.

   Does the gauge now have a good reading at idle? Yes ____ No ____
5. What is the vacuum reading on the gauge? ____________________
6. Is it within specifications? Yes ____ No ____

   Is the reading steady or jumping around? Steady ________ Jumping ________

   If it is not steady and within specifications, what is indicated? ____________________
7. To check cranking vacuum, turn off the engine, plug the breather hose in to the air cleaner, and disable the ignition system.
8. Crank the engine and read the gauge. What is the reading? ______________

   Was the reading steady or jumping? Steady ________ Jumping ________

   What would be indicated if the reading were jumping around? ____________________

9. Check for a restricted exhaust by holding the engine rpm at 2000 rpm and watching the gauge. The reading should be approximately the same as the reading at idle and hold steady. If the reading starts to fall, the exhaust is restricted.
10. Did the gauge maintain its reading or start to fall? Maintained _____ Fell _____
11. Before completing the paperwork, clean your work area, put the tools in their proper places, and wash your hands.
12. List any unusual conditions or recommendations for needed service.

    ____

    ____
13. Record your recommendations for needed service or additional repairs on the repair order.
14. Complete the repair order (R.O.).

Instructor OK ______________ Score ____________

ASE Lab Preparation Worksheet #8-8

READ AN OSCILLOSCOPE

Name______________________________ Class ____________________

OBJECTIVE:

Upon completion of this assignment, you should be able to connect and read an oscilloscope. This task will help prepare you to pass the ASE certification examination in engine performance.

DIRECTIONS:

Before beginning this lab assignment, review the worksheet completely. Fill in the information in the spaces provided as you complete each task.

TOOLS AND EQUIPMENT REQUIRED:

Safety glasses, fender covers, oscilloscope, shop towel

PROCEDURE:

Vehicle year ____________ Make ____________________ Model ____________________

Engine size______________ # of Cylinders ____________

Manufacturer of the scope being used ____.

Model of the scope being used ____.

1. Open the hood and place fender covers on the fenders and front body parts.
2. Attach the shop exhaust hose to the vehicle's exhaust pipe.
3. Connect the timing light lead to the #1 spark plug cable.
4. Connect the positive (+) and negative (–) leads to their respective battery terminals.
5. Connect the coil inductive clamp (pickup) to the coil wire or distributor cap.
6. Connect the tach leads to the primary circuit and a suitable ground.
7. Turn the scope on and adjust the pattern brightness and vertical and horizontal positions.
8. Select the "secondary" pattern position.
9. Adjust the scope for the proper number of cylinders.
10. Select the proper ignition type.

Be sure all scope leads are clear of the engine fan and hot engine parts.

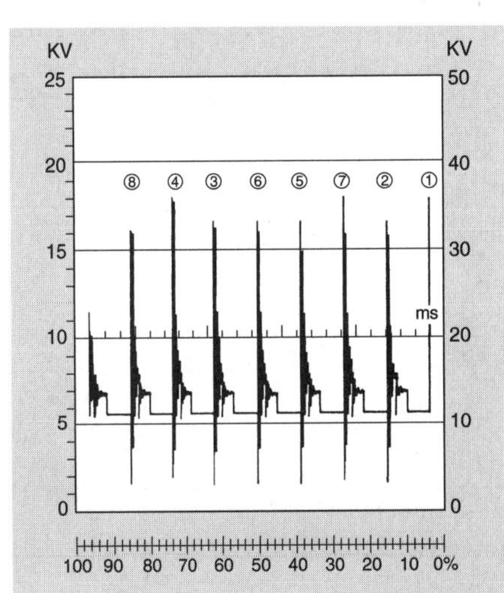

11. Start the engine.
12. Before reading the scope pattern, always check the idle speed and ignition timing.

13. Adjust the scope to display the "parade" pattern.

    a. The parade pattern is displayed from:

       Top to bottom ____ Left to right ____

    b. Where is cylinder number one located on the parade pattern?

       Top ____ Bottom ____ Left ____ Right ____

    c. The primary use of the parade pattern is to look at and compare the:

       Firing lines ____ Spark lines ____ Dwell section ____

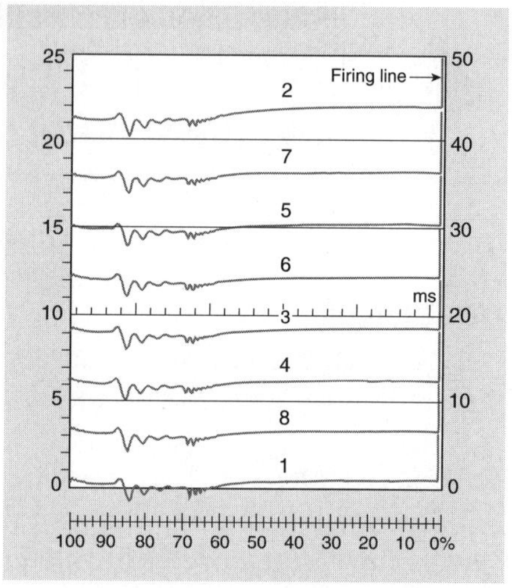

14. Adjust the scope to display the "raster" pattern.

    a. The raster pattern is displayed from:

       Top to bottom ____ Left to right ____

    b. Where is cylinder number one located on the raster pattern?

       Top ____ Bottom ____ Left ____ Right ____

    c. The primary use of the raster pattern is to look at and compare the:

       Firing lines ____ Spark lines ____ Dwell section ____

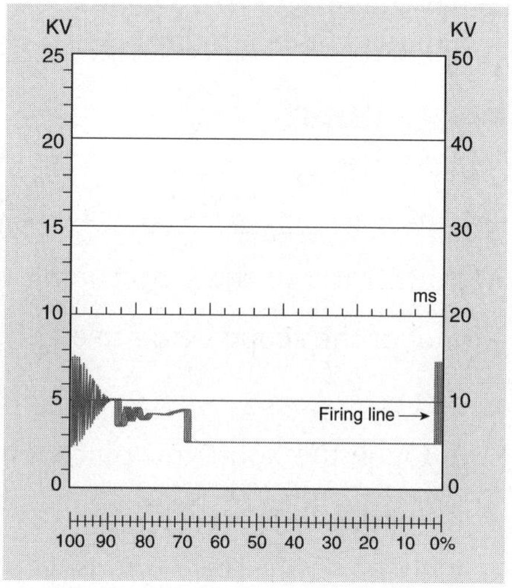

15. Adjust the scope to display the "superimposed" pattern.

    a. The superimposed pattern is displayed from:

       Front to back ____ Left to right ____

    b. The primary use of the superimposed pattern is to look at and compare the:

       Firing lines ____ Spark lines ____ Dwell section ____

16. Switch to the tachometer reading. Read and record the rpm. Idle rpm ____

17. Turn off the engine.

18. Disconnect all connections and neatly coil the wires, returning them to their proper hangers.

19. Before completing the paperwork, clean your work area, put the tools in their proper places, and wash your hands.

20. List any other recommendations for needed service or unusual conditions that you noticed while using the oscilloscope.

Recommendations: ____

____

____

____

____

STOP

Instructor OK ______________ Score ____________

ASE Lab Preparation Worksheet #8-9

POWER BALANCE TESTING

Name______________________________ Class ____________________

OBJECTIVE:

Upon completion of this assignment, you should be able to test the power balance between cylinders in a engine. This task will help prepare you to pass the ASE certification examinations in engine performance and engine repair.

DIRECTIONS:

Before beginning this lab assignment, review the worksheet completely. Fill in the information in the spaces provided as you complete each task.

TOOLS AND EQUIPMENT REQUIRED:

Safety glasses, fender covers, shop towel, oscilloscope or power balance tester

PROCEDURE:

Vehicle year ___________ Make __________________ Model __________________

Repair Order # ________ Engine size_____________ # of Cylinders ___________

Power balance testing, also called load testing, is a quick way to compare the output of an engine's cylinders. The test is best done using an oscilloscope or portable electronic cylinder balance tester. These testers will avoid damage to the ignition module or the catalytic converter.

1. Open the hood and place fender covers on the fenders and front body parts.
2. Connect an oscilloscope or power balance tester to the vehicle.
3. Use the oscilloscope *cylinder kill* feature to cause each cylinder to misfire one at a time. Newer power balance testers automatically short each cylinder for a specified period of time before shorting the next one. They record the amount of rpm drop as each cylinder misfires. If you are shorting the cylinders manually, let the engine stabilize after each cylinder before proceeding to kill the next cylinder.

   Record the rpm drop for each cylinder below.

#1	#2	#3	#4	#5	#6	#7	#8
____	____	____	____	____	____	____	____

4. If a cylinder did not drop in rpm or had an increase in rpm while it was misfiring, a problem is indicated.

   Which cylinders, if any, failed the test?

   Cylinder #s ____________________

   All cylinders tested good ____

5. If the test indicates that there is a problem, there are several follow-up tests that may be done. The tests are listed below. Place a number in front of each test in the order that it should be done.

   _____ Check valve lash

   _____ Compression test

   _____ Vacuum test

   _____ Cylinder leakage test

   _____ Oscilloscope analysis

   _____ Other ____

6. If a problem was noticed during the test, check with your instructor before proceeding with additional tests.
7. Before completing the paperwork, clean your work area, put the tools in their proper places, and wash your hands.
8. Were there any other recommendations for needed service or unusual conditions that you noticed while doing the power balance test?

   Yes _____ No _____

9. Record your recommendations for additional service or repairs on the repair order.
10. Complete the repair order (R.O.).

STOP

Instructor OK ____________ Score ____________

ASE Lab Preparation Worksheet #8-10

COMPRESSION TEST

Name____________________________ Class ____________________

OBJECTIVE:

Upon completion of this assignment, you should be able to check an engine's compression pressure. This task will help prepare you to pass the ASE certification examinations in engine repair and engine performance.

DIRECTIONS:

Before beginning this lab assignment, review the worksheet completely. Fill in the information in the spaces provided as you complete each task.

TOOLS AND EQUIPMENT REQUIRED:

Safety glasses, fender covers, spark plug socket, ratchet, compression gauge, shop towel, oil

PROCEDURE:

Vehicle year __________ Make ________________ Model ________________

Repair Order # ________ Engine size____________ # of Cylinders ___________

1. Open the hood and place fender covers on the fenders and front body parts.
2. Is the engine at normal operating temperature? Yes ____ No ____
3. Is the battery fully charged? Yes ____ No ____
4. Carefully remove the secondary (spark plug) cables. Twist the boots first to avoid internal damage to the cables.
5. Carefully blow around each of the spark plugs with compressed air.

   Why is it a good practice to blow air around the spark plug before removing it?

   ____

6. Remove all of the spark plugs from the engine.

   As each spark plug is removed, check its condition and record your observations below.

Cylinder Number	#1	#2	#3	#4	#5	#6	#7	#8
Normal	____	____	____	____	____	____	____	____
Worn	____	____	____	____	____	____	____	____
Damaged	____	____	____	____	____	____	____	____
Carbon fouled	____	____	____	____	____	____	____	____
Oil fouled	____	____	____	____	____	____	____	____

Why is it a good practice to remove all of the spark plugs when doing a compression test?

____

TURN

7. Use a pedal depressor to block the throttle wide open.

   Note: This is to prevent a carbureted car from drawing fuel into the cylinders during the test. It also allows the engine to breathe more easily.

8. Disable the ignition system in one of the following ways (check the method used):

   ❑ Conventional coil: Use a jumper wire to ground the coil high tension wire at the distributor cap end.

   ❑ Coil wire not accessible: Disconnect the battery power wire to the ignition system.

   ❑ Battery wire not accessible: Remove the ignition fuse. This method works well with newer distributorless ignition systems.

   ❑ On fuel-injected vehicles, disconnect the ignition primary or disable the fuel pump by removing its fuse.

   Ignition system disabled? Yes ____ No ____

9. Select the correct compression gauge adapter and thread it into the cylinder to be tested. Tighten it just enough to compress the O-ring seal. Next install the compression gauge onto the adapter.

10. Crank the engine through at least 4 (four) compression strokes.

    Note: The needle on the gauge should move until it no longer advances. The needle will advance with each compression stroke in response to pressure buildup in the cylinder. Engines with very good compression rings will sometimes reach the highest reading in as little as two compression strokes. With poor rings, six compression strokes might be necessary.

    Record the compression reading for each of the cylinders:

    #1 ____ #2 ____ #3 ____ #4 ____

    #5 ____ #6 ____ #7 ____ #8 ____

    Note: Variations between cylinders should be no more than 20%.

    Are any of the compression readings below acceptable levels? Yes ____ No ____

11. Is a wet test required? Yes ____ No

12. If a wet compression test is required, remove the compression gauge and squirt approximately 1 tablespoon of oil into the cylinder to be tested.

13. Reinstall the compression gauge and crank the engine through at least 4 compression strokes.

    Record the gauge reading and compare it to the dry test reading.

    Dry Reading _____ Wet Reading _____

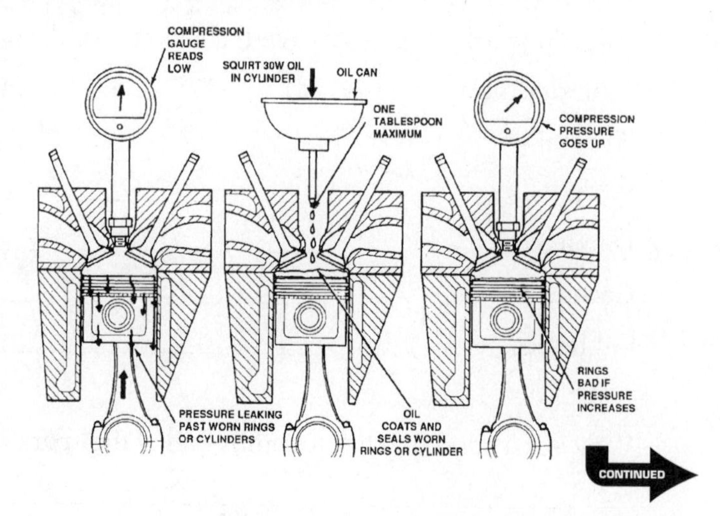

CONTINUED

What might happen if too much oil is squirted into the cylinder?

____

Note: Wet testing is done only on cylinders that may have a problem.

Why would you not wet test all of the cylinders?

____

14. Which of the following diagnoses can be made from the compression test just completed?
    - ❑ Worn piston rings or cylinders
    - ❑ Leaking valves
    - ❑ Blown head gasket
    - ❑ Excessive compression
15. Is the engine's compression within limits (normal)?

    Yes ____ No ____
16. Before completing the paperwork, clean your work area, put the tools in their proper places, and wash your hands.
17. Were there any other recommendations for needed service or unusual conditions that you noticed while doing the compression test?

    Yes ____ No ____
18. Record your recommendations for additional service or repairs on the repair order.
19. Complete the repair order (R.O.).

ASE Lab Preparation Worksheet #8-11

RETRIEVING OBD I TROUBLE CODES

Name______________________________ Class ______________________

OBJECTIVE:

Upon completion of this assignment, you should be able to retrieve OBD I trouble codes. This task will help prepare you to pass the ASE certification examination in engine performance.

DIRECTIONS:

Before beginning this lab assignment, review the worksheet completely. Fill in the information in the spaces provided as you complete each task.

TOOLS AND EQUIPMENT REQUIRED:

Safety glasses, shop towel

PROCEDURE:

Vehicle year ____________ Make ____________________ Model _____________________

Repair Order # _________ Engine size __________________ # of Cylinders ____________

1. Refer to the service information for the suggested method of retrieving the diagnostic trouble codes (DTCs) from the vehicle.

   Information source _________________ Page # _________________

2. What was the recommended tool(s) for retrieving the trouble codes?

   Jumper ____ Test light ____

   Analog voltmeter ____ Scan tool ____

3. Follow the directions as listed in the service information and read the trouble codes.

4. Were any DTCs retrieved from the vehicle?

   Yes ____ No ____

5. List any DTCs that you retrieved and their meaning.

	Code	Meaning
a.	______________________	____
b.	______________________	____
c.	______________________	____

6. Before completing the paperwork, clean your work area, put the tools in their proper places, and wash your hands.

7. List any unusual conditions or recommendations for needed service that you noticed while you were checking OBD I trouble codes.

____

____

8. Record your recommendations for additional service or repairs on the repair order.
9. Complete the repair order (R.O.).

STOP

Instructor OK ____________ Score ____________

ASE Lab Preparation Worksheet #8-12

RETRIEVING OBD I TROUBLE CODES FROM A GENERAL MOTORS VEHICLE

Name____________________ Class ____________

OBJECTIVE:

Upon completion of this assignment, you should be able to retrieve OBD I trouble codes from a typical General Motors vehicle. This task will help prepare you to pass the ASE certification examination in engine performance.

DIRECTIONS:

Before beginning this lab assignment, review the worksheet completely. Fill in the information in the spaces provided as you complete each task.

TOOLS AND EQUIPMENT REQUIRED:

Safety glasses, shop towel

PROCEDURE:

Vehicle year ________ Make ____________ Model ____________

Repair Order # ______ Engine size ____________ # of Cylinders ________

1. Refer to the service information for the suggested method of retrieving the diagnostic trouble codes (DTCs) from the vehicle.

   Information source __________ Page # __________

2. What was the recommended tool(s) for retrieving the trouble codes?

   Jumper ____ Test light ____ Analog voltmeter ____ Scan tool ____

3. With the key on and the engine off (KOEO), use a jumper wire to short between terminals A and B in the diagnostic connector.

4. The check engine light will now flash the codes. Read the codes by counting the number of flashes and pauses.

5. Were any DTCs retrieved from the vehicle?

   Yes ____ No ____

6. List any DTCs that you retrieved and their meaning.

Diagnostic code display

Check engine

Check engine

Pause

Check engine

Check engine

Flash 1

Flash 1 + Flash 1 = 2

1 and 2 = code 12

	Code	Meaning
a.	__________	____________________
b.	__________	____________________
c.	__________	____________________
d.	__________	____________________

7. Before completing the paperwork, clean your work area, put the tools in their proper places, and wash your hands.
8. List any unusual conditions or recommendations for needed service that you noticed.

   ____

   ____

9. Record your recommendations for needed service or additional repairs on the repair order.
10. Complete the repair order (R.O.).

Instructor OK ____________ Score ____________

ASE Lab Preparation Worksheet #8-13

RETRIEVING OBD I TROUBLE CODES FROM A FORD VEHICLE

Name________________________ Class ________________

OBJECTIVE:

Upon completion of this assignment, you should be able to retrieve OBD I trouble codes from a typical Ford vehicle. This task will help prepare you to pass the ASE certification examination in engine performance.

DIRECTIONS:

Before beginning this lab assignment, review the worksheet completely. Fill in the information in the spaces provided as you complete each task.

TOOLS AND EQUIPMENT REQUIRED:

Safety glasses, shop towel

PROCEDURE:

Vehicle year __________ Make ______________ Model ______________

Repair Order # ________ Engine size ______________ # of Cylinders __________

1. Refer to the service information for the suggested method of retrieving the diagnostic trouble codes (DTCs) from the vehicle.

   Information source ____________ Page # ____________

2. What was the recommended tool(s) for retrieving the trouble codes?

   Jumper ____ Test light ____ Analog voltmeter ____ Scan tool ____

3. Locate the diagnostic connector and the self-test input connector. Where are they located?

   ____

4. Connect the voltmeter negative test lead to pin number 4 of the diagnostic connector and the positive test lead of the voltmeter to the positive post of the vehicle's battery.
5. Connect a jumper wire to short between pin number 2 of the diagnostic connector and the self-test input connector.
6. Turn on the ignition key but leave the engine off. The vehicle will run a self-test and then output the codes to the analog voltmeter. Read the codes by counting the number of times the needle sweeps.

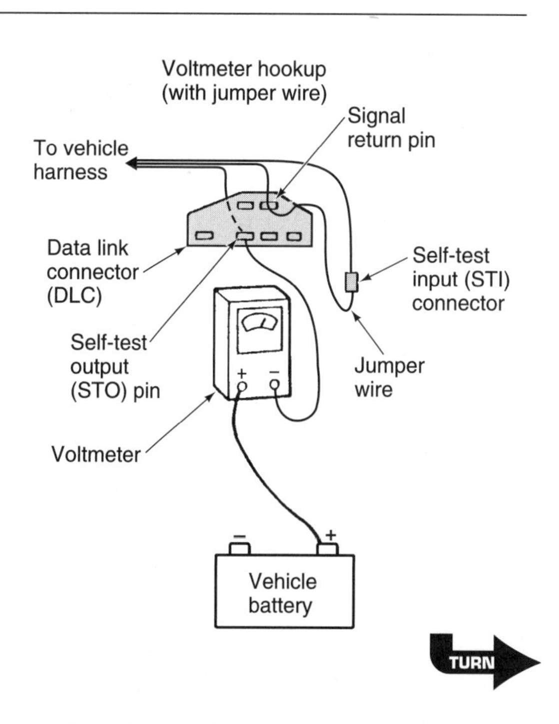

TURN

7. Were any DTCs retrieved from the vehicle? Yes ____ No ____
8. List any DTCs that you retrieved and their meaning.

	Code	Meaning
a.	____________________	____
b.	____________________	____
c.	____________________	____
d.	____________________	____

9. Before completing the paperwork, clean your work area, put the tools in their proper places, and wash your hands.
10. List any unusual conditions or recommendations for needed service that you noticed.

____

____

11. Record your recommendations for needed service or additional repairs on the repair order.
12. Complete the repair order (R.O.).

STOP

Instructor OK ____________ Score ____________

ASE Lab Preparation Worksheet #8-14

RETRIEVING OBD I TROUBLE CODES FROM A CHRYSLER VEHICLE

Name____________________ Class ____________

OBJECTIVE:

Upon completion of this assignment, you should be able to retrieve OBD I trouble codes from a typical Chrysler vehicle. This task will help prepare you to pass the ASE certification examination in engine performance.

DIRECTIONS:

Before beginning this lab assignment, review the worksheet completely. Fill in the information in the spaces provided as you complete each task.

TOOLS AND EQUIPMENT REQUIRED:

Safety glasses, shop towel

PROCEDURE:

Vehicle year ________ Make ____________ Model ____________

Repair Order # ______ Engine size ____________ # of Cylinders ________

1. Refer to the service information for the suggested method of retrieving the diagnostic trouble codes (DTCs) from the vehicle.

   Information source ____________ Page # ____________

2. What was the recommended tool(s) for retrieving the trouble codes?

   Jumper ____ Test light ____ Analog voltmeter ____ Scan tool ____

3. Does the malfunction indicator light go off after the engine is started? Yes ____ No ____

Diagnostic connector

4. Turn off the engine.
5. Cycle (turn on and off) the ignition key three times in 5 seconds without starting the engine.
6. Read the codes from the flashing malfunction indicator light.
   a. The first set of flashes counts as tens. This will be followed by a pause.
   b. The second set of flashes is counted as ones.
   c. Together they make a two-digit code.
   d. Following a long pause, the code will be repeated or another code will be displayed.
7. Were any DTCs retrieved from the vehicle? Yes ____ No ____

8. List any DTCs that you retrieved and their meaning.

	Code	Meaning
a.	____________________	____
b.	____________________	____
c.	____________________	____
d.	____________________	____

9. Before completing the paperwork, clean your work area, put the tools in their proper places, and wash your hands.
10. List any unusual conditions or recommendations for needed service.

____

____

11. Record your recommendations for needed service or additional repairs on the repair order.
12. Complete the repair order (R.O.).

STOP

Instructor OK _______________ Score ___________

ASE Lab Preparation Worksheet #8-15

RETRIEVING OBD II DIAGNOSTIC TROUBLE CODES

Name______________________________ Class ____________________

OBJECTIVE:

Upon completion of this assignment, you should be able to retrieve OBD II diagnostic trouble codes from a vehicle. This task will help prepare you to pass the ASE certification examination in engine performance.

DIRECTIONS:

Before beginning this lab assignment, review the worksheet completely. Fill in the information in the spaces provided as you complete each task.

TOOLS AND EQUIPMENT REQUIRED:

Safety glasses, shop towel

PROCEDURE:

Vehicle year ___________ Make _________________ Model __________________

Repair Order # ________ Engine size _________________ # of Cylinders ____________

1. Refer to the service information for the suggested method of retrieving the diagnostic trouble codes (DTCs) from the vehicle.

   Information source _______________ Page # _______________

2. What was the recommended tool(s) for retrieving the trouble codes?

   Jumper ____ Test light ____ Analog voltmeter ____ Scan tool ____

3. Does the malfunction indicator light go off after the engine is started? Yes ____ No ____
4. Turn off the engine.
5. Locate the OBD II data link connector. Where is it located? _________________
6. Connect the scan tool to the data link connector and turn on the ignition key. The engine should not be running. This is commonly called KOEO (key on, engine off).
7. What is the brand name of the scan tool that you are using? _________________
8. Follow the scan tool manufacturer's directions for displaying the diagnostic trouble codes.
9. Were any DTCs retrieved from the vehicle? Yes ____ No ____

10. List any DTCs that the scan tool displayed and the meaning of the code.

	Code	Meaning
a.	____________	______________________________
b.	____________	______________________________
c.	____________	______________________________
d.	____________	______________________________

11. What does each of the following DTCs indicate?

    a. PO304 ______________________________

    b. PO440 ______________________________

    c. PO133 ______________________________

    d. PO119 ______________________________

12. Before completing the paperwork, clean your work area, put the tools in their proper places, and wash your hands.

13. List any unusual conditions or recommendations for needed service that you noticed while you were checking OBD II trouble codes.

    ____

    ____

14. Record your recommendations for needed service or additional repairs on the repair order.

15. Complete the repair order (R.O.).

STOP

Instructor OK ______________ Score __________

ASE Lab Preparation Worksheet #8-16

CHECK TIMING CHAIN FOR WEAR (PUSHROD-TYPE ENGINE)

Name__________________________ Class __________________

OBJECTIVE:

Upon completion of this assignment, you should be able to check a pushrod engine for excessive timing chain wear. This task will help prepare you to pass the ASE certification examinations in engine repair and engine performance.

DIRECTIONS:

Before beginning this lab assignment, review the worksheet completely. Fill in the information in the spaces provided as you complete each task.

TOOLS AND EQUIPMENT REQUIRED:

Safety glasses, fender covers, shop towel

PROCEDURE:

Vehicle year ___________ Make _________________ Model _________________

Repair Order # ________ Engine size ________________ # of Cylinders ___________

Note: Do *not* perform this test on an overhead cam engine.

1. Open the hood and place fender covers on the fenders and front body parts.
2. Remove the distributor cap.

   How is the distributor cap connected to the distributor?

   Screws ____ Clips ____ Other ____
3. Install a socket on the crankshaft pulley bolt. Turn the crankshaft *counterclockwise* while watching the distributor rotor. Stop turning the crankshaft when the rotor begins to move.
4. Make a mark on the crankshaft pulley in line with the TDC indicator on the timing cover.

   Method used:

   Chalk ____ Pencil ____ Crayon ____

5. Move the crankshaft *clockwise* until the rotor starts to move again.
6. Mark the pulley again in line with the TDC indicator.
7. Estimate the distance in degrees that the crankshaft turned before the rotor moved. (Use the timing mark as an idea of what percentage of an inch is represented by 5°, 10°, etc.)

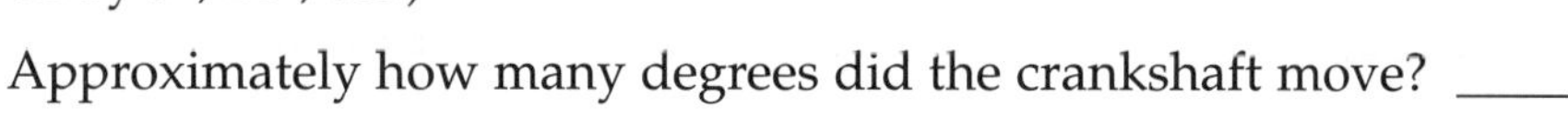

   Approximately how many degrees did the crankshaft move? ____

Note: 8° to 10° of crankshaft movement is allowable.

8. If the maximum allowable movement is exceeded, what repair is required?

____

9. Replace the distributor cap.
10. Before completing the paperwork, clean your work area, put the tools in their proper places, and wash your hands.
11. Were there any other recommendations for needed service or unusual conditions that you noticed while checking the timing chain wear?

    Yes ____ No ____

12. Record your recommendations for additional service or repairs on the repair order.
13. Complete the repair order (R.O.).

Instructor OK ______________ Score __________

ASE Lab Preparation Worksheet #8-17

REPLACE A TIMING BELT

Name____________________________ Class ____________________

OBJECTIVE:

Upon completion of this assignment, you should be able to replace a timing belt. This task will help prepare you to pass the ASE certification examination in engine repair.

DIRECTIONS:

Before beginning this lab assignment, review the worksheet completely. Fill in the information in the spaces provided as you complete each task.

TOOLS AND EQUIPMENT REQUIRED:

Safety glasses, fender covers, shop towel, hand tools, damper puller, torque wrench

PARTS AND SUPPLIES:

Timing belt, tensioner, water pump, idler pulley bearing

Note: Not all vehicles will require all of these parts when replacing the timing belt.

PROCEDURE:

Vehicle year ___________ Make _________________ Model ___________________

Repair Order # ________ Engine size _________________ # of Cylinders ____________

Note: This project should be performed on a shop engine before it is attempted on a live vehicle.

1. Locate a repair manual for the vehicle. Find the instructions for the replacement of the timing belt.

   Which manual was used? ____ Page _______

2. Locate the drawing that shows the valve timing marks for the engine. Draw a sketch in the space below.

3. Open the hood and place fender covers on the fenders and front body parts.
4. Remove the negative battery cable.
5. Remove any accessory drive belts that are in front of the timing cover.

   How many accessory drive belts are there? ______________________

6. Remove the spark plugs.

TURN

7. Turn the engine by hand until the number one cylinder is at TDC on its compression stroke and the timing marks are aligned.

   Note: Always turn the engine in its normal direction of rotation.

8. Use an impact wrench and socket to remove the crankshaft pulley bolt.

   Note: When an impact wrench is used, the crankshaft does not have to be restrained from turning while loosening the bolt.

9. Remove the crankshaft damper (pulley) using the correct type of puller.

   Note: Some crankshaft dampers are press fit to the crankshaft. A three-piece vibration damper can be damaged if a puller is attached to its outer ring rather than to the screw holes in its inner hub.

10. Remove the timing cover.

11. Remove any guides or spacers.

    Spacer ____ Guide ____ None ____

12. Loosen the belt tension adjustment and remove the belt. What kind of adjustment does the belt tensioner use?

    Jackscrew ____ Hydraulic pressure ____

    Spring-loaded lock center ____ Other ____

    Water pump pulley ____

13. Clean both sprockets with a nonpetroleum-based solvent.

    What type of solvent was used? ________________________________

    Note: Petroleum products can be absorbed by the sprockets and cause damage to the belt.

14. Check the bearing in the tensioner or idler pulley for roughness. A rough or loose bearing must be replaced.

    Bearing condition? OK ____ Bad ____

15. Some engines with a timing belt have the water pump located inside the timing cover. If this is the case, now is the time to check the water pump for leakage, bearing wear, or damage, and replace it if necessary.

    Water pump condition?

    OK ____ Leaks ____ Bad bearing ____ N/A ____

16. Install the belt and adjust its belt tension according to manufacturer's recommendations.
17. Check to see that all the sprocket timing marks are correctly located.
18. Reinstall any guides or spacers that were removed.
19. Reinstall the timing cover.
20. Install the crankshaft pulley and torque the crankshaft bolt.

    Torque specification: ________________________________

21. Rotate the engine through two complete revolutions. Do the timing marks still line up as they did previously?

    Yes _____ No _____

Check the valve timing.

Instructor OK before proceeding ____________________

Note: If the timing marks do not line up after rotating the engine two revolutions, there is a problem. Stop, remove the timing belt, and start over at step 16.

22. Install any accessory drive belts that were previously removed and adjust their tension.
23. Install the spark plugs.
24. Install the negative battery cable.
25. Start the engine and adjust the idle speed.

    Idle speed specifications ____________

26. Before completing the paperwork, clean your work area, put the tools in their proper places, and wash your hands.
27. Put a sticker on the valve cover, listing the mileage when the belt was replaced.

    Mileage listed ___________________________

Note: Keeping a record of timing belt replacement is important since the replacement of the timing belt is a required maintenance item for most vehicles.

28. Were there any other recommendations for needed service or unusual conditions that you noticed while replacing the timing belt?

    Yes _____ No _____

29. Record your recommendations for additional service or repairs on the repair order.
30. Complete the repair order (R.O.).

Instructor OK ______________ Score __________

ASE Lab Preparation Worksheet #8-18

CHECKING ENGINE OIL PRESSURE

Name____________________________ Class __________________

OBJECTIVE:

Upon completion of this assignment, you should be able to connect an oil pressure gauge to measure an engine's oil pressure. This task will help prepare you to pass the ASE certification examination in engine repair.

DIRECTIONS:

Before beginning this lab assignment, review the worksheet completely. Fill in the information in the spaces provided as you complete each task.

TOOLS AND EQUIPMENT REQUIRED:

Safety glasses, fender covers, hand tools, flare nut wrenches, oil pressure gauge, shop towel

PROCEDURE:

Vehicle year ___________ Make _________________ Model __________________

Repair Order # ________ Engine size ________________ # of Cylinders ___________

1. Locate the oil pressure specification in the service manual and record the information below.

   Which manual was used? ____

   The specification was located on page number _______

   Minimum: ____ psi at _________ rpm

   Maximum: ____ psi at _________ rpm

2. Open the hood and place fender covers on the fenders and front body parts.
3. What is the engine temperature?

   Cold ____

   Warm ____

4. The vehicle is equipped with an:

   Oil pressure gauge ____

   Indicator light ____

5. Check the operation of the oil pressure light. Turn on the ignition key and leave the engine off. Does the indicator light glow?

   Yes ____ No ____

   Not equipped ____

6. Locate the sending unit on the engine.

   On what part of the engine is it located?

   ____

TURN

Note: If there is a malfunction in the gauge electrical circuit, try unhooking the wire connected to the oil pressure sending unit. When it is connected to ground with the key on, the gauge or light function should change.

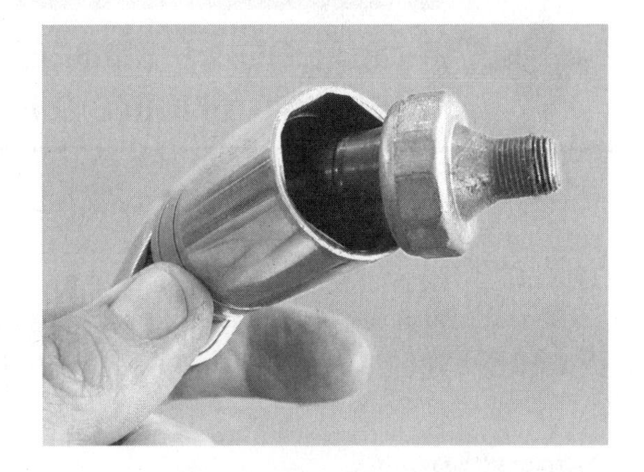

7. Remove the sending unit.

Note: Some sending units require a special socket. In some cases, a twelve-point socket will work.

8. Locate the proper fitting in the oil pressure gauge kit, and install the gauge.

Note: The fitting is national pipe thread (NPT), which is tapered. Tighten it only until it is snug. Do not overtighten it.

9. Start the engine and note the oil pressure while the engine's temperature is cold:

Idle ____ psi

Fast idle ____ psi

10. Run the engine until it reaches operating temperature. Record the oil pressure gauge reading.

Idle ____ psi

Fast idle ____ psi

Note: Low idle oil pressure that goes up with increased engine rpm usually indicates excessive bearing oil clearance.

11. Remove the gauge and reinstall the sending unit.
12. Before completing the paperwork, clean your work area, put the tools in their proper places, and wash your hands.
13. Were there any other recommendations for needed service or unusual conditions that you noticed while checking the oil pressure?

Yes ____ No ____

14. Record your recommendations for additional service or repairs on the repair order.
15. Complete the repair order (R.O.).

Instructor OK ______________ Score __________

ASE Lab Preparation Worksheet #8-19

REPLACE A VALVE COVER GASKET

Name______________________________ Class ____________________

OBJECTIVE:

Upon completion of this assignment, you should be able to replace a valve cover gasket. This task will help prepare you to pass the ASE certification examination in engine repair.

DIRECTIONS:

Before beginning this lab assignment, review the worksheet completely. Fill in the information in the spaces provided as you complete each task.

TOOLS AND EQUIPMENT REQUIRED:

Safety glasses, fender covers, hand tools, shop towel

PARTS AND SUPPLIES:

Valve cover gasket

PROCEDURE:

Vehicle year ___________ Make __________________ Model ___________________

Repair Order # ________ Engine size _________________ # of Cylinders ____________

1. Open the hood and place fender covers on the fenders and front body parts.
2. Remove and label any spark plug wires, vacuum hoses, or electrical wires that interfere with valve cover removal.
3. Remove the screws holding the valve cover in place.
4. Rap the valve cover on its corner with a rubber mallet to loosen it. If it does not come loose easily, see your instructor.
5. Use a gasket scraper to remove the old gasket from the valve cover. In the solvent tank, clean the dirt and oil from the valve cover. Blow it dry with compressed air.

   Note: Remember to blow down and away when using compressed air. Blow the solvent *into* the solvent tank, not onto the floor.

   The solvent was blown: Into the tank ____ On the floor ____

6. If the valve cover is made from sheet metal, flatten the area around the bolt holes using a hammer on a flat surface.
7. Position the new gasket on the valve cover. If there are no locating lugs to hold the gasket in place, use a gasket adhesive or pieces of string to fasten the gasket to the valve cover in four places.

8. Clean any residue from the old valve cover gasket still attached to the cylinder head.

   Note: Do not allow any of the old gasket to fall into the engine.

Area of less clamping force
Valve cover flange
Excessive gasket compression
Cylinder head

9. Before installing the valve cover, check to see that the oil drainback holes in the head are not plugged with sludge.

   Drainback holes clean? Yes ____ No ____

10. Reinstall the valve cover. Fingertighten all the bolts.

    Note: Always start the threads on *all* of the bolts that fasten a part before tightening down *any* of them.

11. Tighten all the screws to 5–6 ft.-lb (60–72 in.-lb) using a torque wrench.

    Which torque wrench was used? Inch-pound ____ Foot-pound ____

    Note: Most foot-pound torque wrenches are not accurate below 15–20 ft.-lb.

12. Reinstall any wires or hoses.
13. Start the engine. Check for smooth engine operation and look for oil leaks around the valve cover gasket.

    Does the engine run smoothly? Yes ____ No ____

    Are there any oil leaks? Yes ____ No ____

14. Before completing the paperwork, clean your work area, put the tools in their proper places, and wash your hands.
15. Were there any other recommendations for needed service or unusual conditions that you noticed while changing the valve cover gasket?

    Yes ____ No ____

16. Record your recommendations for additional service or repairs on the repair order.
17. Complete the repair order (R.O.).

Instructor OK ______________ Score ____________

ASE Lab Preparation Worksheet #8-20

READ A STANDARD VERNIER CALIPER

Name______________________________ Class ____________________

OBJECTIVE:

Upon completion of this assignment, you should be able to read a standard vernier caliper.

DIRECTIONS:

Before beginning this lab assignment, review the worksheet completely.

TOOLS AND EQUIPMENT REQUIRED:

Vernier caliper

PARTS AND SUPPLIES:

Five parts to measure (provided by your instructor)

PROCEDURE:

Example:

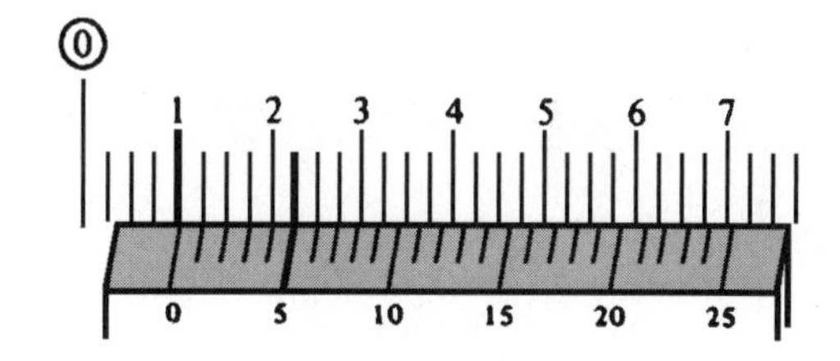

a. Locate the line on the main scale that the zero on the sliding vernier scale is lined up with or is just past.

b. Determine the number of inches by reading the largest number on the main scale before the zero line on the vernier sliding scale.

Inches **0.000**

c. Read the one hundred thousandths (0.100) off the main scale. The first numbered line before the zero line on the vernier sliding scale is the number of one hundred thousandths (0.100).

One hundred thousandths (0.100) **0.100**

d. Next read the twenty-five thousandths (0.025). These are the shorter lines between the one hundred thousandth (0.100) markings on the main scale. Count the number of these lines between the last one hundred thousandth (0.100) mark and the zero line on the vernier sliding scale.

Twenty-five thousandths (0.025) **0.000**

e. Look carefully at the vernier scale; the line on the vernier scale that perfectly lines up with the line on the main scale is the number of one thousandths of an inch. After identifying which lines perfectly match, read the number from the vernier scale.

One thousandths (0.001) **0.005**

f. Add the readings together to determine the final vernier caliper reading.

0.105

b.	Inches	=	**0.000**
c.	0.100	=	**0.100**
d.	0.025	=	**0.000**
e.	0.001	=	**0.005**
f.			**0.105″**

Directions:

Record the readings in the spaces provided from the vernier calipers pictured. Show all of the steps.

1. Inches ______
   .100 ______
   .025 ______
   .001 ______
   Add to determine the vernier caliper reading. ______

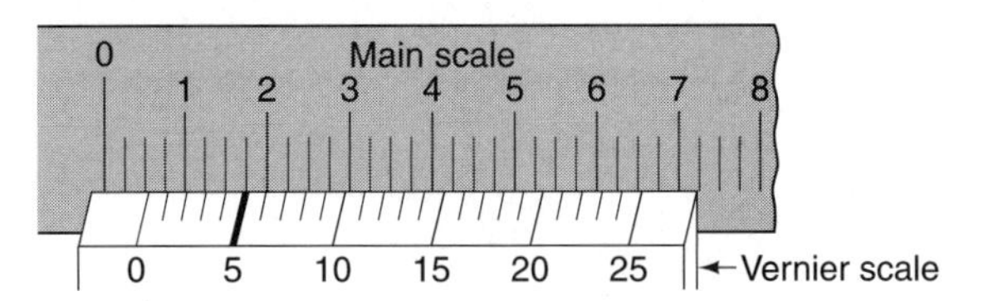

2. Inches ______
   .100 ______
   .025 ______
   .001 ______
   Add to determine the vernier caliper reading. ______

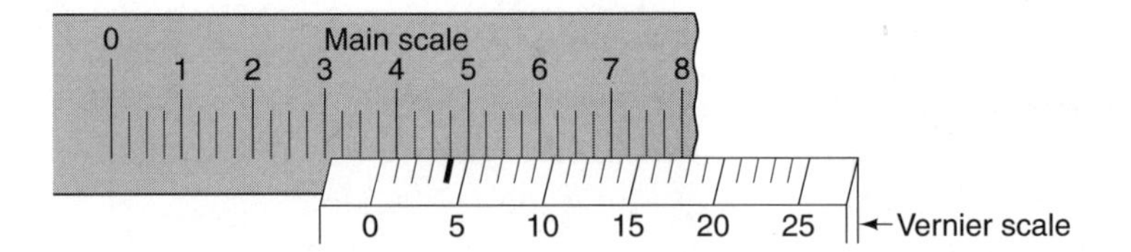

3. Inches ______
   .100 ______
   .025 ______
   .001 ______
   Add to determine the vernier caliper reading. ______

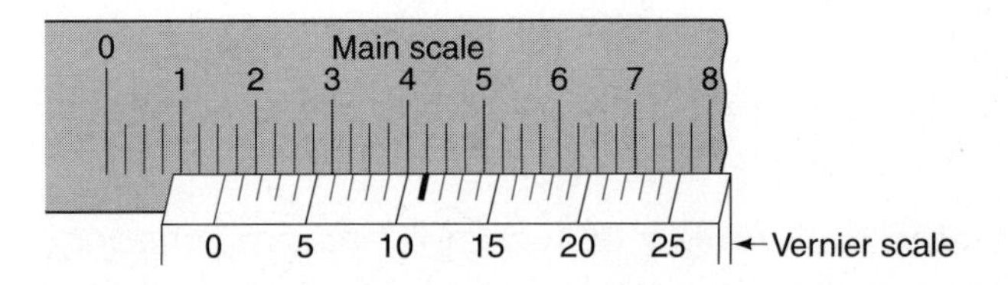

4. Inches ______
   .100 ______
   .025 ______
   .001 ______
   Add to determine the vernier caliper reading. ______

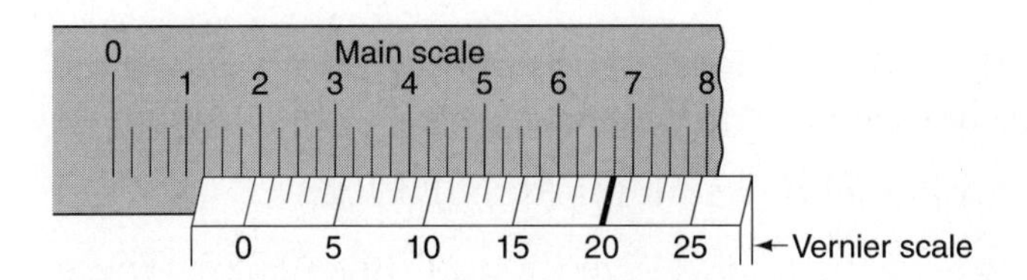

5. Inches ______
   .100 ______
   .025 ______
   .001 ______
   Add to determine the vernier caliper reading. ______

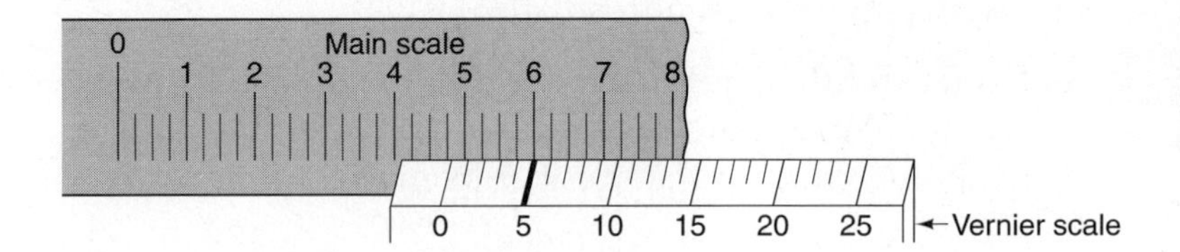

6. Inches ______
   .100 ______
   .025 ______
   .001 ______
   Add to determine the vernier caliper reading. ______

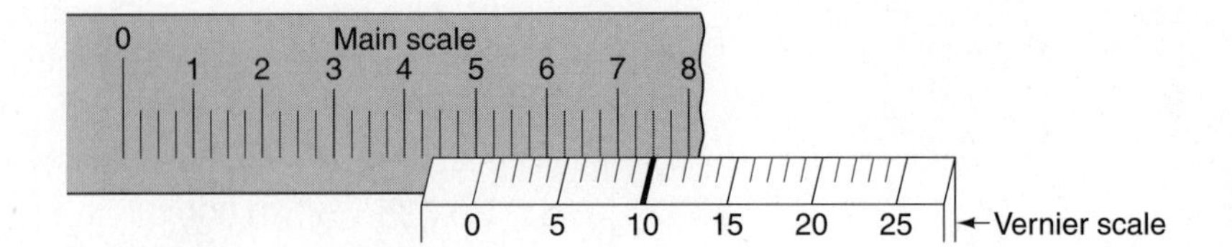

7. Inches ______

   .100 ______

   .025 ______

   .001 ______

   Add to determine the vernier caliper reading. ______

0 Main scale 1 2 3 4 5 6 7 8

0 5 10 15 20 25 ← Vernier scale

8. Inches ______

   .100 ______

   .025 ______

   .001 ______

   Add to determine the vernier caliper reading. ______

0 Main scale 1 2 3 4 5 6 7 8

0 5 10 15 20 25 ← Vernier scale

Directions:

Measure five parts provided by your instructor with a vernier caliper and record the sizes of the parts below.

Example:

Describe the part being measured. **Piston**

Describe where the part is being measured. **Piston skirt**

Record the measurement of the part. **3.875″**

1. Part #1

   Describe the part being measured. ______________________

   Describe where the part is being measured. ______________________

   Record the measurement of the part. ______________________

2. Part #2

   Describe the part being measured. ______________________

   Describe where the part is being measured. ______________________

   Record the measurement of the part. ______________________

3. Part #3

   Describe the part being measured. ______________________

   Describe where the part is being measured. ______________________

   Record the measurement of the part. ______________________

4. Part #4

   Describe the part being measured. ______________________

   Describe where the part is being measured. ______________________

   Record the measurement of the part. ______________________

5. Part #5

   Describe the part being measured. ______________________

   Describe where the part is being measured. ______________________

   Record the measurement of the part. ______________________

ASE Lab Preparation Worksheet #8-21

READ A STANDARD MICROMETER

Name____________________________ Class ____________________

OBJECTIVE:

Upon completion of this assignment, you should be able to read a standard micrometer.

DIRECTIONS:

Before beginning this lab assignment, review the worksheet completely.

TOOLS AND EQUIPMENT REQUIRED:

Micrometer

PARTS AND SUPPLIES:

Six parts to measure (provided by your instructor)

PROCEDURE:

1. Familiarize yourself with the micrometer by identifying its parts. Place the correct letter next to the name of the micrometer part.

   Thimble ____

   Anvil ____

   Sleeve ____

   Spindle ____

   Frame ____

   Ratchet ____

   Lock ____

2. Locate a micrometer.

   What size is it? (Example: 0-1″) ____________________

3. Find the zero (0) line on the micrometer sleeve. Turn the thimble in until the zero on the thimble is aligned with the index line (at the zero line on the sleeve). The distance from the anvil to the end of the spindle is now exactly 0 (or 1″, 2″, 3″, 4″ etc.), depending on the size of the micrometer.

   a. What is the distance from the anvil to the spindle end?

   0 ____ 1″ ____ 2″ ____ 3″ ____ 4″ ____ Other ____

   Turn the thimble exactly forty (40) turns counterclockwise.

   Are any numbers on the sleeve uncovered by the thimble? ____________________________

4. Look at the numbers on the *sleeve*. Each of these numbers is divided into 4 parts.

   How many thousandths of an inch are represented by each of the numbered lines?

   0. ________ ″

   Each of the graduated lines on the sleeve is equal to: 0. ________ ″

5. Each turn of the thimble equals what fraction of an inch? 0. ________ ″

TURN

6. Next, look at the *thimble.* The thimble has numbers and lines around its end.

   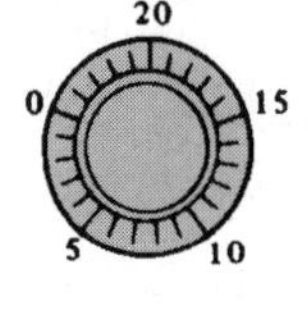

   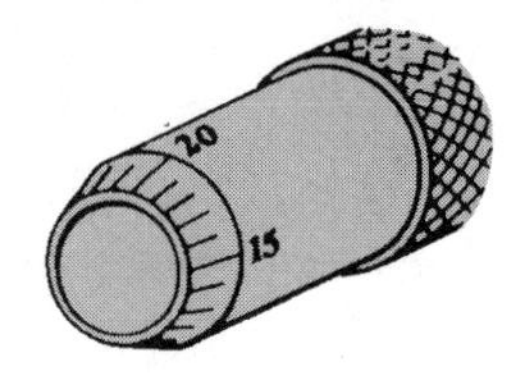

   What is the largest number (not a line) on the thimble?

   ____

   What is the smallest graduation line (not a number) on the thimble?

   ____

   How many graduations are around the thimble? _____

   Each of the graduations on the thimble represents: 0. ________ "

Reading the Micrometer

7. Read the micrometer in the picture. Start by reading the last number on the sleeve that is visible.

   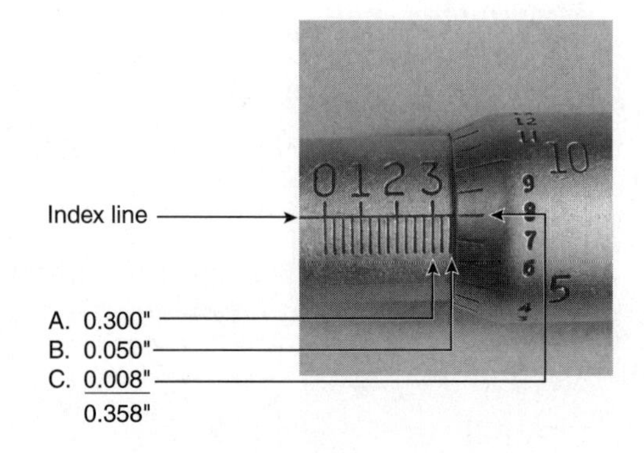

   Record your reading: 0. ________ "

   Next read the number of lines that can be seen from the last number visible on the sleeve to the edge of the thimble.

   Record your reading: 0. ________ "

   Now see which graduation line on the thimble is aligned with the index line.

   Record your reading: 0. ________ "

   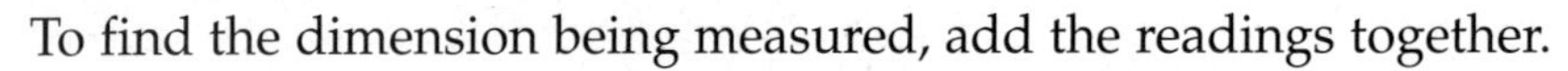

   To find the dimension being measured, add the readings together.

   0. ________ "

   0. ________ "

   0. ________ "

   ____

   Your total should be 0.283"

   If your total was two hundred and eighty-three thousandths, then you are ready to read the micrometers on Worksheet 8-22. Remember to show your work.

STOP

Instructor OK ______________ Score ______________

ASE Lab Preparation Worksheet #8-22

READ A STANDARD MICROMETER PRACTICE

Name______________________________ Class ____________________

DIRECTIONS:

Record the readings in the spaces for micrometers.

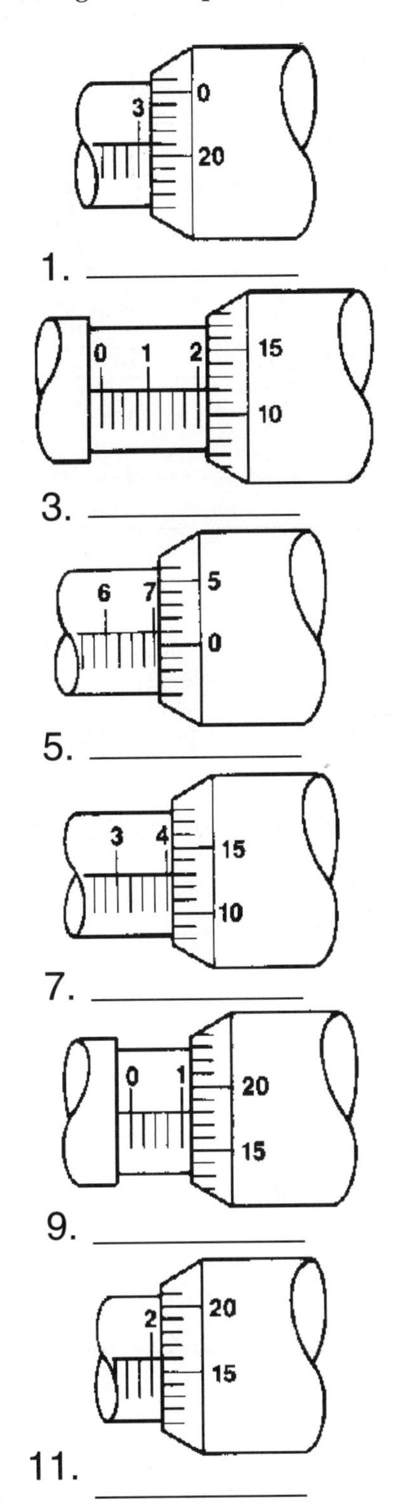

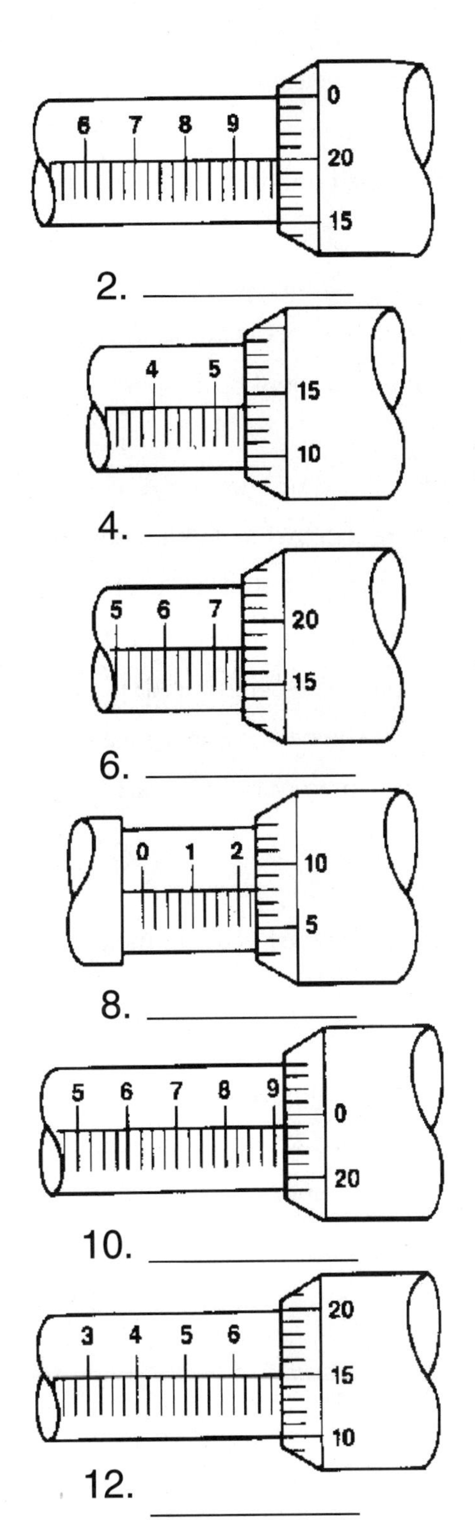

1. ______________
2. ______________
3. ______________
4. ______________
5. ______________
6. ______________
7. ______________
8. ______________
9. ______________
10. ______________
11. ______________
12. ______________

TURN

DIRECTIONS:
Measure six parts provided by your instructor with a micrometer and record the sizes of the parts below.

Example:

Describe the part being measured. **Piston**____________________

Describe where the part is being measured. **Piston skirt**________________

Record the measurement of the part. **3.875″**____________________

1. Part #1

   Describe the part being measured. __________________________

   Describe where the part is being measured. __________________________

   Record the measurement of the part. __________________________

2. Part #2

   Describe the part being measured. __________________________

   Describe where the part is being measured. __________________________

   Record the measurement of the part. __________________________

3. Part #3

   Describe the part being measured. __________________________

   Describe where the part is being measured. __________________________

   Record the measurement of the part. __________________________

4. Part #4

   Describe the part being measured. __________________________

   Describe where the part is being measured. __________________________

   Record the measurement of the part. __________________________

5. Part #5

   Describe the part being measured. __________________________

   Describe where the part is being measured. __________________________

   Record the measurement of the part. __________________________

6. Part #6

   Describe the part being measured. __________________________

   Describe where the part is being measured. __________________________

   Record the measurement of the part. __________________________

Instructor OK ____________ Score ____________

ASE Lab Preparation Worksheet #8-23

VALVE LASH ADJUSTMENT (OVERHEAD CAM ENGINE WITH SHIM-TYPE ADJUSTMENT)

Name____________________ Class ____________

OBJECTIVE:

Upon completion of this assignment, you should be able to adjust the valves on an overhead cam engine that has shim-type adjustment. This task will help prepare you to pass the ASE certification examinations in engine performance and engine repair.

DIRECTIONS:

Before beginning this lab assignment, review the worksheet completely. Fill in the information in the spaces provided as you complete each task.

TOOLS AND EQUIPMENT REQUIRED:

Safety glasses, fender covers, hand tools, feeler gauges, special adjustment tools, shop towel

PROCEDURE:

Vehicle year __________ Make ______________ Model ______________

Repair Order # ________ Engine size ______________ # of Cylinders __________

1. Open the hood and place fender covers on the fenders and front body parts.

Measuring the Valve Clearance

2. Locate the valve lash specifications in the repair manual and record them below.

	Hot		Cold
Intake	0. _______ "	Intake	0. _______ "
Exhaust	0. _______ "	Exhaust	0. _______ "

3. Record the engine's firing order below.

____ ____ ____ ____ ____ ____ ____ ____

4. List the companion cylinders below.

____ ____ ____ ____

____ ____ ____ ____

5. Remove the valve cover(s).

   The valve cover is made of:

   Cast aluminum ____ Stamped steel ____ Plastic ____

Camshaft

Lash pad adjuster

Cam follower (bucket)

TURN

6. Inspect the runners on the intake and exhaust manifolds to determine which valves are intake and which are exhaust. Draw a sketch below showing which valves are intake and which are exhaust.

   Note: The service manual will also identify which valves are intake and which are exhaust.

   Show your drawing to the instructor before proceeding. **Instructor OK** ____________________

7. Locate the large bolt on the front of the crankshaft pulley. What size socket fits it? __________

8. Use a large (1/2"-drive) ratchet to turn the bolt (and crankshaft) in its normal direction of rotation.

   Note: Always turn an engine in its normal direction of rotation. Turning an engine backward can result in serious engine damage.

   Which way does the crankshaft on this engine normally turn?

   Clockwise ____ Counterclockwise ____

9. As the crankshaft is turned, watch the number one cylinder's exhaust valve. As it begins to close, continue turning the crankshaft until the number one intake valve just begins to open. The intake and exhaust valves for that cylinder are now in a rocking motion. Show your instructor.

   Instructor OK ______________________________

   Note: At this point, the valves for the *companion* cylinder to number one can be checked or adjusted.

   Which cylinder is the companion to cylinder #1? ____

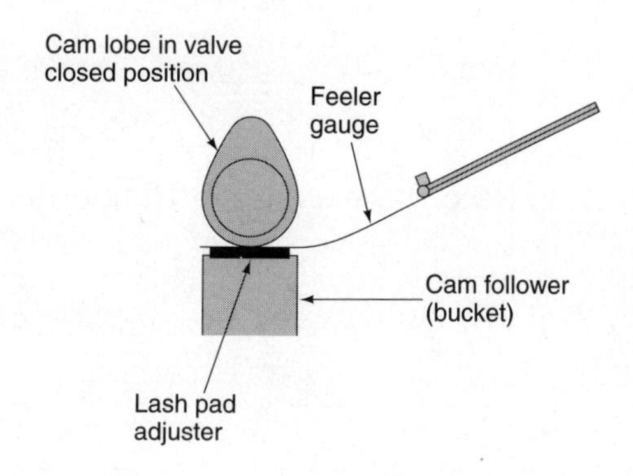

10. Measure the clearance between the valve shims and cam lobes on the companion cylinder with a thickness (feeler) gauge. Record your measurements in the chart on the next page.

11. Turn the crankshaft until the next cylinder in the firing order has its valves rocking.

    Which cylinder is this? ____

    How many degrees did you turn the crankshaft? ____

12. Measure the clearance for its companion.

13. Following the firing order, repeat the process for the remaining cylinders.

    Record your measurements on the following chart.

Cylinder	#1	#2	#3	#4	#5	#6
Intake	_______	_______	_______	_______	_______	_______
Exhaust	_______	_______	_______	_______	_______	_______

14. In the chart above, circle the valves that need to be adjusted.

Adjusting the Valve Clearance

Note: The method for adjusting valve clearance (lash) varies among manufacturers. Check the service manual for the proper adjustment method for this engine. The method presented here is typical of many manufacturers.

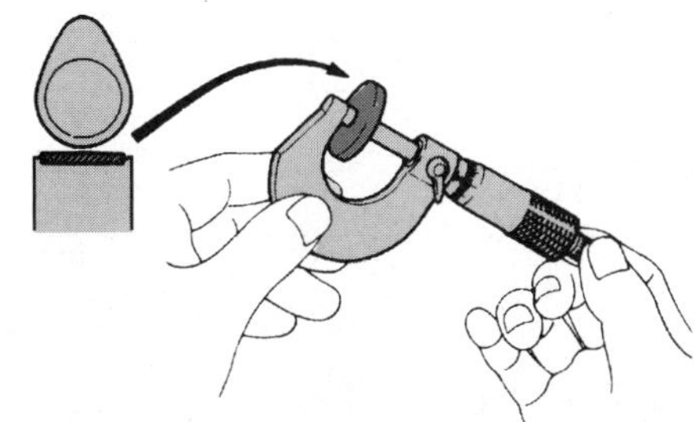

15. Turn the crankshaft to position the cam lobe for the valve to be adjusted so that the cam lobe faces away from the valve.

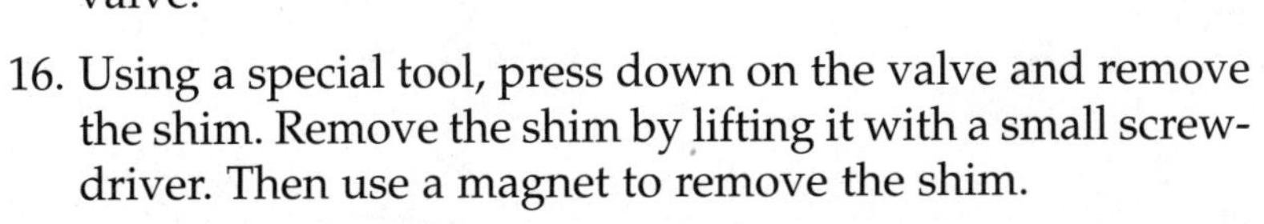

16. Using a special tool, press down on the valve and remove the shim. Remove the shim by lifting it with a small screwdriver. Then use a magnet to remove the shim.

Note: Compressed air from a rubber-tipped blow gun carefully directed under the shim will help to break the seal formed by the engine oil.

17. Determine the thickness of the correct replacement shim by measuring the thickness of the removed shim. Then add the measured clearance, minus the clearance specification.

    R + [C – S] = N

    R = Thickness of the removed shim

    C = Measured valve clearance

    S = Valve lash specification

    N = New shim thickness

18. Select the new shim and install it by reversing the removal procedure.

    Replacement shim thickness: 0. _______ "

19. Recheck the valve clearance. Is it now correct? Yes _____ No _____

20. Have the instructor check your work before continuing.

    Instructor OK ________________________________

21. If the clearance is within the specified tolerance, continue adjusting the valves as necessary.

    How many cylinders does the engine have? _____

    How many valves are there per cylinder? _____

TURN

22. Record the shims replaced below:

Cylinder	Valve	Old Shim Thickness	Measured Clearance	Specification	New Shim Thickness
____	____	________	________	________	________
____	____	________	________	________	________
____	____	________	________	________	________
____	____	________	________	________	________
____	____	________	________	________	________
____	____	________	________	________	________
____	____	________	________	________	________
____	____	________	________	________	________
____	____	________	________	________	________
____	____	________	________	________	________
____	____	________	________	________	________
____	____	________	________	________	________
____	____	________	________	________	________
____	____	________	________	________	________
____	____	________	________	________	________
____	____	________	________	________	________

23. Install the valve cover and any other components that were removed. Start the engine.

    Did the engine start? Yes ____ No ____

24. Before completing the paperwork, clean your work area, put the tools in their proper places, and wash your hands.

25. Were there any other recommendations for needed service or unusual conditions that you noticed while checking and adjusting the valves?

    Yes ____ No ____

26. Record your recommendations for additional service or repairs on the repair order.

27. Complete the repair order (R.O.).

STOP

Instructor OK ______________ Score __________

ASE Lab Preparation Worksheet #8-24

MECHANICAL VALVE LASH ADJUSTMENT (ROCKER ARM)

Name__________________________ Class ________________

OBJECTIVE:

Upon completion of this assignment, you should be able to check and adjust the valves on a cylinder head with rocker arms. This task will help prepare you to pass the ASE certification examinations in engine performance and engine repair.

DIRECTIONS:

Before beginning this lab assignment, review the worksheet completely. Fill in the information in the spaces provided as you complete each task.

TOOLS AND EQUIPMENT REQUIRED:

Safety glasses, fender covers, hand tools, feeler gauges, shop towel

PROCEDURE:

Vehicle year __________ Make ________________ Model ________________

Repair Order # _______ Engine size _______________ # of Cylinders __________

1. Locate the valve lash specifications in the repair manual and record them below.

	Hot		Cold
Intake	0. _______ "	Intake	0. _______ "
Exhaust	0. _______ "	Exhaust	0. _______ "

2. Record the engine's firing order below.

   ____ ____ ____ ____ ____ ____ ____ ____

3. List the companion cylinders below.

   ____ ____ ____ ____

   ____ ____ ____ ____

4. Open the hood and place fender covers on the fenders and front body parts.
5. Remove the valve cover(s).

   The valve cover is made of: Cast aluminum ____ Stamped steel ____ Plastic ____
6. Inspect the runners on the intake and exhaust manifolds to determine which valves are intake and which are exhaust. Draw a sketch below showing which valves are intake and which are exhaust.

Note: The service manual will also identify which valves are intake and which are exhaust.

Show your drawing to the instructor before proceeding. **Instructor OK** ____________________

7. Locate the large bolt on the front of the crankshaft pulley. What size socket fits it? __________
8. Use a large (1/2"-drive) ratchet to turn the bolt (and crankshaft) in its normal direction of rotation.

Note: Always turn an engine in its normal direction of rotation. Turning an engine backward can result in serious engine damage.

Which way does the crankshaft on this engine normally turn?

Clockwise ____ Counterclockwise ____

9. As the crankshaft is turned, watch the number one cylinder's exhaust valve. As it begins to close, continue turning the crankshaft until the number one intake valve just begins to open. The intake and exhaust valves for that cylinder are now in a rocking motion. Show your instructor.

Instructor OK ______________________________

Note: At this point, the valves for the *companion* cylinder to number one can be checked or adjusted.

Which cylinder is the companion to cylinder #1? __________________

10. On the companion cylinder, measure the clearance between the valve and the rocker arm with a thickness (feeler) gauge.
11. Adjust the valve clearance as necessary.
12. Turn the crankshaft until the next cylinder in the firing order has its valves rocking.

Which cylinder is this? ____

How many degrees did you turn the crankshaft? ____

13. Measure the clearance for its companion.
14. Following the firing order, repeat the process for the remaining cylinders.

Record your measurements here.

Cylinder	#1	#2	#3	#4	#5	#6
Intake	______	______	______	______	______	______
Exhaust	______	______	______	______	______	______

15. In the chart above, circle the valves that require adjustment.
16. Replace the valve cover.

Note: Overtightening the valve cover fasteners can damage the gasket.

17. Start the engine. Did it start? Yes ____ No ____
18. Before completing the paperwork, clean your work area, put the tools in their proper places, and wash your hands.
19. Were there any other recommendations for needed service or unusual conditions that you noticed while checking and adjusting the valves? Yes ____ No ____
20. Record your recommendations for additional service or repairs on the repair order.
21. Complete the repair order (R.O.).

STOP

Instructor OK ________________ Score ____________

ASE Lab Preparation Worksheet #8-25

VALVE LASH ADJUSTMENT (HYDRAULIC LIFTER) (ENGINE NOT RUNNING)

Name_____________________________ Class ____________________

OBJECTIVE:

Upon completion of this assignment, you should be able to adjust the lifters on an engine with hydraulic lifters. This task will help prepare you to pass the ASE certification examinations in engine repair and engine performance.

DIRECTIONS:

Before beginning this lab assignment, review the worksheet completely. Fill in the information in the spaces provided as you complete each task.

TOOLS AND EQUIPMENT REQUIRED:

Safety glasses, fender covers, shop towel, hand tools

PROCEDURE:

Vehicle year ____________ Make __________________ Model __________________

Repair Order # ________ Engine size _________________ # of Cylinders ____________

1. Locate the correct specification in the service manual or computer program.

   Which manual or computer program was used? ______________________________

   On which page of the service manual is the valve adjustment specification located? __________

   Turn the adjusting nut ____ turns after zero clearance (zero lash).

2. Find the instructions for the adjustment of the valves.

   Page # ___________

3. Record the following information:

   Which valves are to be adjusted with cylinder #1 at TDC?

   Intake valves for cylinder numbers ____ ____ ____ ____

   Exhaust valves for cylinder numbers ____ ____ ____ ____

   Which valves are to be adjusted with cylinder # ____ at TDC?

   Intake valves for cylinder numbers ____ ____ ____ ____

   Exhaust valves for cylinder numbers ____ ____ ____ ____

4. Open the hood and place fender covers on the fenders and front body parts.

5. Remove the valve covers.

6. Rotate the crankshaft until the number one cylinder is at top dead center on the compression stroke and the timing marks are aligned.

7. Adjust the valves identified in step 3 by backing off the adjusting nut until clearance is felt in the pushrod.

TURN

8. Rotate the pushrod while tightening the adjuster nut until all the clearance is removed.
9. Turn the adjusting nut the specified number of turns.
10. Adjust the remaining valves to be adjusted in this crankshaft position.
11. Rotate the crankshaft one revolution and align the timing marks.

    Which cylinder is not at TDC? ______
12. Adjust the remaining valves.
13. Reinstall the valve covers.
14. Start engine and check and adjust the idle speed as required.
15. Check for leaks. Are there any leaks? Yes ____ No ____
16. Before completing the paperwork, clean your work area, put the tools in their proper places, and wash your hands.
17. Were there any other recommendations for needed service or unusual conditions that you noticed while adjusting the valves?

    Yes ____ No ____
18. Record your recommendations for additional service or repairs on the repair order.
19. Complete the repair order (R.O.).

STOP

Instructor OK ______________ Score __________

ASE Lab Preparation Worksheet #8-26

RESTORE A SCREW THREAD

Name__________________________ Class ________________

OBJECTIVE:

Upon completion of this assignment, you should be able to remove a broken fastener and restore the damaged threads. This task will help prepare you to pass the ASE certification examination in engine repair.

DIRECTIONS:

Before beginning this lab assignment, review the worksheet completely. Fill in the information in the spaces provided as you complete each task.

TOOLS AND EQUIPMENT REQUIRED:

Safety glasses, shop towel, drill motor, drills, taps

PROCEDURE:

1. What caused the bolt to break?

   ____

2. What size is the broken bolt?

   ____

3. Carefully file the broken bolt flat.
4. Center punch the top of the broken bolt.

   Note: Be absolutely certain that the center punch mark is exactly on center. If not, pound the center punch mark deeper until it is on the center.

5. Use a sharp, small drill to drill a pilot hole exactly in the center of the bolt.

   Note: If the pilot hole is off-center, the restored hole will be off-center.

6. What size drill bit and drill motor chuck are being used?

   Drill bit size _____

   Drill motor chuck size _____

   Remember: Bolts are made of steel, so cutting oil is *required.*

7. After starting to drill the hole, double-check to see that the hole is being drilled *exactly* in the center of the bolt. If not, use a file or a die grinder and burr to remove the small amount of hole already drilled. Again, center punch the hole and drill exactly in the center of the bolt.

8. Was the hole drilled on-center? Yes ____ No ____
9. If possible, drill the pilot hole all the way through the broken bolt.

   Note: This will relieve some of the internal tension in the bolt. Many times after the bolt has been drilled through, it is easily removed.

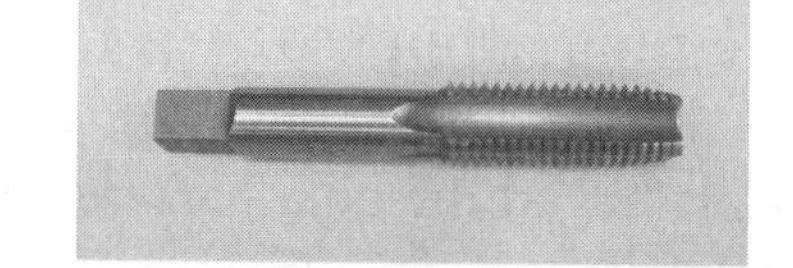

10. Finish drilling the hole with the largest size drill bit that can be used without damaging the original threads.
11. Run a tap through the hole to clean out the threads. Tap size ____
12. If the threads are not in good condition after chasing them with the tap, install a thread insert.

    Thread condition: Good ____ Bad ____

    Is a thread insert required? Yes ____ No ____
13. Before completing the paperwork, clean your work area, put the tools in their proper places, and wash your hands.

STOP

ASE Lab Preparation Worksheet #8-27

INSTALL A HELI-COIL® THREAD INSERT

Name____________________________ Class ____________________

OBJECTIVE:

Upon completion of this assignment, you should be able to repair a damaged thread by installing a Heli-coil® insert. This task will help prepare you to pass the ASE certification examination in engine repair.

DIRECTIONS:

Before beginning this lab assignment, review the worksheet completely. Fill in the information in the spaces provided as you complete each task.

TOOLS AND EQUIPMENT REQUIRED:

Safety glasses, shop towel, Heli-coil® tool set

PARTS AND SUPPLIES:

Heli-coil® insert

PROCEDURE:

1. What part is being repaired? ________________________
2. What is the size of the threads being repaired?

   ____

3. Drill the damaged hole with a drill bit of the specified size.

   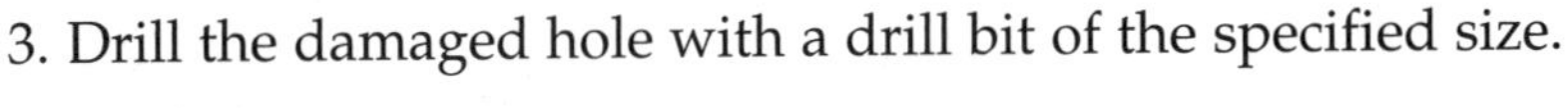

   Drill size? ____

   Note: The success of the job depends on the drill being held straight (perpendicular to the hole).

4. Tap the hole with the proper size tap.

   Tap size ____

   Note: Turn the tap counterclockwise after each revolution to break off the cuttings.

5. Install the thread insert on the mandrel.
6. Put a *small* amount of thread locking adhesive on the thread.

   Note: Thread sealer is sometimes provided with the insert kit. Use only a small amount. This is an *anaerobic* sealer (which hardens only without the presence of air). It will not work properly if too much is used.

7. Thread the insert into the hole until it is just below the surface of the part.

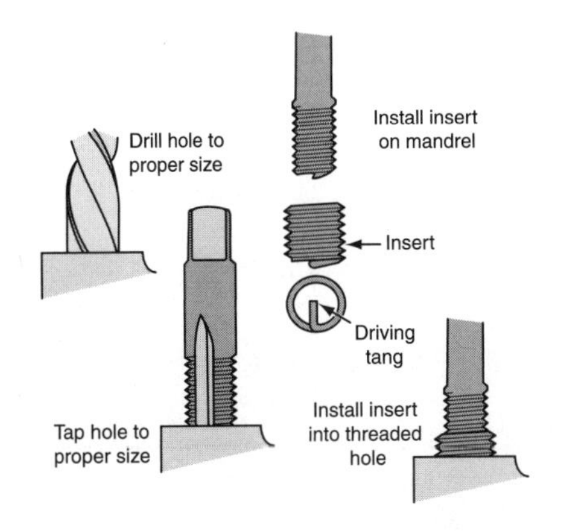

8. The bottom of the insert has a tang used to assist in turning the insert into the hole. Break the tang off with a punch to complete the job.

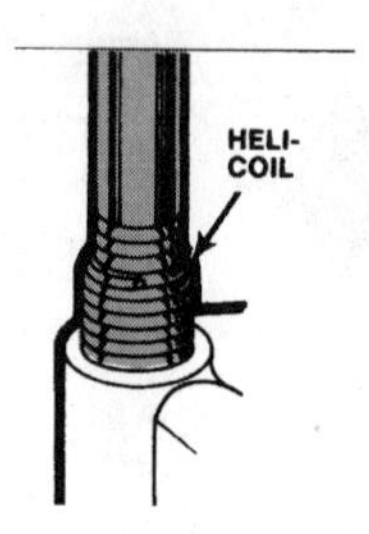

Install insert into newly threaded hole

9. Thread a new bolt into the repair hole. Does the fastener turn in easily at least three full turns?

   Yes _____ No _____

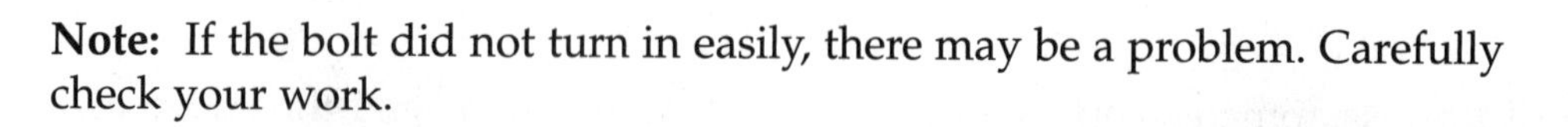

   Note: If the bolt did not turn in easily, there may be a problem. Carefully check your work.

10. Before completing the paperwork, clean your work area, put the tools in their proper places, and wash your hands.

STOP

Instructor OK ______________ Score __________

ASE Lab Preparation Worksheet #8-28

DRILL AND TAP A HOLE

Name______________________________ Class ____________________

OBJECTIVE:

Upon completion of this assignment, you should be able to drill a hole and tap it to accept a fastener. This task will help prepare you to pass the ASE certification examinations in all areas.

DIRECTIONS:

Before beginning this lab assignment, review the worksheet completely. Fill in the information in the spaces provided as you complete each task.

TOOLS AND EQUIPMENT REQUIRED:

Safety glasses, fender cover, shop towel, drill index, tap and die set, drill motor, lubricant, center punch, countersink or burr

PROCEDURE:

What part is being repaired? ______________________

1. If the part is on a vehicle, open the hood and place fender covers on the fenders and front body parts.
2. Determine the size of the hole to be threaded by identifying the bolt to be used.

   What is the size of its screw thread? _____
3. Select the correct drill from the tap/drill size chart.

   What is the correct drill size? _____

   What is its decimal equivalent? _____

   What is the closest fractional drill size? _____

   Note: It is important that the correct drill be used. A hole that is drilled too large will not leave enough material to make good threads. A hole that is too small could bind the tap and possibly break it.
4. Use a center punch to put a mark where the hole is to be drilled.

   Is the mark exactly on center? Yes _____ No _____
5. Before drilling the finished hole, use a small drill to make a pilot hole. This will make the drilling easier.

   Size of pilot drill used: ________
6. Drill the hole to its finished size.

Size	Threads per inch			Outside Diameter Inches	Tap Drill Approx. 75% Full Thread	Decimal Equivalent of Tap Drill
	NC	NF	NS			
0	...	80	...	.0600	3/64	.0469
1	...	...	56	.0730	54	.0550
1	64	...	...	.0730	53	.0595
1	...	72	...	.0730	53	.0595
2	56	...	...	.0860	50	.0700
2	...	64	...	.0860	50	.0700
3	48	...	...	.0990	47	.0785
3	...	56	...	.0990	45	.0820
4	...	...	32	.1120	45	.0820
4	...	...	36	.1120	44	.0860
4	40	...	...	.1120	43	.0890
4	...	48	...	.1120	42	.0935
5	...	...	36	.1250	40	.0980
5	40	...	...	.1250	38	.1015
5	...	44	...	.1250	37	.1040
6	32	...	...	.1380	36	.1065
6	...	...	36	.1380	34	.1110
6	...	40	...	.1380	33	.1130
8	...	...	30	.1640	30	.1285
8	32	...	...	.1640	29	.1360
8	...	36	...	.1640	29	.1360
8	...	...	40	.1640	28	.1405
10	24	...	...	.1900	25	.1495
10	...	...	28	.1900	23	.1540
10	...	...	30	.1900	22	.1570
10	...	32	...	.1900	21	.1590
12	24	...	...	.2160	16	.1770
12	...	28	...	.2160	14	.1820
12	...	...	32	.2160	13	.1850
1/4	20	...	...	.2500	7	.2010
1/4	...	28	...	.2500	3	.2130
5/16	18	...	...	.3125	F	.2570
5/16	...	24	...	.3125	I	.2720
3/8	16	...	...	.3750	5/16	.3125
3/8	...	24	...	.3750	Q	.3320
7/16	14	...	...	.4375	U	.3680
7/16	...	20	...	.4375	25/64	.3906
1/2	13	...	...	.5000	27/64	.4219
1/2	...	20	...	.5000	29/64	.4531
9/16	12	...	...	.5625	31/64	.4844
9/16	...	18	...	.5625	33/64	.5156
5/8	11	...	...	.6250	17/32	.5312
5/8	...	18	...	.6250	37/64	.5781
3/4	10	...	...	.7500	21/32	.6562
3/4	...	16	...	.7500	11/16	.6875
7/8	9	...	...	.8750	49/64	.7656

Note: If the hole is being drilled in cast iron, no lubricant is needed. Remember, the hole must be drilled perpendicular to the surface of the part. If the hole is not drilled all the way through the part, be sure to leave space at the bottom of the hole for bolt clearance.

7. Use a countersink or burr to chamfer the top of the hole.
8. Select the proper size and type of tap.

   Type of tap selected: ____________________

   Bottoming tap ____ Plug tap ____ Taper tap ____ Pipe tap ____

   Tap size? _________

9. Begin tapping by carefully turning the tap clockwise while gently pushing down. Keep the tap perpendicular to the part.

   Tap started straight? Yes ____ No ____

10. Once the tap is started correctly, continue tapping using both hands to turn the tap. This helps to ensure that the tap will continue perpendicular into the hole.

    Note: Advance the tap 1/2 turn clockwise and then turn it back 1/4 turn. Do this until the thread is completely cut. If the tap binds, stop immediately and check with your instructor.

11. Clean the completed thread with compressed air. Test the thread by turning a new bolt into the hole. It should turn easily into the hole a minimum of three complete turns without the use of tools.

    Does the bolt turn into the new threads easily? Yes ____ No ____

    Note: If the bolt does not turn in easily, there may be a problem. Carefully check the new threads and the bolt size. Repair the problem as necessary.

12. Before completing the paperwork, clean your work area, put the tools in their proper places, and wash your hands.

STOP

Part II

ASE Lab Preparation Worksheets

Service Area 9

Chassis Service

Instructor OK ______________ Score ___________

ASE Lab Preparation Worksheet #9-1

MANUALLY BLEED BRAKES AND FLUSH THE SYSTEM

Name______________________________ Class ____________________

OBJECTIVE:

Upon completion of this assignment, you should be able to manually bleed brakes and flush the system. This task will help prepare you to pass the ASE certification examination in brakes.

DIRECTIONS:

Before beginning this lab assignment, review the worksheet completely. Fill in the information in the spaces provided as you complete each task.

TOOLS AND EQUIPMENT REQUIRED:

Safety glasses, fender cover, jack stands or vehicle lift, shop towel, bleeder wrench, siphon, hose, container

PARTS AND SUPPLIES:

Brake fluid

This worksheet is not intended for use with vehicles that have antilock brake systems. Consult a service manual for the proper procedures before bleeding antilock brake systems.

PROCEDURE:

Vehicle year ___________ Make _________________ Model __________________

Repair Order # ________ Engine size ________________ # of Cylinders ___________

1. In the service manual or computer program, locate the proper bleeding sequence for the vehicle being serviced. The proper bleeding sequence is important. Vehicle manufacturers publish brake bleeding sequence charts.

   Which manual or computer program was used? ______________________________ Page _____

2. On the sketch below, place numbers next to the wheels in the order that they are to be bled.

   Note: Traditional systems are bled beginning with the farthest cylinder from the master cylinder. Many front-wheel-drive vehicles use a diagonally split-brake system. These systems are bled by bleeding the right rear first, followed by the left front, left rear, and right front.

3. Open the hood and place fender covers on the fenders and front body parts.

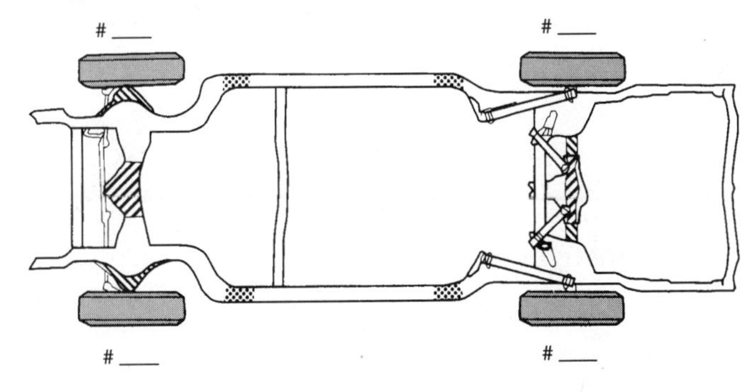

Brake fluid will damage vehicle paint. Be sure to *always* use fender covers and clean spills immediately.

Note: Water will clean up brake fluid spills.

4. Open the master cylinder and use a siphon to remove the old fluid.
5. Refill the master cylinder with clean brake fluid of the correct DOT specification.

   DOT specification: ____
6. Raise and support the vehicle.
7. Inspect the bleed screws. Free from dirt ____ Turn freely ____ Frozen ____
8. Place a small hose over the bleed screw and direct it into a container to catch the old brake fluid.
9. Have an assistant push down on the brake pedal while the bleed screw is loose. He or she should let up on the pedal only when the bleed screw is closed. Good communication is important to successfully complete this procedure.

   Does your helper understand the bleed procedure?

   Yes ____ No ____

   Bleeding method used: 1 person ____ 2 person ____

Loosen 1/4 turn

No air bubbles

Note: An alternate "one-person" method is to loosen a bleed screw and run a hose from it into a partially filled container of brake fluid. The old fluid will go into the container when the brakes are applied and air will not be able to enter when the pedal is released. The brake pedal should never be pumped. Pumping the pedal will cause the air in the system to break down into tiny bubbles that are difficult to bleed from the system.

10. Bleed the cylinders in the correct order until the fluid is clean, clear, and contains no sign of air.

    Was all of the air removed? Yes ____ No ____

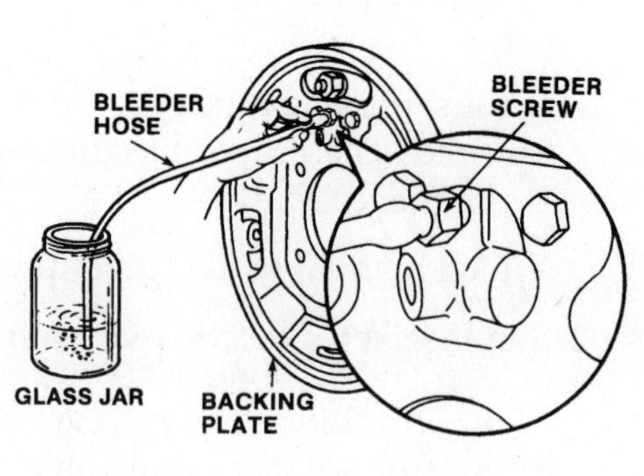

11. Check and refill the master cylinder after each wheel cylinder is bled.

Note: Never reuse brake fluid that has been siphoned or bled from the system.

Note: Some master cylinders have smaller reservoirs. Check the fluid level more often when servicing these systems.

12. Apply the brake pedal and check pedal feel and height.

    OK ____ Spongy ____

    Firm ____ Low ____

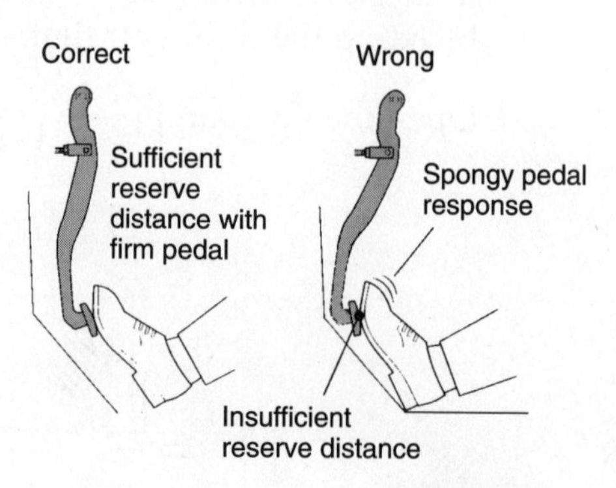

13. Check that all the bleed screws are tight.
14. Recheck the master cylinder fluid level.

15. Before completing the paperwork, clean your work area, put the tools in their proper places, and wash your hands.
16. Were there any other recommendations for needed service or unusual conditions that you noticed while bleeding the brakes?

    Yes _____ No _____
17. Record your recommendations for additional service or repairs on the repair order.
18. Complete the repair order (R.O.).

STOP

Instructor OK ______________ Score __________

ASE Lab Preparation Worksheet #9-2

REMOVE A BRAKE DRUM USING AN IMPACT WRENCH

Name______________________________ Class ______________________

OBJECTIVE:

Upon completion of this assignment, you should be able to remove and replace a brake drum using an impact wrench. This task will help prepare you to pass the ASE certification examinations in brakes, suspension, and steering.

DIRECTIONS:

Before beginning this lab assignment, review the worksheet completely. Fill in the information in the spaces provided as you complete each task.

TOOLS AND EQUIPMENT REQUIRED:

Safety glasses, jack and jack stands or vehicle lift, shop towel, 1/2″ impact wrench, impact sockets, hammer, torque wrench, rubber mallet

PROCEDURE:

Vehicle year __________ Make ________________ Model __________________

Repair Order # ________ Engine size ______________ # of Cylinders __________

1. Raise the rear of the vehicle and place it on safety stands or raise it on a lift.

   How is the vehicle supported? Safety stands ____ Lift ____

2. Remove the wheel cover or hubcap.

   Wheel cover ____

   Hubcap ____

   N/A ____

   Note: Some cars have lug nuts that extend through holes in the wheel cover or hubcap. On these cars it is not necessary to remove the hubcap.

3. Select the correct size *impact* socket. Socket size ____

4. Thread loosening direction:

   Clockwise (left-hand) ____

   Counterclockwise (right-hand) ____

   Note: Check to see that the impact wrench turns the proper direction before loosening the lug nuts.

5. Loosen the lug nuts and remove the wheel.

6. Sometimes special fasteners may retain the brake drum.

   Were any fasteners retaining the brake drum? Yes ____ No ____

   Were the fasteners removed? Yes ____ No ____

TURN

7. Use a large hammer to rap sharply on the brake drum *between* the lug bolts. The drum should pop free. If not, seek instructor assistance.

   Note: The dust around the brake sometimes contains asbestos residue. It is recommended that a respirator be worn when working around brakes for protection from breathing asbestos.

8. Reinstall the drum and wheel.
9. Tighten all of the lug nuts by hand.
10. Tighten the lug nuts in the proper order with a torque wrench.

    What is the wheel torque specification for the vehicle being serviced?

    ____

    Draw a sketch of the proper order in which to tighten wheel lug nuts in the box to the right.
11. Install the wheel cover or hubcap using a rubber mallet.
12. Lower the vehicle.
13. Before completing the paperwork, clean your work area, put the tools in their proper places, and wash your hands.
14. Were there any other recommendations for needed service or unusual conditions that you noticed while removing the brake drum?

    Yes ____ No ____
15. Record your recommendations for additional service or repairs on the repair order.
16. Complete the repair order (R.O.).

Instructor OK _______________ Score ___________

ASE Lab Preparation Worksheet #9-3

INSPECT DRUM BRAKES

Name_____________________________ Class ____________________

OBJECTIVE:

Upon completion of this assignment, you should be able to inspect drum brakes for needed service. This task will help prepare you to pass the ASE certification examination in brakes.

DIRECTIONS:

Before beginning this lab assignment, review the worksheet completely. Fill in the information in the spaces provided as you complete each task.

TOOLS AND EQUIPMENT REQUIRED:

Safety glasses, jack and jack stands or vehicle lift, shop towel, 1/2" impact wrench, impact sockets, hammer, drum gauge, torque wrench

CAUTION

- ❑ If a vehicle is found to have unsafe brakes, it should not be driven from the shop. It should be towed either to a repair facility or to the owner's home.
- ❑ When brake work is performed, wheels should not be reinstalled on the vehicle until all brake work has been satisfactorily completed.
- ❑ For liability reasons, an owner may only remove an unsafe vehicle if the state police or highway patrol has been notified and issues an equipment repair citation.

PROCEDURE:

Vehicle year ___________ Make ___________________ Model ____________________

Repair Order # ________ Engine size _________________ # of Cylinders ___________

Note: Step 1 does not apply to antilock braking systems (ABS). If the vehicle you are inspecting has ABS, start at step 2.

1. Sit in the vehicle and apply the brake pedal.

   Does it feel firm when applied? Yes ____ No ____

   A spongy pedal indicates:

   A normal condition ____

   That the brakes need bleeding ____

   That the brakes need to be adjusted ____

   Pump the pedal twice in rapid succession. Does the pedal height change? Yes ____ No ____

   If the pedal height is higher on the second application, this indicates:

   A brake adjustment is needed ____

   The brakes need to be bled ____

   A normal condition ____

2. Check the brake master cylinder reservoir fluid level. Full ____ Low ____

Note: Low fluid level in the disc brake reservoir is an indication that disc pads may be worn. Use Worksheet 9-4 (Inspect Front Disc Brakes) to check the disc brakes.

3. Raise the rear of the vehicle and place it on safety stands or raise it on a lift.

   How is the vehicle supported? Safety stands ____ Lift ____

4. Remove the wheel cover or hubcap.

   Note: Some cars have lug nuts that extend through holes in the wheel cover or hubcap. On these cars it is not necessary to remove the hubcap.

5. Select the correct size *impact* socket. Socket size ____

6. In which direction do the lug nuts need to be turned to loosen them?

   Clockwise (left-hand) ____ Counterclockwise (right-hand) ____

   Note: Check that the impact wrench turns in the proper direction before loosening the lug nuts.

7. Loosen the lug nuts and remove the wheel.

   Note: The dust around the brake sometimes contains asbestos residue. It is recommended that a respirator be worn when working around brakes for protection from asbestos.

8. Sometimes special fasteners may retain the brake drum. Were any special fasteners retaining the brake drum?

   Yes ____ No ____

   Were the fasteners removed? Yes ____ No ____

9. Is the parking brake released? Yes ____ No ____

10. Slide the drum off the axle flange. If it will not come off, use a large hammer to rap sharply on the brake drum *between* the lug bolts. The drum should pop free. If not, seek instructor assistance.

    Came off easily ____ Tight, needed assistance ____

11. Inspect the inner surface of the brake drum. Smooth ____ Scored ____ Other ____

12. Measure the drum size with a drum gauge.

    Standard drum specification ___ . ___ "

    What size is the drum? ___ . ___ "

    Is the drum oversize? Yes ____ No ____

    What service does the drum need?

    None ____ Machine oversize ____ Replace ____

13. Carefully pull back the rubber dust boots on the wheel cylinder to check for excessive leakage.

    Slightly moist (OK) ____

    Wet or soaked (need service) ____

CONTINUED

14. Inspect the rubber brake hose leading from the axle housing to the metal tube on the chassis.

    Cracked/weathered ____ OK ____

15. Inspect the linings.

    Lining construction: Bonded ____ Riveted ____

    Grease or brake fluid contamination ____

    Lining thickness: Thicker than the metal shoe ____ Thinner than the metal shoe ____

16. Have an assistant operate the parking brake while you check its operation.

    Works properly ____ Needs service ____

17. Reinstall the drum and wheel.
18. Tighten all of the lug nuts by hand.
19. Tighten the lug nuts in the proper order with a torque wrench.

    What is the wheel torque specification for the vehicle? ________

    Draw a sketch of the proper order in which to tighten wheel lug nuts in the box to the right.

20. Install wheel cover or hubcap using a rubber mallet.
21. Lower the vehicle.
22. Before completing the paperwork, clean your work area, put the tools in their proper places, and wash your hands.
23. Were there any other recommendations for needed service or unusual conditions that you noticed while inspecting the drum brakes?

    Yes ____ No ____

24. Record your recommendations for additional service or repairs on the repair order.
25. Complete the repair order (R.O.).

STOP

Instructor OK ______________ Score __________

ASE Lab Preparation Worksheet #9-4

INSPECT FRONT DISC BRAKES

Name____________________________ Class ____________________

OBJECTIVE:

Upon completion of this assignment, you should be able to inspect the front disc brakes for needed repairs. This task will help prepare you to pass the ASE certification examination in brakes.

DIRECTIONS:

Before beginning this lab assignment, review the worksheet completely. Fill in the information in the spaces provided as you complete each task.

TOOLS AND EQUIPMENT REQUIRED:

Safety glasses, jack and jack stands or vehicle lift, shop towel, 1/2″ impact wrench, impact sockets, flashlight, torque wrench

CAUTION

❑ If a vehicle is found to have unsafe brakes, it should not be driven from the shop. It should be towed either to a repair facility or to the owner's home.

❑ When brake work is performed, wheels should not be reinstalled on the vehicle until all brake work has been satisfactorily completed.

❑ For liability reasons, an owner may only remove an unsafe vehicle if the state police or highway patrol has been notified and issues an equipment repair citation.

PROCEDURE:

Vehicle year ___________ Make _________________ Model ________________

Repair Order # ________ Engine size ________________ # of Cylinders ___________

Note: Step 1 does not apply to antilock braking systems (ABS). If the vehicle you are inspecting has ABS, start at step 2.

1. Sit in the vehicle and apply the brake pedal.

   Does it feel firm when applied? Yes ____ No ____

   If the pedal feels spongy, this indicates:

   A normal condition ____

   That the brakes need bleeding ____

   That the brakes need to be adjusted ____

   Pump the pedal twice in rapid succession. Does the pedal height change? Yes ____ No ____

   If the pedal height is higher on the second application, this indicates:

   A brake adjustment is needed ____

   The brakes need to be bled ____

   A normal condition ____

2. Check the fluid level in the brake master cylinder disc brake reservoir. Full ____ Low ____

Note: Low fluid level in the disc brake reservoir indicates that the disc pads may be worn.

3. Raise the rear of the vehicle and place it on safety stands or raise it on a lift.

How was the vehicle supported? Safety stands ____ Lift ____

4. Remove the wheel cover or hubcap.

Note: Some cars have lug nuts that extend through holes in the wheel cover or hubcap. On these cars it is not necessary to remove the hubcap.

5. Select the correct size *impact* socket. Socket size ____

6. In which direction do the lug nuts need to be turned to loosen them?

Clockwise (left-hand) ____ Counterclockwise (right-hand) ____

Note: Check that the impact wrench turns in the proper direction before loosening the lug nuts.

7. Loosen the lug nuts and remove the wheel.

8. Use a flashlight to inspect both the leading and trailing edges of both linings. Linings sometimes wear unevenly.

Lining thickness at thinnest point:

Thicker than metal shoe ____ Worn evenly ____

Thinner than metal shoe ____ Worn unevenly ____

Do the linings need to be replaced?

Yes ____ No ____

Direction of rotation
Rotor
Piston
1/8″ (3.18-mm) maximum taper
Outer pad and plate
Inner pad and plate

Note: The dust around the brake assembly sometimes contains asbestos residue. Wear a respirator for protection from asbestos when working around brakes.

9. Inspect the condition of the rotor. Smooth ____ Wavy ____ Scored ____

Is there a lip at the outer edge? Yes ____ No ____

Rotate the rotor. Is any warpage (runout) evident?

Yes ____ No ____

10. Check the rotor for parallelism. Are the rotor faces parallel?

Yes ____ No ____

11. Check rotor for runout. Is there any runout?

Yes ____ No ____

How much ________

12. Inspect the condition of the rubber brake hose leading to each brake caliper.

Cracked/weathered ____ Twisted ____ OK ____

13. Reinstall the wheel.
14. Tighten all of the lug nuts by hand.
15. Tighten the lug nuts in the proper order with a torque wrench.

    What is the proper wheel torque for the vehicle being serviced? _______________

    Draw a sketch of the proper order in which to tighten wheel lug nuts in the box to the right.

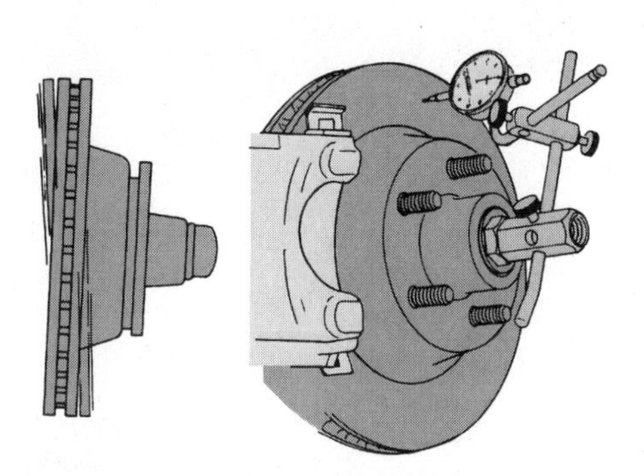

16. Install the wheel covers or hubcaps using a rubber mallet.
17. Lower the vehicle.
18. Before completing the paperwork, clean your work area, put the tools in their proper places, and wash your hands.
19. Were there any other recommendations for needed service or unusual conditions that you noticed while inspecting the disc brakes?

    Yes _____ No _____

20. Record your recommendations for additional service or repairs on the repair order.
21. Complete the repair order (R.O.).

STOP

Instructor OK ______________ Score __________

ASE Lab Preparation Worksheet #9-5

REPLACE FRONT DISC BRAKE PADS

Name______________________________ Class ____________________

OBJECTIVE:

Upon completion of this assignment, you should be able to replace disc brake pads. This task will help prepare you to pass the ASE certification examination in brakes.

DIRECTIONS:

Before beginning this lab assignment, review the worksheet completely. Fill in the information in the spaces provided as you complete each task.

TOOLS AND EQUIPMENT REQUIRED:

Safety glasses

PROCEDURE:

Vehicle year ___________ Make _________________ Model ___________________

Repair Order # ________ Engine size ________________ # of Cylinders ___________

1. Check the level of the brake fluid in the master cylinder. Full ____ Low ____
2. Raise the vehicle on a lift or place it on jack stands.
3. Remove the front wheel and tire assemblies.
4. Remove the caliper guide pins or disc pad retainers, and remove the disc pads or caliper.

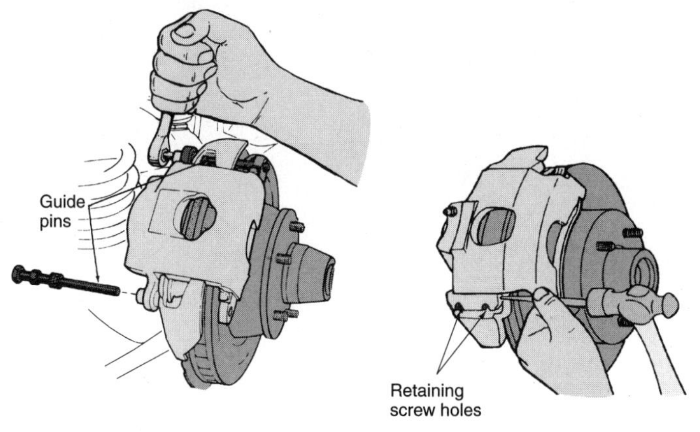

5. Do not let the caliper hang from its hose. Support it by hanging it on a wire from the suspension or the frame of the vehicle.
6. Inspect the condition of the rotors. Smooth ____ Wavy ____ Scored ____
7. Measure the rotor parallelism. Measurement ___________ Specification ___________
8. Measure the rotor for runout. Measurement ___________ Specification ___________
9. Is the rotor in reusable shape or does it need to be serviced or possibly replaced?

   Reusable _________ Service _________ Replace _________

TURN

Note: If the rotors need to be serviced or replaced, this will need to be done before proceeding. Consult with your instructor.

10. Open the caliper bleeder screw and push the piston back into the caliper bore using a C-clamp or other suitable tool.

Note: Tighten the bleeder screw when you are finished so that it is not accidentally left open.

11. Install the caliper without the disc pads and check that it moves freely in its mount. Does it move freely? Yes ____ No ____
12. Remove the caliper and install the disc pads. Then reinstall the caliper on its mount. Remember to torque all bolts as required.
13. Install the tires and wheel assemblies.
14. Lower the vehicle.
15. Torque the lug nuts and install the wheel covers. Torque specification: ______________
16. Before completing the paperwork, clean your work area, put the tools in their proper places, and wash your hands.
17. Were there any recommendations for needed service or unusual conditions that you noticed while you were replacing the disc pads? Yes ____ No ____
18. Record your recommendations for needed service or additional repairs on the repair order.
19. Complete the repair order (R.O.).

Instructor OK ______________ Score __________

ASE Lab Preparation Worksheet #9-6

BRAKE ADJUSTMENT (SELF-ADJUSTING BENDIX BRAKES)

Name________________________ Class ________________

OBJECTIVE:

Upon completion of this assignment, you should be able to adjust the brakes on a vehicle that has self-adjusting Bendix brakes. This task will help prepare you to pass the ASE certification examination in brakes.

DIRECTIONS:

Before beginning this lab assignment, review the worksheet completely. Fill in the information in the spaces provided as you complete each task.

TOOLS AND EQUIPMENT REQUIRED:

Safety glasses, fender cover, jack and jack stands or vehicle lift, shop towel, hammer, punch, screwdriver, brake spoon

PROCEDURE:

Vehicle year __________ Make ________________ Model ________________

Repair Order # ________ Engine size ______________ # of Cylinders __________

1. Locate the brakes section in a service manual or computer program for the vehicle. Determine if the vehicle is equipped with Bendix (dual servo) or leading trailing brakes. Bendix-type brakes are typically found on rear-wheel-drive vehicles.

   Service manual or computer program used: ______________________________ Page #____

2. Raise and support the vehicle.
3. Locate the adjusting slots (check any that apply):

   Rear of brake backing plates ____

   In the brake drum ____

   There are none ____

   Note: Many vehicles do not have an open adjusting slot. There is a knockout that must be removed by drilling or knocking it out with a hammer and punch.

   Adjusting slot ____

   Knockout ____

4. Remove the dust cover from the slot, *or* remove the knockout to form an adjusting slot.

   Note: If the knockout was removed, a replacement rubber dust cover will need to be installed in the slot to prevent dirt and moisture from entering the brakes.

5. Use a brake spoon to tighten the adjuster (star wheel). The adjuster will only turn in one direction (tightening). Tighten until a heavy drag is felt while rotating the wheel by hand.

6. Insert a thin screwdriver into the hole and push the self-adjuster arm away from the starwheel.
7. Loosen the adjuster until the wheel turns freely.
8. Install the dust cover into the adjusting hole.
9. Repeat the procedure on the remaining drum brake.
10. Lower the vehicle to the ground.
11. Check brake pedal height. On the second application of the pedal within one second, the pedal height should remain in the same position as on the first application.

    Does the pedal height remain the same?

    Yes ____ No ____

    Does the pedal have a firm feel?

    Yes ____ No ____

    Do the brakes require further adjustment? Yes ____ No ____

    Note: If the self-adjuster is operating properly, the brakes can be adjusted by backing up the vehicle and applying the brakes firmly several times. Speed of the vehicle is not a factor in adjusting the brakes. Drive slowly and carefully as you back the vehicle to adjust the brakes.

12. Was it necessary to back the vehicle to adjust the brakes? Yes ____ No ____
13. Do the brakes feel firm and is the pedal height consistent after several applications?

    Yes ____ No ____

14. Before completing the paperwork, clean your work area, put the tools in their proper places, and wash your hands.
15. Were there any other recommendations for needed service or unusual conditions that you noticed while adjusting the brakes?

    Yes ____ No ____

16. Record your recommendations for additional service or repairs on the repair order.
17. Complete the repair order (R.O.).

Instructor OK ______________ Score __________

ASE Lab Preparation Worksheet #9-7

PARKING BRAKE ADJUSTMENT

Name_____________________________ Class ________________

OBJECTIVE:

Upon completion of this assignment, you should be able to adjust a vehicle's parking brake. This task will help prepare you to pass the ASE certification examination in brakes.

DIRECTIONS:

Before beginning this lab assignment, review the worksheet completely. Fill in the information in the spaces provided as you complete each task.

TOOLS AND EQUIPMENT REQUIRED:

Safety glasses, jack and jack stands or vehicle lift, hand tools, shop towel

PROCEDURE:

Vehicle year __________ Make ________________ Model _________________

Repair Order # ________ Engine size _______________ # of Cylinders ___________

1. Complete a rear drum brake adjustment before making any adjustment to the parking brake. Refer to Worksheet 9-6.

   Are the rear brakes correctly adjusted?

   Yes ____ No ____

2. Apply the parking brake to 1/4 of its full travel. 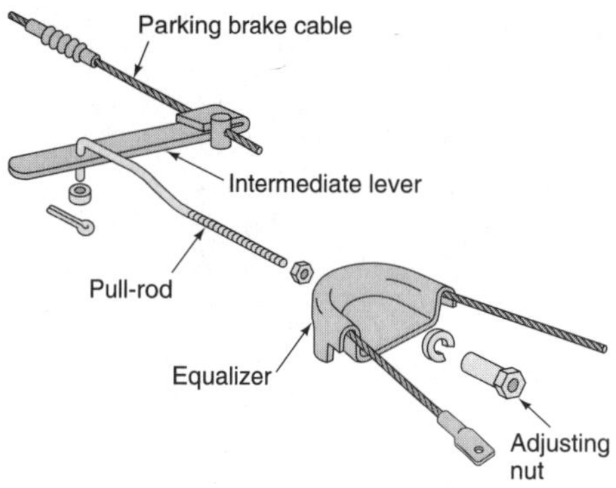
3. Put the transmission shift selector in neutral.
4. Raise and support the vehicle.
5. Loosen the adjusting locknut on the parking brake equalizer bar.
6. Tighten the adjusting nut at the equalizer bar until the rear wheels will no longer turn when rotated by hand.
7. Release the parking brake lever.
8. Do both rear wheels turn freely?

   Yes ____ No ____

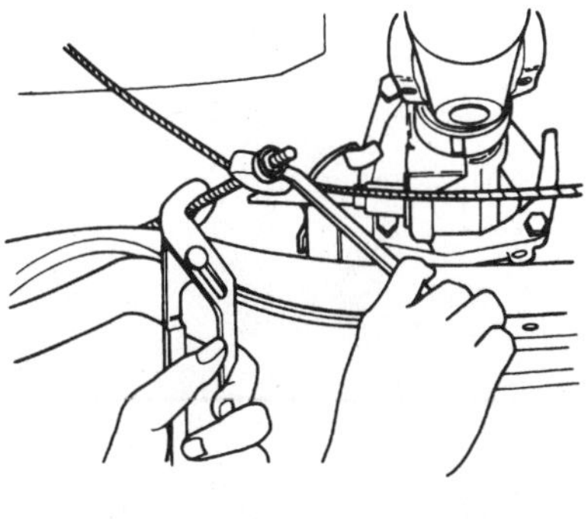

9. Tighten the locknut on the adjuster.
10. Lower the vehicle.
11. Check the operation of the parking brake. Will it hold the vehicle on an incline?

    Yes ____ No ____

12. Before completing the paperwork, clean your work area, put the tools in their proper places, and wash your hands.

Parking brake cable
Intermediate lever
Pull-rod
Equalizer
Adjusting nut

13. Were there any other recommendations for needed service or unusual conditions that you noticed while adjusting the parking brake?

    Yes ____ No ____

14. Record your recommendations for additional service or repairs on the repair order.
15. Complete the repair order (R.O.).

STOP

Instructor OK ______________ Score __________

ASE Lab Preparation Worksheet #9-8

COMPLETE BRAKE INSPECTION WORKSHEET

Name____________________________ Class ____________________

OBJECTIVE:

Upon completion of this assignment, you should be able to inspect a vehicle's brake system. This task will help prepare you to pass the ASE certification examination in brakes.

DIRECTIONS:

Before beginning this lab assignment, review the worksheet completely. Fill in the information in the spaces provided as you complete each task.

TOOLS AND EQUIPMENT REQUIRED:

Safety glasses, shop towel

PROCEDURE:

Vehicle year ___________ Make _________________ Model __________________

Repair Order # ________ Engine size ________________ # of Cylinders ___________

1. Master cylinder fluid level: OK _______ Low _______ Overfilled _______
2. Parking brake: OK _______ Needs adjustment _______
3. Raise the vehicle and place it on safety stands or raise it on a lift.
4. Loosen the lug nuts and remove the wheels.
5. Inspect the front brakes.
   a. Approximate brake pad thickness: _________________
   b. Rotor condition: Smooth _______ Rough _______ Grooved _______
   c. Rotor thickness: __________ (in./mm).
   d. Rotor parallelism: __________ (in./mm).
   e. Rotor runout: __________ (in./mm). Specification: __________ (in./mm).
   f. Caliper condition: OK _______ Leaking _______ Stuck _______
   g. Condition of the brake lines: OK _______ Damaged __________
   h. Leaks? Yes ______ No ______ Location of leak (if any) ________________________
6. Inspect the rear brakes.
   a. What type of rear brakes is used on this vehicle? Drum ______ Disc ______

   Rear drum brakes

   b. If the vehicle has rear drum brakes, answer the following:

      Brake shoe thickness: _______________

      Drum condition: Smooth _______ Rough _______ Grooved _______

      Drum diameter: __________ (in./mm). Specification: __________ (in./mm).

Drum out of round: Yes _______ No _______

Wheel cylinder condition: OK _______ Leaking _______

Condition of the brake lines: OK _______ Damaged ___________

Leaks: Yes _______ No _______ Location of leak (if any) ___________________________

Rear Disc brakes

c. If the vehicle has rear disc brakes, answer the following:

Brake pad thickness: _________________

Rotor condition: Smooth _______ Rough _______ Grooved _______

Rotor thickness: ___________ (in./mm). Specification: ___________ (in./mm).

Rotor parallelism: ___________ (in./mm).

Rotor runout: ___________ (in./mm).

Caliper condition: OK _______ Leaking _______ Stuck _______

Condition of the brake lines: OK _______ Damaged ___________

Leaks: Yes _______ No _______ Location of leak (if any) ___________________________

7. Install the wheels and lower the vehicle.
8. Before completing the paperwork, clean your work area, put the tools in their proper places, and wash your hands.
9. List any unusual conditions or recommendations for needed service.

____

____

10. Record your recommendations for additional service or repairs on the repair order.
11. Complete the repair order (R.O.).

STOP

Instructor OK ______________ Score __________

ASE Lab Preparation Worksheet #9-9
ADJUST A TAPERED ROLLER WHEEL BEARING

Name___________________________ Class ___________________

OBJECTIVE:
Upon completion of this assignment, you should be able to adjust a tapered roller wheel bearing. This task will help prepare you to pass the ASE certification examination in suspension and steering.

DIRECTIONS:
Before beginning this lab assignment, review the worksheet completely. Fill in the information in the spaces provided as you complete each task.

TOOLS AND EQUIPMENT REQUIRED:
Safety glasses, jack and jack stands or vehicle lift, shop towel, adjustable pliers, diagonal cutters

PARTS AND SUPPLIES:
Cotter pins

PROCEDURE:
Vehicle year ___________ Make _________________ Model _________________

Repair Order # ________ Engine size ________________ # of Cylinders ___________

1. Raise and support the vehicle.

   Note: Tapered roller bearings are found on the front of rear-wheel-drive vehicles and the rear of most front-wheel-drive vehicles. Consult the service manual for specific adjusting information.

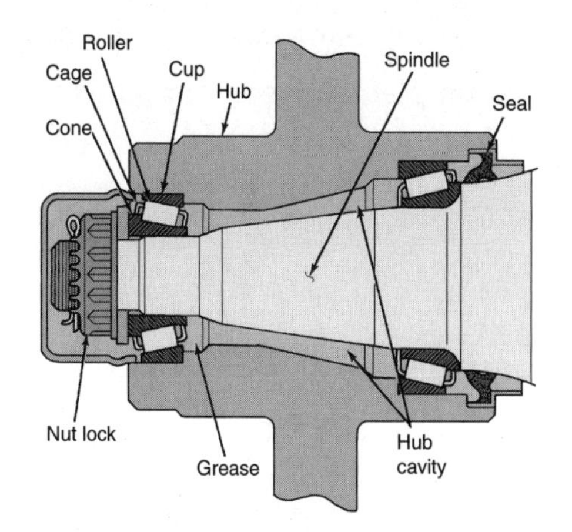

2. The vehicle being serviced has:

   Front-wheel drive ____ Rear-wheel drive ____

3. Remove the wheel cover or hubcap.

   Wheel cover ____ Hubcap ____

4. Remove the grease cap.

   Which tool was used?

   Grease cap pliers ____ Adjustable joint pliers ____

5. Remove the cotter pin. Which tool was used?

   Diagonal cutters ____ Cotter pin tool ____

6. Use an adjustable wrench or adjustable joint pliers to tighten the spindle nut while turning the wheel. Tighten the nut until it is "snug" (25 ft.-lb).

7. Back off the spindle nut and retighten until *"zero-lash"* (1 ft.-lb)is reached. Rock the wheel as you tighten the nut until all looseness disappears.

Note: It is difficult to use a torque wrench when adjusting wheel bearings. The following is a convenient way to check for proper adjustment. If the tabbed washer under the spindle nut can be moved easily with a screwdriver, the bearing is not too tight. Endplay can be 0.001″ to 0.010″. This is from 1/16 to 1/8 turn backed off from snug (zero-lash).

8. Install the largest diameter cotter pin that will fit into the spindle hole. What size cotter pin is needed?

   5/32″ ____ 1/8″ ____ 3/16″ ____

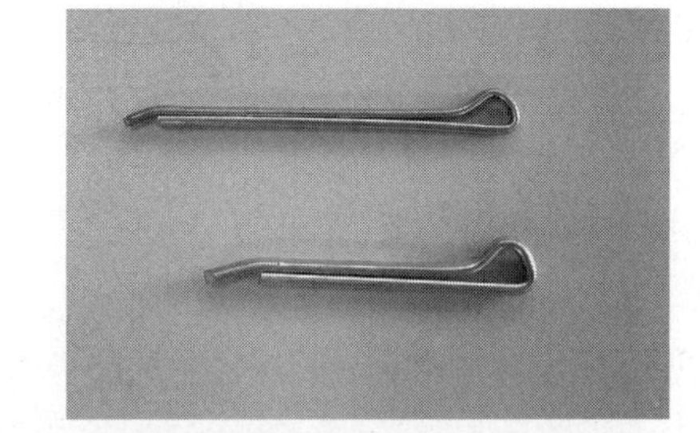

9. If the hole does not line up, see if there is another hole 90° from it that does line up.

   If no hole lines up, *loosen* the nut until the cotter pin can be installed.

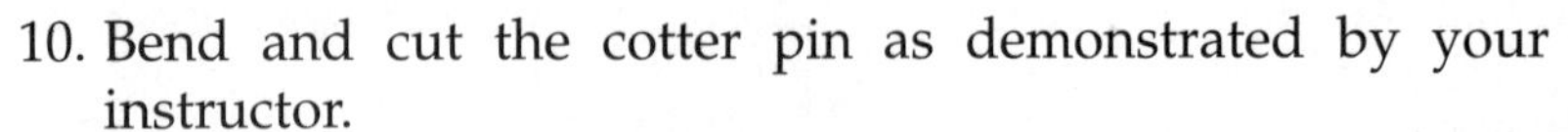

10. Bend and cut the cotter pin as demonstrated by your instructor.

11. Reinstall the grease cap.
12. Install the wheel cover or hubcap.
13. Repeat steps 2 through 12 to adjust the other tapered wheel bearing.
14. After both wheel bearings are adjusted, lower the vehicle.
15. Before completing the paperwork, clean your work area, put the tools in their proper places, and wash your hands.
16. Were there any other recommendations for needed service or unusual conditions that you noticed while adjusting the wheel bearings?

    Yes ____ No ____

17. Record your recommendations for additional service or repairs on the repair order.
18. Complete the repair order (R.O.).

Instructor OK ______________ Score __________

ASE Lab Preparation Worksheet #9-10

REPACK WHEEL BEARINGS

Name______________________________ Class ____________________

OBJECTIVE:

Upon completion of this assignment, you should be able to repack wheel bearings on front- and rear-wheel-drive vehicles that use tapered wheel bearings. This task will help prepare you to pass the ASE certification examination in suspension and steering.

DIRECTIONS:

Before beginning this lab assignment, review the worksheet completely. Fill in the information in the spaces provided as you complete each task.

TOOLS AND EQUIPMENT REQUIRED:

Safety glasses, jack and jack stands or vehicle lift, hand tools, seal drivers, shop towel

PARTS AND SUPPLIES:

Cotter pin, seal, grease

Note: Complete Worksheet 9-9, Adjust a Tapered Roller Wheel Bearing, before attempting to complete this worksheet.

PROCEDURE:

Vehicle year ___________ Make __________________ Model __________________

Repair Order # ________ Engine size _________________ # of Cylinders ___________

Note: During a laboratory period, there may not be enough time to complete repacking both wheel bearings. Complete the bearing pack on one side of the vehicle at a time. If both sides are being packed at the same time, remember, *do not interchange* the parts. Bearings "wear-mate" to the bearing cups. Interchanging the parts leads to premature bearing failure. When a bearing is replaced, its cup must be replaced also.

1. Raise and support the car.
2. Remove the wheel cover or hubcap.
3. Unless the vehicle has disc brakes, it is not necessary to remove the wheel lug nuts.

   Were the lug nuts left tight? Yes ____ No ____

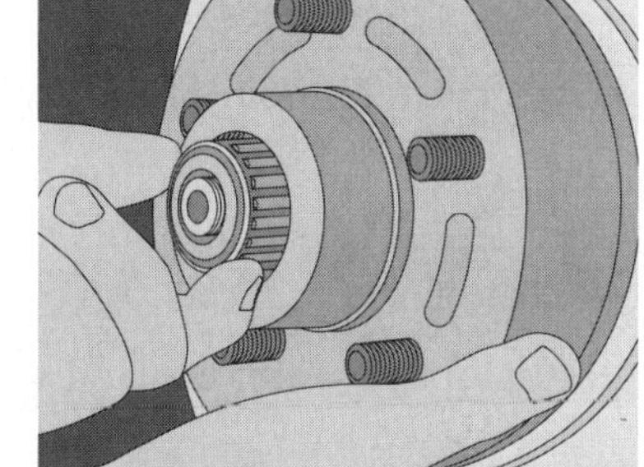

4. Remove the grease cover, cotter pin, and spindle nut.
5. Hold the tire at the top and bottom and gently rock it. The washer and outer bearing will slide outward on the spindle for easy removal.
6. Rotate the tire as you slide the drum/wheel assembly off of the spindle. If it does not come off easily, seek help from your instructor.

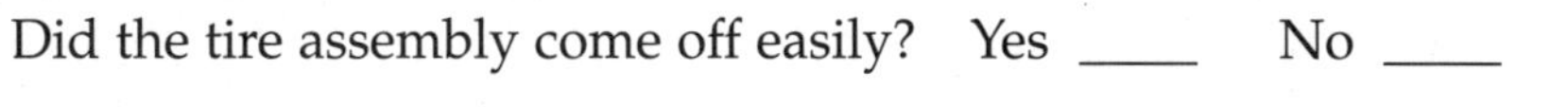

   Did the tire assembly come off easily? Yes ____ No ____

7. Reach through the hub with a punch. Gently tap the punch with a hammer to remove the inner bearing and seal.

   Note: On many cars, the spindle nut can be reinstalled and used to *gently* pull the bearing and seal loose from the hub.

8. Wipe excess grease off the bearings with a paper towel before washing the bearing in solvent.
9. Thoroughly clean the bearings in solvent.
10. Completely dry the bearings with compressed air.

When using compressed air to dry a bearing, hold the bearing cage so the bearing cannot spin. Spinning an unlubricated bearing will ruin the bearing, and it is dangerous.

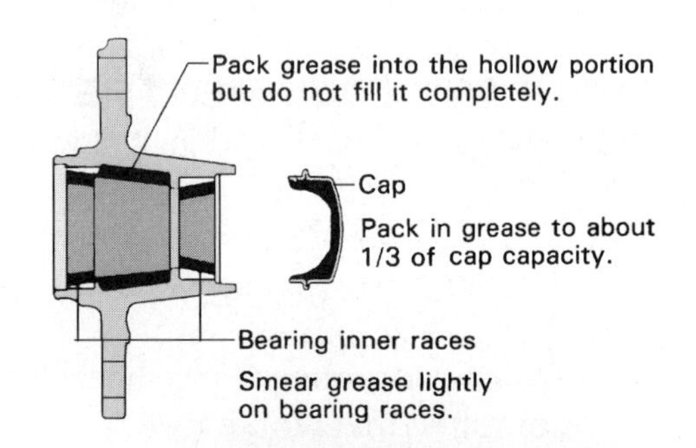

11. Wipe the old grease out of the hub with a paper towel.
12. Install a small amount of new grease into the hub.
13. Inspect the condition of the bearings and cups (races). Record your findings below.

    Reusable ____ Stained ____

    Pitted ____ Scored ____

    Cage damage____

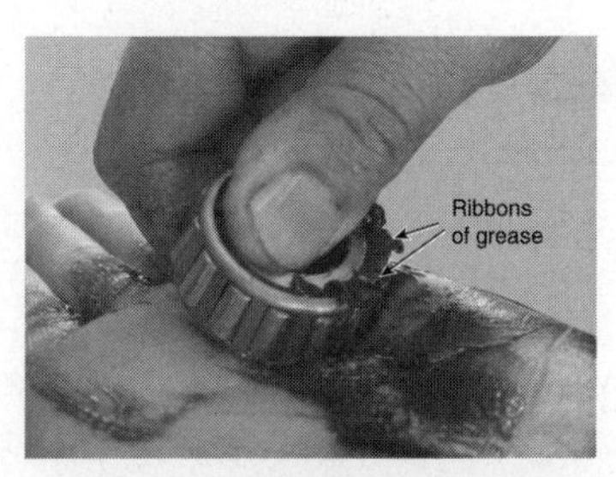

14. Thoroughly repack the bearings with high temperature wheel bearing grease.

    Which method was used to repack the bearings?

    Hand ____ Bearing packer ____

15. Install the inner (large) bearing into its cup and carefully install the seal. Is the old seal or a new seal being used?

    New seal ____ Old seal ____

16. What method was used to install the seal?

    Seal installing tool ____

    Hammer and block of wood ____

    Hammer only ____

    Other (describe) ____

17. Lubricate the lip of the seal with grease.
18. Carefully center the drum as you rotate it over the brake assembly. Be careful not to damage the grease seal during installation.
19. Install the outer bearing *and washer.*

    Note: The washer is indexed to the spindle, preventing the rotating assembly from accidentally tightening or loosening the spindle nut.

20. Adjust the bearing.
21. Have your instructor inspect the adjustment before lowering the vehicle to the ground.

    Instructor's OK ____________________
22. Before completing the paperwork, clean your work area, put the tools in their proper places, and wash your hands.
23. Were there any other recommendations for needed service or unusual conditions that you noticed while packing the wheel bearings?
24. Record your recommendations for additional service or repairs on the repair order.
25. Complete the repair order (R.O.).

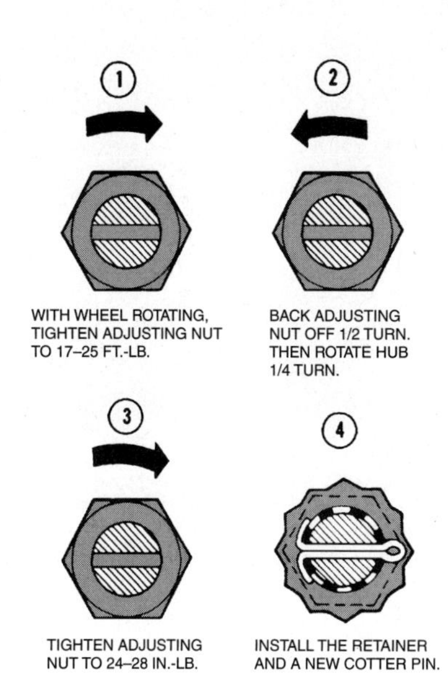

STOP

Instructor OK ______________ Score __________

ASE Lab Preparation Worksheet #9-11

REPACK WHEEL BEARINGS (DISC BRAKE)

Name__________________________ Class ___________________

OBJECTIVE:

Upon completion of this assignment, you should be able to repack tapered wheel bearings. This task will help prepare you to pass the ASE certification examination in suspension and steering.

DIRECTIONS:

Before beginning this lab assignment, review the worksheet completely. Fill in the information in the spaces provided as you complete each task.

TOOLS AND EQUIPMENT REQUIRED:

Safety glasses, jack and jack stands or vehicle lift, impact wrench, impact sockets, hand tools, hammer, punch, bearing packer, shop towel

PARTS AND SUPPLIES:

High temperature grease, cotter pins

PROCEDURE:

Vehicle year ___________ Make _________________ Model __________________

Repair Order # ________ Engine size ________________ # of Cylinders ___________

Note: Complete Worksheet 9-10, Repack Wheel Bearings, before or along with this worksheet.

Note: During a laboratory period there may not be enough time to complete repacking both wheel bearings. Complete the bearing pack on one side of the vehicle at a time. If both sides are being packed at the same time, remember: *do not interchange* the parts. Bearings "wear-mate" to the bearing cups. Interchanging the parts leads to premature bearing failure. When a bearing is replaced, its cup must be replaced also.

1. Check the service manual or computer program for the correct procedure for disc brake caliper removal.

   Which manual or computer program was used for this information?

   ____

   On what page did you find the procedure for removing the disc brake caliper?

   ____

2. Raise and support the car.
3. Remove the wheel cover or hubcap.
4. Use an impact wrench and socket to remove the lug nuts. What size socket was needed? ______
5. Remove the tire and wheel assembly.
6. Remove the grease cap and spindle nut.

Remove the Disc Brake Caliper

7. Loosen the bleed fitting on the brake caliper.
8. Hold the rotor at the top and bottom and gently rock it.

   Note: This procedure moves the caliper pistons back into their bores so that the caliper is more easily removed.

9. Retighten the bleed screw.
10. Remove the caliper and support it while repacking the wheel bearing. Do *not* let it hang by the brake hose.

    How did you support the caliper?

    Supported with wire ____

    Used a special tool ____

    Set it on the steering linkage ____

11. Remove the grease cover, cotter pin, and spindle nut.
12. Hold the rotor/disc at the top and bottom and gently rock it. The washer and outer bearing will slide outward on the spindle for easy removal.
13. Rotate the rotor/disc as you slide it off the spindle. If it does not come off easily, seek help from your instructor.

    Did the rotor and hub come off easily? Yes ____ No ____

14. Reach through the hub with a punch and gently tap the punch to remove the inner bearing and seal.
15. Wipe excess grease off the bearings with a paper towel before washing them in solvent.
16. Thoroughly clean the bearings in solvent.
17. Dry the bearings with compressed air.

    Note: While using compressed air to dry a bearing, hold the bearing cage so the bearing cannot spin. Spinning an unlubricated bearing will ruin the bearing, and it is dangerous.

18. Wipe old grease out of the hub with a paper towel.
19. Inspect the condition of the bearings and cups (races). Record your findings below:

    Reusable? ____ Pitted? ____ Cage damaged? ____

    Stained? ____ Scored? ____ Other (describe) ________________________

20. Thoroughly repack the bearings with high temperature wheel bearing grease.

    What method was used to repack the bearings? Hand ____ Bearing packer ____

21. Install the inner (large) bearing into its cup and carefully install the seal. Is the old seal or a new seal being used?

    New seal ____ Old seal ____

22. What method was used to install the seal?

    Hammer only ____ Seal installing tool ____

    Hammer and block of wood ____ Other (describe) ____________________

23. Carefully center the rotor as you rotate it over the spindle. Be careful not to damage the grease seal during installation.
24. Install the outer bearing and washer.

    Note: The washer is indexed to the spindle, preventing the rotating assembly from accidentally tightening or loosening the spindle nut.

25. Adjust the bearing as described in Worksheet 9-9, Adjust a Tapered Roller Wheel Bearing.
26. Reinstall the caliper. Torque the caliper bolts to specifications.

    Caliper bolt torque specification: ____ ft.-lb

27. Install the wheel, tighten the lug nuts, lower the vehicle to the ground, and retorque the lug nuts.

    Lugnut torque specification: ____ ft.-lb

28. Install the wheel covers or hubcaps.
29. Apply the foot brake until a firm pedal is felt.

    Does the brake pedal feel normal and firm? Yes ____ No ____

CAUTION It is not unusual for the brake pedal to move all the way to the floor on the first application after the caliper has been removed. On vehicles with disc brakes, the piston is usually moved back in its bore when the caliper is removed. The pedal must be depressed to readjust the brakes. Always be sure to apply the brake pedal several times before moving the vehicle.

30. Before completing the paperwork, clean your work area, put the tools in their proper places, and wash your hands.
31. Were there any other recommendations for needed service or unusual conditions that you noticed while repacking the wheel bearings?

    Yes ____ No ____

32. Record your recommendations for additional service or repairs on the repair order.
33. Complete the repair order (R.O.).

Instructor OK ______________ Score ___________

ASE Lab Preparation Worksheet #9-12

REPLACE A TAPERED WHEEL BEARING

Name______________________________ Class ____________________

OBJECTIVE:

Upon completion of this assignment, you should be able to replace a tapered wheel bearing. This task will help prepare you to pass the ASE certification examination in suspension and steering.

DIRECTIONS:

Before beginning this lab assignment, review the worksheet completely. Fill in the information in the spaces provided as you complete each task.

TOOLS AND EQUIPMENT REQUIRED:

Safety glasses, jack and jack stands or vehicle lift, hand tools, hammer, wheel bearing punch, shop towel

PARTS AND SUPPLIES:

Wheel bearing, axle seal, grease

> **Note:** When replacing the wheel bearings, it is always necessary to pack them with grease. Use the appropriate worksheets for help with the packing and adjustment procedures.

PROCEDURE:

Vehicle year ___________ Make _________________ Model __________________

Repair Order # ________ Engine size _________________ # of Cylinders ___________

1. Raise and support the vehicle.
2. Remove the tire and hub. Remove the wheel bearings from the hub.
3. Type of eye protection worn:

   Goggles ____

   Safety glasses ____

   Face shield ____
4. Locate the recesses in the hub on the inside of the bearing race.
5. Using a hammer and wheel bearing punch, drive the cup from the hub. Tap first on one side of the cup at the recess, then on the other side, until the race is removed.
6. Clean and inspect the hub.

   Is the hub damaged? Yes ____ No ____

   Will the hub need to be replaced? Yes ____ No ____

7. Install the new bearing cup.

   Note: If a tapered bearing cup driver is not available, grind a small amount off the outside of the old cup. Then use it as a driver to install the new cup in the hub. Tap the cup into the hub until it is bottomed in its bore.

   What tool was used to install the new cup?

   Old cup ____

   Bearing cup driver ____

   Does the bearing cup fit tightly into the hub?

   Yes ____ No ____

   Was the cup bottomed in its bore?

   Yes ____ No ____

8. Pack the wheel bearing and reassemble it to the hub using Worksheet 9-10, Repack Wheel Bearings, as a guide.
9. Before completing the paperwork, clean your work area, put the tools in their proper places, and wash your hands.
10. Were there any other recommendations for needed service or unusual conditions that you noticed while replacing the wheel bearing?

    Yes ____ No ____

11. Record your recommendations for additional service or repairs on the repair order.
12. Complete the repair order (R.O.).

Instructor OK ______________ Score __________

ASE Lab Preparation Worksheet #9-13

REPLACE A REAR AXLE WITH A PRESS-FIT BEARING

Name__________________________ Class __________________

OBJECTIVE:

Upon completion of this assignment, you should be able to remove and replace a rear axle with a press-fit bearing. This task will help prepare you to pass the ASE certification examinations in manual transmission and differentials.

DIRECTIONS:

Before beginning this lab assignment, review the worksheet completely. Fill in the information in the spaces provided as you complete each task.

TOOLS AND EQUIPMENT REQUIRED:

Safety glasses, jack and jack stands or vehicle lift, drain pan, hand tools, seal puller, slide hammer, shop towel

PARTS AND SUPPLIES:

Lubricant (possibly a bearing and/or seal)

PROCEDURE:

Vehicle year __________ Make ________________ Model ________________

Repair Order # ________ Engine size _______________ # of Cylinders ___________

1. Raise and support the vehicle.
2. Remove the rear wheel.
3. Remove the brake drum or the caliper and rotor.

   Brake drum ____ Caliper and rotor ____
4. Use a ratchet, extension, and socket to remove the retainer flange nuts and bolts.

   Number of retainer flange bolts ____

Retainer nuts

Hole in axle flange

Note: There is a hole provided in the axle flange for easy access to the bolts. Hold the nuts from the rear of the backing plate with a combination wrench.

5. Remove the axle. Use a slide hammer if necessary.

   Slide hammer needed? Yes ____ No ____

Note: When using a slide hammer, if it feels really solid, either all of the bolts might not have been removed or the axle is of the C-lock type. In either case, damage could result if the slide hammer is used.

6. Feel the axle bearing for roughness or damage. Does it need to be replaced?

   Yes ____ No ____

Note: Check with your instructor to see if the equipment is available to replace the bearing. If not, the axle will need to be sent to an automotive machine shop for bearing replacement.

Was the bearing replaced? Yes ____ No ____

7. If there is a separate seal, replace it now.

   Does the axle have a separate seal? Yes ____ No ____

   Was the seal replaced? Yes ____ No ____

   Note: When replacing a seal, always put a little grease on the sealing lip to protect it during initial startup.

   Lip of the seal lubricated? Yes ____ No ____

8. Install the axle assembly. Slide the axle in until the splines align with the splines in the differential side gears. Be careful not to damage the axle seal.
9. Install the retainer flange nuts and bolts and torque them to specifications.

   Torque specification: ________

10. Replace the brake drum and tire/wheel assembly.
11. Check the lubricant level in the differential and add as needed.

    How much lubricant was added? ____ pints None ____

12. Before completing the paperwork, clean your work area, put the tools in their proper places, and wash your hands.
13. Were there any other recommendations for needed service or unusual conditions that you noticed while removing and replacing the axle?

    Yes ____ No ____

14. Record your recommendations for additional service or repairs on the repair order.
15. Complete the repair order (R.O.).

STOP

Instructor OK ______________ Score __________

ASE Lab Preparation Worksheet #9-14

REPLACE A C-LOCK-TYPE REAR AXLE

Name______________________________ Class ____________________

OBJECTIVE:

Upon completion of this assignment, you should be able to remove and replace a C-lock-type rear axle. This task will help prepare you to pass the ASE certification examinations in manual transmissions and differentials.

DIRECTIONS:

Before beginning this lab assignment, review the worksheet completely. Fill in the information in the spaces provided as you complete each task.

TOOLS AND EQUIPMENT REQUIRED:

Safety glasses, jack and jack stands or vehicle lift, drain pan, impact wrench, impact sockets, hand tools, slide hammer, prybar, seal puller, shop towel, torque wrench

PARTS AND SUPPLIES:

Differential cover gasket, possibly an axle seal and/or axle bearing

This worksheet is to be used to remove the axles on a C-lock rear axle. Refer to Worksheet 9-13, Replace a Rear Axle with a Press-Fit Bearing, for other types of axles.

PROCEDURE:

Vehicle year ___________ Make _________________ Model ________________

Repair Order # ________ Engine size _________________ # of Cylinders ___________

1. Raise and support the vehicle.
2. Remove the wheel. Right ____ or Left ____
3. Remove the brake drum or the caliper and rotor.

   Brake drum ____ or Caliper and rotor ____
4. Clean dirt and grease from the cover on the back of the differential.
5. Place a drain pan under the differential and drain the differential lubricant into it.
6. Remove the differential cover bolts and remove the cover.

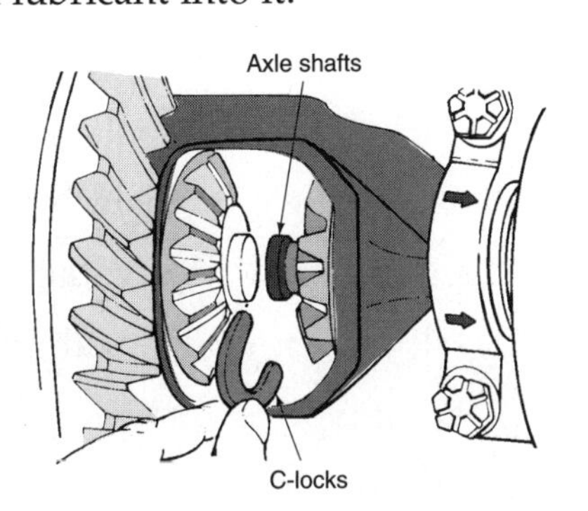

7. Remove the pinion shaft lock bolt and the pinion shaft. The shaft will slide out.

   Note: Once the pinion shaft has been removed, do not turn the axle shafts.
8. Push the axle shaft inward to the center of the vehicle until the C-lock is visible. Remove the C-lock.
9. Slide the axle completely out of the housing, being careful not to damage the seal.

Note: Do not allow the axle to hang in the housing. The axle seal will be damaged, resulting in a leak.

Seal and Bearing Replacement

10. If the seal requires replacement, remove it from the housing.

    Is the seal being replaced? Yes ____ No ____

    Tool used to remove the seal:

    Puller ____

    Prybar ____

    Screwdriver ____

11. If the bearing needs to be replaced, pull it from the housing using a slide hammer.

    Axle shaft bearing remover

    Is the bearing being replaced? Yes ____ No ____

12. Install the new bearing until it is fully seated into the housing.

    Fully seated? Yes ____ No ____

13. Install the new seal in the housing. Tap it into place until it is fully seated or flush with the housing.

    Fully seated ____ Flush with housing ____

14. Coat the sealing lip with grease or gear lubricant.

Reinstall Axles

15. Slide the axles carefully into the axle housing. Be careful not to damage the axle seal.
16. Slide the axles in until their splines engage the splines in the differential side gears.
17. Install the C-locks. Pull the axles outward to seat the C-locks into the side gears.
18. Install the pinion shaft. Align the bolt hole and install the lock bolt. Torque it to specifications.

CAUTION Do not overtighten the lock bolt! These bolts break easily.

    Pinion shaft lock bolt torque specification: ____ ft.-lb

19. Clean *all* gasket material from the gasket surfaces.
20. Install a new differential cover gasket. Install the cover.
21. Fill the differential with the correct type of lubricant.

    What type of gear lubricant was used? ________________

    Does this differential require a friction modifier (limited slip) additive?

    Yes ____ No ____

CONTINUED

22. Replace the brake drum or rotor/caliper and reinstall the wheel.
23. Torque the lug nuts. Lug nut torque specification: _____ ft.-lb
24. Install the wheel cover or hubcap.
25. Lower the vehicle.
26. Before completing the paperwork, clean your work area, put the tools in their proper places, and wash your hands.
27. Were there any other recommendations for needed service or unusual conditions that you noticed while removing and replacing the axle?

    Yes _____ No _____
28. Record your recommendations for additional service or repairs on the repair order.
29. Complete the repair order (R.O.).

STOP

Instructor OK ______________ Score __________

ASE Lab Preparation Worksheet #9-15

REPLACE SHOCK ABSORBERS

Name______________________________ Class ____________________

OBJECTIVE:

Upon completion of this assignment, you should be able to replace the shock absorbers on a vehicle. This task will help prepare you to pass the ASE certification examination in suspension and steering.

DIRECTIONS:

Before beginning this lab assignment, review the worksheet completely. Fill in the information in the spaces provided as you complete each task.

TOOLS AND EQUIPMENT REQUIRED:

Safety glasses, fender cover, jack and jack stands or vehicle lift, hand tools, shock absorber wrench, shop towel

PARTS AND SUPPLIES:

Shock absorbers

Note: The following instructions are for standard shock absorbers. To replace MacPherson strut-type shock absorbers, consult a service manual.

PROCEDURE:

Vehicle year ___________ Make __________________ Model ___________________

Repair Order # ________ Engine size _________________ # of Cylinders ____________

1. Purchase the replacement shock absorbers.
2. When shocks are being replaced, the car should be raised on a lift that supports the wheels or suspension.

   Type of lift: Frame-contact ____ Wheel-contact ____ Other ____
3. Position the vehicle on a wheel-contact lift or safety stands so that the suspension does *not* hang free.

   Note: It is safer and easier to change the shocks when the wheels are not hanging free.
4. Which shocks are being replaced? Front ____ Rear ____ All four ____

   Note: Shocks should always be replaced in pairs, both fronts or both rears. It is not advisable to replace only one shock.
5. Locate the upper and lower shock mount bolts.

   Note: Sometimes the rear shock's upper mounting is located inside the trunk, or on hatchbacks, inside the passenger compartment. If working inside the vehicle, use fender covers to protect the paint.
6. Is a special shock tool needed to hold one end of the shock? Yes ____ No ____

TURN

7. Spray the shock absorber fastener threads with penetrating oil. 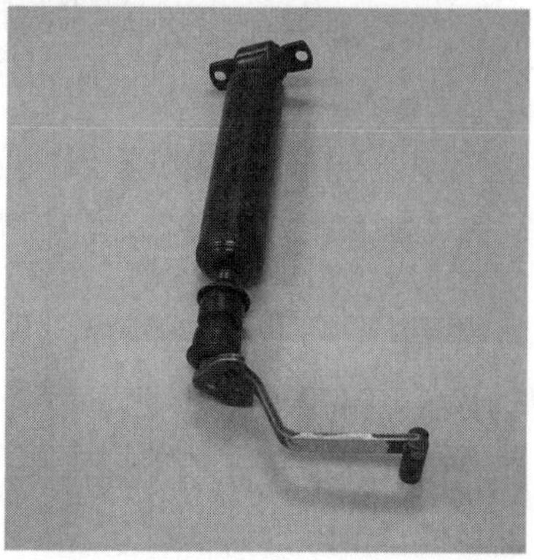
8. Are the new shocks gas charged? Yes ____ No ____

   Note: Gas shocks contain pressurized gas and do not require bleeding. They are shipped with a strap around them because they will expand to full travel when unrestrained. The strap should not be removed until after the shock is installed on the vehicle.

9. Remove the shock absorber and compare it to a new one. Are they the same length when fully compressed?

   Yes ____ No ____

   If the new shock is not gas charged, compare the shocks when they are fully extended. Are they the same length when fully extended?

   Yes ____ No ____

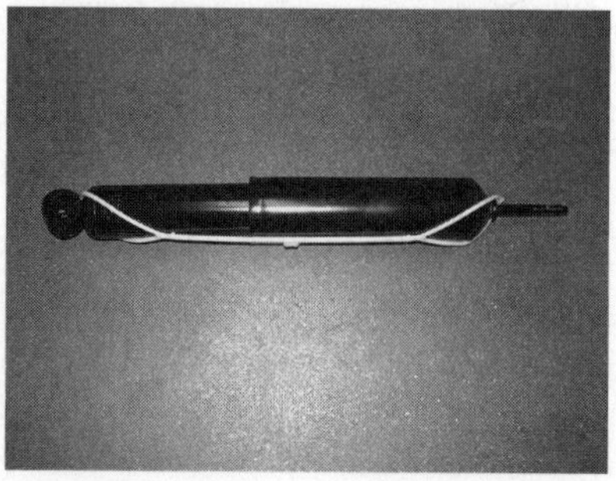

10. Does the old shock still resist movement when extended and compressed?

    Yes ____ No ____

    If the shocks are gas charged, it is not necessary to bleed them; proceed to step 15.

    If they are not gas charged, then it will be necessary to bleed the shocks; proceed to step 11.

11. Extend the shock while it is in its normal vertical position.
12. Turn the shock over so that its top is down. Fully collapse the shock.
13. Repeat the process four or five times to work out any air trapped in the shock.
14. When does the new shock give more resistance?

    Compression ____ Extension ____ Equal resistance ____

15. Install the new shocks. Do not overtighten the nuts! Tighten them until the rubber bushings are just barely compressed.
16. Cut the retaining strap from gas-charged shocks. Yes ____ No ____ N/A ____
17. Lower the vehicle and bounce test the shock absorbers.

    Do the shocks resist spring oscillation? Yes ____ No ____

    Did you hear any unusual noise? Yes ____ No ____

18. Before completing the paperwork, clean your work area, put the tools in their proper places, and wash your hands.
19. Were there any other recommendations for needed service or unusual conditions that you noticed while replacing the shock absorbers?

    Yes ____ No ____

20. Record your recommendations for additional service or repairs on the repair order.
21. Complete the repair order (R.O.).

STOP

Instructor OK ______________ Score __________

ASE Lab Preparation Worksheet #9-16

CHECK BALL JOINT WEAR

Name____________________________ Class ____________________

OBJECTIVE:

Upon completion of this assignment, you should be able to check a vehicle's ball joints for wear. This task will help prepare you to pass the ASE certification examination in suspension and steering.

DIRECTIONS:

Before beginning this lab assignment, review the worksheet completely. Fill in the information in the spaces provided as you complete each task.

TOOLS AND EQUIPMENT REQUIRED:

Safety glasses, fender covers, jack and jack stands or vehicle lift, prybar, dial indicator, ball joint checker, shop towel

PROCEDURE:

Vehicle year ___________ Make __________________ Model ___________________

Repair Order # ________ Engine size _________________ # of Cylinders ____________

Locate the ball joint specifications in the service manual or computer program.

What service manual or computer program was used? ____________________________________

On what page are the specifications located? ___________

What is the radial play specification? ___________

What is the axial play specification? ___________

1. Raise the car on a wheel ramp-type lift. If one is not available, it will be necessary to work on the ground.
2. Check the wheel bearing adjustment before testing ball joint looseness.

   Needs adjustment ____ OK ____
3. Which is the load-carrying ball joint? Location: Upper ____ Lower ____

   Note: When there are two control arms, the load-carrying joint is the one on the control arm that supports the spring.

SINTERED IRON BEARING

WEAR SURFACES

.050"

PRELOAD

WEAR INDICATOR

4. Does the vehicle have wear indicator ball joints?

   Yes ____ No ____

   Note: Late-model ball joints usually have a wear indicator on the grease fitting. Wear on these joints is inspected with the weight of the vehicle on the tires.

   Is any wear indicated? Yes ____ No ____

TURN

5. To check a ball joint without a wear indicator, remove the load from the ball joint. Raise the vehicle with a jack placed on the *frame* or *lower control arm* as specified.

   The jack is on the: Frame ____ Lower control arm ____

Check Ball Joint Axial Play

6. Raise the vehicle until the tire is off the ground just enough to be able to fit a prybar under it.
7. Pry the tire up and down while checking for vertical movement at the ball joint.

   Was any vertical movement noticed? Yes ____ No ____
8. Measure and record the vertical movement:

   Measurement ____

   What method was used to check the vertical movement?

   Dial indicator ____ Ball joint checker ____

Check Ball Joint Radial Play

9. Raise the vehicle until the tire is off the ground.
10. Grasp the tire at the top and bottom and try to move it alternately in and out at the top and bottom.

    Was any radial movement noticed? Yes ____ No ____
11. Measure and record the radial movement.

    Measurement ____

    What method was used to check the radial movement?

    Dial indicator ____

    Ball joint checker____
12. Before completing the paperwork, clean your work area, put the tools in their proper places, and wash your hands.
13. Were there any other recommendations for needed service or unusual conditions that you noticed while checking the ball joints?

    Yes ____ No ____
14. Record your recommendations for additional service or repairs on the repair order.
15. Complete the repair order (R.O.).

Instructor OK ______________ Score __________

ASE Lab Preparation Worksheet #9-17

CENTERING THE STEERING WHEEL

Name______________________________ Class ____________________

OBJECTIVE:

Upon completion of this assignment, you should be able to center a steering wheel. This task will help prepare you to pass the ASE certification examination in suspension and steering.

DIRECTIONS:

Before beginning this lab assignment, review the worksheet completely. Fill in the information in the spaces provided as you complete each task.

TOOLS AND EQUIPMENT REQUIRED:

Safety glasses, fender covers, jack and jack stands or vehicle lift, hand tools, steering wheel lock, shop towel

PROCEDURE:

Vehicle year ___________ Make __________________ Model ___________________

Repair Order # ________ Engine size _________________ # of Cylinders ____________

The steering wheel is usually centered using the tie-rods, not by removing the steering wheel and putting it back on straight. When a toe-in adjustment is done, adjusting one of the tie-rods more than the other will cause the steering wheel to be off-center. If the steering wheel is not straight ahead when driving, follow this procedure to straighten it.

1. Count the number of turns the steering wheel makes as it is turned from lock to lock (include fractions.)

   Number of turns lock to lock ____

2. Position the steering wheel one-half (1/2) way between the locks. It should be centered. If not, remove it and put it back on straight.

   Steering wheel centered? Yes ____ No ____

3. Use a steering wheel lock to hold the steering wheel in the centered position.

4. From the front of the car, sight down the tires on each side. The front tire should align with the rear one. This will give a rough estimate of the direction the tie-rods need to be turned.

   From which side can more of one rear tire than the other be seen? Right ____ Left ____

   If the right rear tire is more visible when sighting down the tires, the steering needs to be adjusted to the left.

   If the left rear tire is more visible when sighting down the tires, the steering needs to be adjusted to the right.

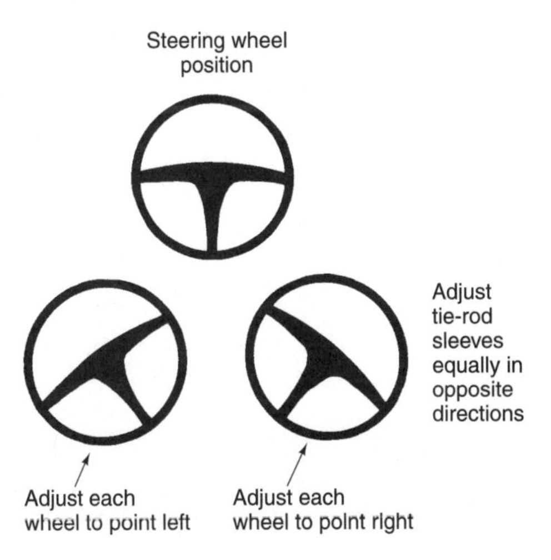

The steering linkage needs to be adjusted to the: Right ____ Left ____

5. Turn each tie-rod an equal amount in opposite directions. This maintains the current toe setting but moves the steering wheel position.

   Note: Whenever a tie-rod clamp is loosened, the clamp must be positioned so that it does not bind before retightening. See your instructor.

6. Test drive the car on a straight, level road.

   Is the steering wheel centered? Yes ____ No ____

   If the steering wheel is off-center and points to the right, readjust the tie rods so that the tires also are pointed to the right.

7. Before completing the paperwork, clean your work area, put the tools in their proper places, and wash your hands.

8. Were there any other recommendations for needed service or unusual conditions that you noticed while centering the steering wheel?

   Yes ____ No ____

9. Record your recommendations for additional service or repairs on the repair order.

10. Complete the repair order (R.O.).

STOP

Part II

ASE Lab Preparation Worksheets

Service Area 10

Miscellaneous

Instructor OK ______________ Score ___________

ASE Lab Preparation Worksheet #10-1

PREPARE FOR A TEST DRIVE

Name______________________________ Class ____________________

OBJECTIVE:

Upon completion of this assignment, you should know how to prepare a vehicle for a test drive.

DIRECTIONS:

Before beginning this lab assignment, review the worksheet completely. Fill in the information in the spaces provided as you complete each task.

PROCEDURE:

Vehicle year ___________ Make __________________ Model ____________________

Unless a vehicle is unsafe to drive, a test drive should always be done before performing repairs, and again after repairs are completed.

Before a test drive, check the following items:

1. Open the hood and place fender covers on the fenders and front body parts.
   a. Check the oil level. Full ____ Low ____
   b. Are there any coolant leaks? Yes ____ No ____
   c. Check the power steering for drive belt tension and proper fluid level. OK ____ Low ____
   d. Check the belt tension. OK ____ Loose ____
2. Walk around the vehicle and inspect the condition of the tires:
   a. Damage to the sidewalls or tread area? Yes ____ No ____
   b. Signs of impact damage/bent rim? Yes ____ No ____
   c. Condition of valve stems? Good ____ Bad ____
   d. Adjust tire pressures to specifications. Specification: ____ psi
   e. Are the tires of the correct size? Yes ____ No ____
   f. Are both radial and bias tires used? Yes ____ No ____
   g. Are front tires of the same brand and tread pattern? Yes ____ No ____
3. Sit in the vehicle and inspect the following:
   a. Depress the brake pedal. How does it feel? Spongy ____ Firm ____
   b. Turn the key to the on position. Is the malfunction indicator light working properly?
   Yes ____ No ____
   c. Does the vehicle have enough fuel for the test drive? Yes ____ No ____
   d. Start the engine. Is there adequate oil pressure? Yes ____ No ____

e. Are any of the driver warning lights on? Yes ____ No ____

f. Does the power steering assist work with equal ease in both directions? Yes ____ No ___

4. Describe any problems you noticed during the pretest drive inspection.

____

____

Upon completion of the pretest drive inspection, get your instructor's approval before taking a vehicle on a test drive.

Before leaving on a test drive, adjust the seat and mirrors.
Never drive a vehicle without using the seat belt.

STOP

Instructor OK ______________ Score __________

ASE Lab Preparation Worksheet #10-2

PERFORM A TEST DRIVE

Name_________________________ Class ________________

OBJECTIVE:

Upon completion of this assignment, you should be able to perform a test drive to check vehicle condition and locate problems. This task will help prepare you to pass the ASE certification examinations in all areas.

DIRECTIONS:

Before beginning this lab assignment, review the worksheet completely. Fill in the information in the spaces provided as you complete each task.

PROCEDURE:

Vehicle year __________ Make ________________ Model _______________

Repair Order # _______ Engine size _______________ # of Cylinders __________

Do not attempt this worksheet without the knowledge and permission of your instructor.

During the test drive, a professional technician will check a vehicle for many driving conditions. The following are a few of the important checks that should be made. Read through them completely and make sure you understand them before attempting to test drive a vehicle. Complete Worksheet 10-1, Prepare for a Test Drive, before the test drive:

1. Engine Performance
   a. The vehicle should accelerate smoothly.
   b. The vehicle should have the necessary power to pull hills.

   Note: Problems here could be due to an engine or transmission problem or just the need for a maintenance tune-up.

2. Transmission Operation
   a. Standard transmission—smooth engagement of the clutch, shifts smoothly, runs quietly
   b. Automatic transmission—shifts smoothly at the appropriate time on acceleration and deceleration

3. Brakes
   a. Brake pedal is firm and will hold pressure.
   b. Vehicle stops straight.

4. Hard Steering
   a. The vehicle should be easier to steer at higher speeds.

Note: This can be caused by binding parts, incorrect alignment, low tires, or a failure in the power steering system.

b. The tires should not squeal on turns.

Note: Squealing tires could be due to a bent steering arm or low tire pressures.

5. Noises

a. There should not be any squeaks and clunks when going over bumps.

Note: This could be caused by bad bushings, which can also cause changes in camber, resulting in brake pull.

b. There should not be any noise during turns that changes in pitch as the car weaves to the left and then to the right.

Note: The outer wheel bearing turns faster during a turn. It will make more noise when turning to one side than to the other.

6. Shimmy or Tramp

The steering wheel should not shake from side to side.

Note: This could indicate a bent or out-of-balance wheel or excessive caster.

7. Wanders

The vehicle should not drift and require constant steering.

Note: The vehicle could have an incorrect caster angle setting.

8. Pull

The vehicle should not pull to one side or the other while accelerating or at a cruise or under braking.

Note: This could be due either to alignment or brakes.

9. Rough Ride

The vehicle should not have a rougher than normal ride.

Note: This could be due to tire pressures that are too high, a bent or frozen shock, or to the installation of radial tires on an older vehicle.

10. Excess Body Roll

The vehicle should not lean excessively to one side or the other during fast turns.

Note: The shocks could be worn out.

Upon completion of the pretest drive inspection, get your instructor's approval before taking a vehicle on a test drive.

CONTINUED

Before leaving on a test drive, adjust the seat and mirrors. Never drive a vehicle without using the seat belt.

Test Drive Checklist

a. Does the vehicle go straight when you let go of the steering wheel? OK ____ Problem ____

b. Accelerates smoothly? OK ____ Problem ____

c. Good acceleration hot and cold, climbs hills easily? OK ____ Problem ____

d. Standard transmission shifts smoothly. OK ____ Problem ____

e. Automatic transmission shifts smoothly at the correct speed. OK ____ Problem ____

f. Stops straight and fast? OK ____ Problem ____

g. Rides smooth on the highway (tire balance, shocks, suspension)? OK ____ Problem ____

h. No clunks when going over speed bumps? OK ____ Problem ____

i. Quiet on highway (no wind leaks, no differential noise)? OK ____ Problem ____

j. No smoke from exhaust on heavy acceleration or deceleration? OK ____ Problem ____

k. Gauges appear accurate after warmup? OK ____ Problem ____

11. Before completing the paperwork, clean your work area, put the tools in their proper places, and wash your hands.
12. Were there any other recommendations for needed service or unusual conditions that you noticed while test driving the vehicle?

    Yes ____ No ____
13. Record your recommendations for additional service or repairs on the repair order.
14. Complete the repair order (R.O.).

STOP

Instructor OK ______________ Score __________

ASE Lab Preparation Worksheet #10-3

USED CAR CONDITION APPRAISAL CHECKLIST

Name____________________________ Class __________________

OBJECTIVE:

Upon completion of this assignment, you should be able to inspect the condition of a used vehicle with the intent to purchase.

DIRECTIONS:

Fill in the information in the spaces provided as you inspect the vehicle.

Vehicle year __________ Make ________________ Model ________________

Engine size _______________ # of Cylinders ___________

Documents:

	OK	Bad	Cost
Registration documents			
Identification of seller			
Identification of vehicle			
Emission certification			

Blue book value: High _______ Low _______

Is the seller the registered owner? Yes ____ No ____

Comments: ____

Body/Paint Condition:

	OK	Bad	Cost
Rust: roof, bottoms of doors, and fenders			
Paint: shines?			
Faded evenly (no new repairs)			
Door jambs and underhood same color?			
Paint on chrome strips or weatherstripping?			
Vinyl top in good condition?			
Chrome in good condition?			
Dents: evidence of repair?			
Body fit			
Hidden welds on body joints			

Comments: ____

Wear and Tear:

Odometer mileage________________

Owner has file of maintenance records?

Yes ___No___

Wear and tear consistent with mileage? Yes ____ No ____

Comments: ____

	OK	Bad	Cost
Tires:			
All the same brand?			
All the same size?			
Radials?			
Proper type spare			
Condition of spare			

Comments: ____

Electrical System:	OK	Bad	Cost
Charging system working?			
Starting system working?			
Battery condition			
Gauges			
Radio			
Lights			
Heater			
Air conditioning			
Clock			
Horn			

Comments: ____

Doors:	OK	Bad	Cost
Shut tightly?			
Driver's door handle loose?			
Windows tight, roll easily?			
Electric windows working?			
Weatherstripping condition			

Comments: ____

	OK	Bad	Cost
Interior Condition:			
Upholstery			
Headliner			
Dash board			
Carpets			
Gas, brake pedals worn			

Comments: ____

	OK	Bad	Cost
Windshield:			
Signs of leakage			
Frost around edges			
Rock chips			
Cracks			
Wipers:			
Condition			
Operation			
Washers:			

Comments: ____

	OK	Bad	Cost
Mechanical:			
Condition of Chassis:			
Evidence of frame or collision damage			
Tire wear/alignment			
Tires worn?			
Brake pedal height			
Brake pedal firmness			
Front brake linings			
Rear brake linings			
Steering wheel freeplay (less than 3")?			
Steering linkage			
Suspension parts			
Springs (ride height)			
Suspension bushings			
Shock absorbers			
Rubber snubbers			

Comments: ____

	OK	Bad	Cost
Leaks:			
Engine oil			
Transmission fluid			
Differential oil			
Power steering fluid			
Brake fluid			
Coolant			

Comments: ______________________________

	OK	Bad	Cost
Idles smooth?			
Engine hot			
Engine cold			

Comments: ______________________________

	OK	Bad	Cost
Transmission:			
Stall test (5 sec.)			
Fluid level			
Fluid condition			

Comments: ______________________________

	OK	Bad	Cost
Powertrain:			
U-joints, CV joint boots			
Clutch slippage			
Clutch freeplay			
Gear engagement			
Transmission noise			

Comments: ______________________________

	OK	Bad	Cost
Engine:			
Oil appearance			
Last oil change, mileage			
Oil pressure idling (warm)			
Lifter noise when cold?			
Vacuum, reading at idle			
Compression			
Radiator condition			
Hoses and belts			
Blowby (check PCV and air cleaner)			
Exhaust system condition			

	OK	Bad	Cost
Oil or soot in tailpipe			
Choke operation			
Exhaust smoke			
Motor mounts			

Comments: ______________________________

Emission Controls:

	OK	Bad	Cost
Underhood label			
Air injection system			
Exhaust recirculation			
Catalytic converter			
Crankcase vent, PCV			
Evaporative controls			
Thermostatic air cleaner			
Modifications?			

Other ______________________________

Test Drive:

Comments:______________________________

____

____

STOP

Instructor OK ______________ Score __________

ASE Lab Preparation Worksheet #10-4

PERSONAL TOOL INVENTORY

Name______________________________ Class ___________________

OBJECTIVE:

The following list includes those tools recommended for an entry-level technician. Upon completion of this assignment, you should have a complete inventory of your personal tool set. This inventory will help you decide which additional tools to add to your set.

DIRECTIONS:

Place a check (✓) in the space next to the tools that you own.

TOOLS AND EQUIPMENT REQUIRED:

Personal tool set

Toolbox

Top chest ______________________________

Rollaway ______________________________

Utility cart______________________________

Portable box______________________________

Screwdrivers

Slot screwdrivers:

3″___4″___6″___8″___10″___12″ _________

Phillips screwdrivers:

Standard length

#1___#2 ___#3 ____________________

Short #2___Long #2 _________________

Magnetic screwdriver ________________

Torx® drivers:

T8 ___T10___T15___T20___T25___T30 ____

Other______________________________

Pliers

10″ Multipurpose pliers ________________

Large multipurpose pliers ________________

7″ Locking pliers ______________________

10″ Locking pliers _____________________

Needle nose pliers _____________________

Slip joint pliers _______________________

Diagonal cutters _______________________

Terminal pliers (wire/crimper) ___________

Other______________________________

Snap Ring Pliers

Inside___Outside___Convertible __________

Other______________________________

Hammers

16 oz. Ball peen hammer _________________

Plastic hammer ________________________

Brass hammer__________________________

Rubber mallet _________________________

Dead blow hammer ____________________

Other______________________________

Socket Drive Tools

3/8″ Drive________________________________

Ratchet________________________________

Wobble extensions (15°) 3″___6″___12″ ___

Speed handle ______________________

Impact driver ______________________

1/2″ Drive________________________________

Ratchet________________________________

6″ extension ______________________

12″ extension ______________________

Flex handle (breaker bar) ____________

Torque Wrenches

1/2″ Drive ft.-lb ____________________

3/8″ Drive ft.-lb ____________________

3/8″ Drive in.-lb ____________________

Metric Sockets

3/8″ Drive Socket Set

6 mm___7 mm___ 8 mm___9 mm ________

10 mm___11 mm___12 mm ____________

13 mm___14 mm___15 mm ____________

16 mm___17 mm___18 mm___19 mm ______

Other________________________________

Deep socket set

6 mm___7 mm___ 8 mm___9 mm ________

10 mm___11 mm___12 mm ____________

13 mm___14 mm___15 mm ____________

16 mm___17 mm___18 mm___19 mm ______

Other________________________________

Impact socket set

8 mm___9 mm___10 mm___11 mm _______

12 mm___13 mm___14 mm___15 mm ______

16 mm___17 mm___18 mm___19 mm ______

Other________________________________

Universal impact socket set

8 mm___9 mm___10 mm___11 mm _______

12 mm___13 mm___14 mm___15 mm ______

16 mm___17 mm___18 mm___19 mm ______

Other________________________________

1/2″ Drive Socket Set

10 mm___11 mm___12 mm___13 mm ______

14 mm___15 mm___16 mm___17 mm ______

18 mm___19 mm___20 mm___21 mm ______

22 mm___23 mm___24 mm___25 mm ______

Other________________________________

Deep socket set

10 mm___11 mm___12 mm___13 mm ______

14 mm___15 mm___16 mm___17 mm ______

18 mm___19 mm___20 mm___21 mm ______

22 mm___23 mm___24 mm___25 mm ______

Other________________________________

Impact socket set

10 mm___11 mm___12 mm___13 mm ______

14 mm___15 mm___16 mm___17 mm ______

18 mm___19 mm___20 mm___21 mm ______

22 mm___23 mm___24 mm___25 mm ______

Other________________________________

Metric Wrenches

Combination wrenches (standard length)

7 mm___8 mm___9 mm___10 mm ________

11 mm___12 mm___13 mm___14 mm ______

15 mm___16 mm___17 mm___18 mm ______

19 mm___Other________________________

Flare-nut wrenches

8 mm___9 mm___10 mm___11 mm _______

12 mm___13 mm___14 mm___15 mm ______

16 mm___17 mm___18 mm___19 mm ______

Other________________________________

Allen wrench set

0.7 mm___ 0.9 mm___ 2 mm___ 3 mm _____

4 mm___5 mm___6 mm___7 mm ________

8 mm___10 mm___12 mm___14 mm ______

17 mm___19 mm___

Other________________________________

Standard Sockets

Note: All vehicles currently manufactured use metric fasteners.

3/8" Drive Socket Set

1/4___5/16 ___3/8___ 7/16 ___1/2_______

9/16 ___5/8___11/16___3/4___Other______

Deep socket set

1/4___5/16 ___3/8___ 7/16 ___1/2_______

9/16 ___5/8___11/16___3/4___Other______

Impact Socket Set

5/16 ___3/8___7/16 ___1/2 ___9/16 ______

5/8___11/16___3/4___Other _____________

1/2" Drive Socket Set

3/8 ___7/16 ___1/2___9/16___ 5/8 ______

11/16___3/4___13/16___7/8 ___15/16 _____

1___1 1/16___1 1/4___1 5/16 ___________

Other ______________________________

Deep socket set

7/16 ___1/2___9/16___ 5/8 _____________

11/16___3/4___13/16___7/8 ___15/16 _____

1___1 1/16___1 1/4___1 5/16 ___________

Other ______________________________

Impact socket set

3/8 ___7/16 ___1/2___9/16___ 5/8 ______

11/16___3/4___13/16___7/8 ___15/16 _____

1___1 1/16___1 1/4___1 5/16 ___________

Other ______________________________

Standard Wrenches

Note: All vehicles currently manufactured use metric fasteners.

Combination wrenches (standard length)

1/4___5/16 ___3/8___ 7/16 ___1/2_______

9/16 ___5/8___11/16___3/4 ____________

13/16___7/8 ___15/16 ________________

1___1 1/16___1 1/4___1 5/16 ___________

Other ______________________________

Flare-nut wrenches

5/16 ___3/8___7/16 ___1/2 ___9/16 ______

5/8___11/16___3/4___ 13/16___ 7/8 ______

15/16 ___ 1___ Other __________________

Allen wrench set

1/16 ___3/32 ___ 7/64___1/8___9/64 _____

5/32___3/16 ___7/32___1/4___5/16 ______

3/8___ 7/16 ___ 1/2___9/16 ___5/8 ______

Other_______________________________

Punch and Chisel Set

Flat tip chisel___ Center punch ___________

Brass punch ________________________

Pin punches: 1/16"___1/8"___1/4" _______

Other ______________________________

Files

Mill___Half round___Round ____________

Other_______________________________

Drill Index (Fractional sizes 1/16–1/2")

Every 1/64?___ Every 1/32? __ Every 1/16?__

TURN

Air Tools

1/2″ drive impact wrench ________________

3/8″ drive impact wrench ________________

3/8″ drive ratchet ________________

Drill motor ________________

Blowgun ________________

Grinder ________________

Spark Plug Tools

Spark plug gauge ________________

5/8″ spark plug socket ________________

13/16″ spark plug socket ________________

Insulated pliers ________________

Battery Tool Set

Battery pliers ________________

Battery terminal puller ________________

Battery post cleaner ________________

Battery clamp spreader ________________

Side terminal cleaner ________________

5/16 battery terminal wrench ________________

Brake Tools

Brake spoon ________________

Brake spring tool ________________

Brake retaining spring tool ________________

Wheel cylinder hone ________________

Disc brake piston removal tool ________________

Miscellaneous

Safety glasses ________________

Safety goggles ________________

Hacksaw ________________

Prybar ________________

Flexible magnetic pickup ________________

Scraper (putty knife) ________________

Wire brush ________________

Tire air pressure gauge ________________

Circuit tester (continuity tester) ________________

Feeler gauge set ________________

Adjustable long-handled mirror ________________

Flashlight ________________

Digital multimeter (DMM) ________________

Remote starter switch ________________

Compression gauge ________________

Vacuum fuel pressure gauge ________________

Oil filter wrench ________________

Creeper ________________

Tape measure ________________

Additional Tools: List any additional tools that you have in your toolbox.

____

____

____

____

____

____

____

____

____

____

____

____

____

____

Part II

ASE Lab Preparation Worksheets

Appendix

Repair Orders

THOMSON

DELMAR LEARNING™

DATE	TIME REQUESTED	WRITTEN BY
NAME		
ADDRESS		
CITY		ZIP
HOME PHONE	BUSINESS PHONE	

YEAR	MAKE	MODEL	LICENSE NO.	MILEAGE
VEHICLE ID #				

Part or Lubricant Description:

TOTAL PARTS		

Labor Description | **Labor Time:** | **HRS.** | **MINS.**

Labor Description	Labor Time	HRS.	MINS.

I authorize the listed labor and materials required for this repair in an amount not to exceed
You are authorized to operate this vehicle on streets for testing purposes

$ ____________
Original estimate

Customer Signature ______________________________

ALL PARTS WILL BE DISCARDED UNLESS INSTRUCTED OTHERWISE SAVE ☐ DISCARD ☐

Revised Estimate | Date | Time | Person Contacted | Contacting Person | BY PERSON ☐ BY PHONE ☐

I acknowledge being notified and approving an increase in the original estimate.

Customer Signature ______________________________

DISCLAIMER STATEMENT

The School District does not assume responsibility or accept liability for vehicles and vehicle parts or work performed by students. The proper performance of tasks by students is a slow, methodical process which cannot be accelerated. Student work is done as an educational experience and is observed by the instructor to see that it is done properly. However, instances do occur where work has to be re done because the student failed to follow instructions.

Parts installed are not warranted beyond that given by respective manufacturers. No other warranties are made.

I understand that all jobs done in this laboratory are for student learning experience, with no expressed warranty or specific completion time.

I hereby authorize the repair work listed hereon, including sublet work to be done along with the purchase of necessary materials. The automotive staff and/or students may operate the described vehicle for testing and inspection at owner's risk. An express lien is acknowledged on said vehicle to secure the amount of repairs thereto. I hereby agree to hold the School District, the Board of Trustees, and all District offices, agents, and employees free from any loss, damages, liability, cost or expense caused in any way that may arise during or as a result of this vehicle being repaired by or used in the school Auto Shop.

RECOMMENDED SERVICE & COMMENTS	EST. COST

TOTAL LABOR TIME		
x SHOP RATE		
LABOR TOTAL		
TOTAL LABOR AND PARTS		
LABOR		
PARTS		
SALES TAX		
TOTAL		

STUDENT TECHNICIANS

THOMSON
DELMAR LEARNING™

DATE	TIME REQUESTED	WRITTEN BY
NAME		
ADDRESS		
CITY		ZIP
HOME PHONE	BUSINESS PHONE	

YEAR	MAKE	MODEL	LICENSE NO.	MILEAGE
VEHICLE ID #				

Part or Lubricant Description:

TOTAL PARTS		

Labor Description | **Labor Time:** | **HRS.** | **MINS.**

I authorize the listed labor and materials required for this repair in an amount not to exceed
You are authorized to operate this vehicle on streets for testing purposes

$ ____________
Original estimate

Customer Signature ______________________________

ALL PARTS WILL BE DISCARDED UNLESS INSTRUCTED OTHERWISE SAVE ☐ DISCARD ☐

Revised Estimate	Date	Time	Person Contacted	Contacting Person	
					BY PERSON ☐ BY PHONE ☐

I acknowledge being notified and approving an increase in the original estimate.

Customer Signature ______________________________

RECOMMENDED SERVICE & COMMENTS	EST. COST

TOTAL LABOR TIME		
x SHOP RATE		
LABOR TOTAL		
TOTAL LABOR AND PARTS		
LABOR		
PARTS		
SALES TAX		
TOTAL		

STUDENT TECHNICIANS

DISCLAIMER STATEMENT

The School District does not assume responsibility or accept liability for vehicles and vehicle parts or work performed by students. The proper performance of tasks by students is a slow, methodical process which cannot be accelerated. Student work is done as an educational experience and is observed by the instructor to see that it is done properly. However, instances do occur where work has to be re done because the student failed to follow instructions.

Parts installed are not warranted beyond that given by respective manufacturers. No other warranties are made.

I understand that all jobs done in this laboratory are for student learning experience, with no expressed warranty or specific completion time.

I hereby authorize the repair work listed hereon, including sublet work to be done along with the purchase of necessary materials. The automotive staff and/or students may operate the described vehicle for testing and inspection at owner's risk. An express lien is acknowledged on said vehicle to secure the amount of repairs thereto. I hereby agree to hold the School District, the Board of Trustees, and all District offices, agents, and employees free from any loss, damages, liability, cost or expense caused in any way that may arise during or as a result of this vehicle being repaired by or used in the school Auto Shop.

THOMSON
DELMAR LEARNING™

Part or Lubricant Description:

TOTAL PARTS		

DATE	TIME REQUESTED	WRITTEN BY
NAME		
ADDRESS		
CITY		ZIP
HOME PHONE	BUSINESS PHONE	
YEAR / MAKE	MODEL	LICENSE NO. / MILEAGE
VEHICLE ID #		

Labor Description	Labor Time:	HRS.	MINS.

I authorize the listed labor and materials required for this repair in an amount not to exceed
You are authorized to operate this vehicle on streets for testing purposes

$ ____________ Original estimate

Customer Signature ____________

ALL PARTS WILL BE DISCARDED UNLESS INSTRUCTED OTHERWISE SAVE ☐ DISCARD ☐

Revised Estimate ____ Date ____ Time ____ Person Contacted ____ Contacting Person ____ BY PERSON ☐ BY PHONE ☐

I acknowledge being notified and approving an increase in the original estimate.

Customer Signature ____________

RECOMMENDED SERVICE & COMMENTS	EST. COST

TOTAL LABOR TIME		
x SHOP RATE		
LABOR TOTAL		
TOTAL LABOR AND PARTS		
LABOR		
PARTS		
SALES TAX		
TOTAL		

STUDENT TECHNICIANS

DISCLAIMER STATEMENT

The School District does not assume responsibility or accept liability for vehicles and vehicle parts or work performed by students. The proper performance of tasks by students is a slow, methodical process which cannot be accelerated. Student work is done as an educational experience and is observed by the instructor to see that it is done properly. However, instances do occur where work has to be re done because the student failed to follow instructions.

Parts installed are not warranted beyond that given by respective manufacturers. No other warranties are made.

I understand that all jobs done in this laboratory are for student learning experience, with no expressed warranty or specific completion time.

I hereby authorize the repair work listed hereon, including sublet work to be done along with the purchase of necessary materials. The automotive staff and/or students may operate the described vehicle for testing and inspection at owner's risk. An express lien is acknowledged on scid vehicle to secure the amount of repairs thereto. I hereby agree to hold the School District, the Board ofTrustees, and all District offices, agents, and employees free from any loss, damages, liability, cost or expense caused in any way that may arise during or as a result of this vehicle being repaired by or used in the school Auto Shop.

THOMSON

DELMAR LEARNING™

DATE	TIME REQUESTED	WRITTEN BY
NAME		
ADDRESS		
CITY	ZIP	
HOME PHONE	BUSINESS PHONE	
YEAR / MAKE	MODEL / LICENSE NO.	MILEAGE
VEHICLE ID #		

Part or Lubricant Description:

TOTAL PARTS		

Labor Description **Labor Time:** **HRS.** **MINS.**

I authorize the listed labor and materials required for this repair in an amount not to exceed
You are authorized to operate this vehicle on streets for testing purposes

$ ________ Original estimate

Customer Signature ________________________

ALL PARTS WILL BE DISCARDED UNLESS INSTRUCTED OTHERWISE SAVE ☐ DISCARD ☐

Revised Estimate Date Time Person Contacted Contacting Person BY PERSON ☐ BY PHONE ☐

I acknowledge being notified and approving an increase in the original estimate.

Customer Signature ________________________

RECOMMENDED SERVICE & COMMENTS	EST. COST

TOTAL LABOR TIME		
x SHOP RATE		
LABOR TOTAL		
TOTAL LABOR AND PARTS		
LABOR		
PARTS		
SALES TAX		
TOTAL		

STUDENT TECHNICIANS

DISCLAIMER STATEMENT

The School District does not assume responsibility or accept liability for vehicles and vehicle parts or work performed by students. The proper performance of tasks by students is a slow, methodical process which cannot be accelerated. Student work is done as an educational experience and is observed by the instructor to see that it is done properly. However, instances do occur where work has to be re done because the student failed to follow instructions.

Parts installed are not warranted beyond that given by respective manufacturers. No other warranties are made.

I understand that all jobs done in this laboratory are for student learning experience, with no expressed warranty or specific completion time.

I hereby authorize the repair work listed hereon, including sublet work to be done along with the purchase of necessary materials. The automotive staff and/or students may operate the described vehicle for testing and inspection at owner's risk. An express lien is acknowledged on said vehicle to secure the amount of repairs thereto. I hereby agree to hold the School District, the Board of Trustees, and all District offices, agents, and employees free from any loss, damages, liability, cost or expense caused in any way that may arise during or as a result of this vehicle being repaired by or used in the school Auto Shop.

THOMSON

DELMAR LEARNING™

DATE	TIME REQUESTED	WRITTEN BY
NAME		
ADDRESS		
CITY	ZIP	
HOME PHONE	BUSINESS PHONE	
YEAR / MAKE	MODEL	LICENSE NO. / MILEAGE
VEHICLE ID #		

Part or Lubricant Description:

TOTAL PARTS		

Labor Description **Labor Time:** **HRS.** **MINS.**

I authorize the listed labor and materials required for this repair in an amount not to exceed
You are authorized to operate this vehicle on streets for testing purposes

$ ____________
Original estimate

Customer Signature ____________________

ALL PARTS WILL BE DISCARDED UNLESS INSTRUCTED OTHERWISE SAVE ☐ DISCARD ☐

Revised Estimate ______ Date ______ Time ______ Person Contacted ______ Contacting Person ______ BY PERSON ☐ BY PHONE ☐

I acknowledge being notified and approving an increase in the original estimate.

Customer Signature ____________________

DISCLAIMER STATEMENT

The School District does not assume responsibility or accept liability for vehicles and vehicle parts or work performed by students. The proper performance of tasks by students is a slow, methodical process which cannot be accelerated. Student work is done as an educational experience and is observed by the instructor to see that it is done properly. However, instances do occur where work has to be re done because the student failed to follow instructions.

Parts installed are not warranted beyond that given by respective manufacturers. No other warranties are made.

I understand that all jobs done in this laboratory are for student learning experience, with no expressed warranty or specific completion time.

I hereby authorize the repair work listed hereon, including sublet work to be done along with the purchase of necessary materials. The automotive staff and/or students may operate the described vehicle for testing and inspection at owner's risk. An express lien is acknowledged on said vehicle to secure the amount of repairs thereto. I hereby agree to hold the School District, the Board of Trustees, and all District offices, agents, and employees free from any loss, damages, liability, cost or expense caused in any way that may arise during or as a result of this vehicle being repaired by or used in the school Auto Shop.

RECOMMENDED SERVICE & COMMENTS	EST. COST

TOTAL LABOR TIME		
x SHOP RATE		
LABOR TOTAL		
TOTAL LABOR AND PARTS		
LABOR		
PARTS		
SALES TAX		
TOTAL		

STUDENT TECHNICIANS

THOMSON

DELMAR LEARNING™

DATE	TIME REQUESTED	WRITTEN BY
NAME		
ADDRESS		
CITY	ZIP	
HOME PHONE	BUSINESS PHONE	
YEAR / MAKE / MODEL	LICENSE NO.	MILEAGE
VEHICLE ID #		

Part or Lubricant Description:

TOTAL PARTS		

Labor Description | **Labor Time:** **HRS.** **MINS.**

Labor Description	Labor Time	HRS.	MINS.

I authorize the listed labor and materials required for this repair in an amount not to exceed
You are authorized to operate this vehicle on streets for testing purposes

$ ______ Original estimate

Customer Signature ______

ALL PARTS WILL BE DISCARDED UNLESS INSTRUCTED OTHERWISE SAVE ☐ DISCARD ☐

Revised Estimate Date Time Person Contacted Contacting Person BY PERSON ☐ BY PHONE ☐

I acknowledge being notified and approving an increase in the original estimate.

Customer Signature ______

DISCLAIMER STATEMENT

The School District does not assume responsibility or accept liability for vehicles and vehicle parts or work performed by students. The proper performance of tasks by students is a slow, methodical process which cannot be accelerated. Student work is done as an educational experience and is observed by the instructor to see that it is done properly. However, instances do occur where work has to be re done because the student failed to follow instructions.

Parts installed are not warranted beyond that given by respective manufacturers. No other warranties are made.

I understand that all jobs done in this laboratory are for student learning experience, with no expressed warranty or specific completion time.

I hereby authorize the repair work listed hereon, including sublet work to be done along with the purchase of necessary materials. The automotive staff and/or students may operate the described vehicle for testing and inspection at owner's risk. An express lien is acknowledged on said vehicle to secure the amount of repairs thereto. I hereby agree to hold the School District, the Board ofTrustees, and all District offices, agents, and employees free from any loss, damages, liability, cost or expense caused in any way that may arise during or as a result of this vehicle being repaired by or used in the school Auto Shop.

RECOMMENDED SERVICE & COMMENTS	EST. COST

TOTAL LABOR TIME	
x SHOP RATE	
LABOR TOTAL	

TOTAL LABOR AND PARTS		
LABOR		
PARTS		
SALES TAX		
TOTAL		

STUDENT TECHNICIANS

THOMSON
DELMAR LEARNING™

DATE | TIME REQUESTED | WRITTEN BY

NAME

ADDRESS

CITY | ZIP

HOME PHONE | BUSINESS PHONE

YEAR | MAKE | MODEL | LICENSE NO. | MILEAGE

VEHICLE ID #

Part or Lubricant Description:

TOTAL PARTS

Labor Description | **Labor Time:** | **HRS.** | **MINS.**

I authorize the listed labor and materials required for this repair in an amount not to exceed $ ________ Original estimate
You are authorized to operate this vehicle on streets for testing purposes

Customer Signature ____________________

ALL PARTS WILL BE DISCARDED UNLESS INSTRUCTED OTHERWISE SAVE ☐ DISCARD ☐

Revised Estimate | Date | Time | Person Contacted | Contacting Person | BY PERSON ☐ BY PHONE ☐

I acknowledge being notified and approving an increase in the original estimate.

Customer Signature ____________________

DISCLAIMER STATEMENT

The School District does not assume responsibility or accept liability for vehicles and vehicle parts or work performed by students. The proper performance of tasks by students is a slow, methodical process which cannot be accelerated. Student work is done as an educational experience and is observed by the instructor to see that it is done properly. However, instances do occur where work has to be re done because the student failed to follow instructions.
Parts installed are not warranted beyond that given by respective manufacturers. No other warranties are made.
I understand that all jobs done in this laboratory are for student learning experience, with no expressed warranty or specific completion time.
I hereby authorize the repair work listed hereon, including sublet work to be done along with the purchase of necessary materials. The automotive staff and/or students may operate the described vehicle for testing and inspection at owner's risk. An express lien is acknowledged on said vehicle to secure the amount of repairs thereto. I hereby agree to hold the School District, the Board of Trustees, and all District offices, agents, and employees free from any loss, damages, liability, cost or expense caused in any way that may arise during or as a result of this vehicle being repaired by or used in the school Auto Shop.

RECOMMENDED SERVICE & COMMENTS | EST. COST

TOTAL LABOR TIME
x SHOP RATE
LABOR TOTAL

TOTAL LABOR AND PARTS
LABOR
PARTS
SALES TAX
TOTAL

STUDENT TECHNICIANS

THOMSON
DELMAR LEARNING™

DATE	TIME REQUESTED	WRITTEN BY
NAME		
ADDRESS		
CITY	ZIP	
HOME PHONE	BUSINESS PHONE	

YEAR	MAKE	MODEL	LICENSE NO.	MILEAGE
VEHICLE ID #				

Part or Lubricant Description:

TOTAL PARTS		

Labor Description **Labor Time:** **HRS.** **MINS.**

Labor Description	Labor Time	HRS.	MINS.

I authorize the listed labor and materials required for this repair in an amount not to exceed
You are authorized to operate this vehicle on streets for testing purposes

$ ________ Original estimate

Customer Signature ________

ALL PARTS WILL BE DISCARDED UNLESS INSTRUCTED OTHERWISE SAVE ☐ DISCARD ☐

Revised Estimate Date Time Person Contacted Contacting Person BY PERSON ☐ BY PHONE ☐

I acknowledge being notified and approving an increase in the original estimate.

Customer Signature ________

RECOMMENDED SERVICE & COMMENTS	EST. COST

TOTAL LABOR TIME		
x SHOP RATE		
LABOR TOTAL		
TOTAL LABOR AND PARTS		
LABOR		
PARTS		
SALES TAX		
TOTAL		

STUDENT TECHNICIANS

DISCLAIMER STATEMENT

The School District does not assume responsibility or accept liability for vehicles and vehicle parts or work performed by students. The proper performance of tasks by students is a slow, methodical process which cannot be accelerated. Student work is done as an educational experience and is observed by the instructor to see that it is done properly. However, instances do occur where work has to be re done because the student failed to follow instructions.

Parts installed are not warranted beyond that given by respective manufacturers. No other warranties are made.

I understand that all jobs done in this laboratory are for student learning experience, with no expressed warranty or specific completion time.

I hereby authorize the repair work listed hereon, including sublet work to be done along with the purchase of necessary materials. The automotive staff and/or students may operate the described vehicle for testing and inspection at owner's risk. An express lien is acknowledged on said vehicle to secure the amount of repairs thereto. I hereby agree to hold the School District, the Board ofTrustees, and all District offices, agents, and employees free from any loss, damages, liability, cost or expense caused in any way that may arise during or as a result of this vehicle being repaired by or used in the school Auto Shop.

THOMSON
DELMAR LEARNING™

DATE	TIME REQUESTED	WRITTEN BY
NAME		
ADDRESS		
CITY		ZIP
HOME PHONE	BUSINESS PHONE	

YEAR	MAKE	MODEL	LICENSE NO.	MILEAGE
VEHICLE ID #				

Part or Lubricant Description:

TOTAL PARTS		

Labor Description | **Labor Time:** | **HRS.** | **MINS.**

I authorize the listed labor and materials required for this repair in an amount not to exceed $______ Original estimate
You are authorized to operate this vehicle on streets for testing purposes

Customer Signature______

ALL PARTS WILL BE DISCARDED UNLESS INSTRUCTED OTHERWISE SAVE ☐ DISCARD ☐

Revised Estimate Date Time Person Contacted Contacting Person BY PERSON ☐ BY PHONE ☐

I acknowledge being notified and approving an increase in the original estimate.

Customer Signature ______

RECOMMENDED SERVICE & COMMENTS	EST. COST

TOTAL LABOR TIME		
x SHOP RATE		
LABOR TOTAL		
TOTAL LABOR AND PARTS		
LABOR		
PARTS		
SALES TAX		
TOTAL		

STUDENT TECHNICIANS

DISCLAIMER STATEMENT

The School District does not assume responsibility or accept liability for vehicles and vehicle parts or work performed by students. The proper performance of tasks by students is a slow, methodical process which cannot be accelerated. Student work is done as an educational experience and is observed by the instructor to see that it is done properly. However, instances do occur where work has to be re done because the student failed to follow instructions.

Parts installed are not warranted beyond that given by respective manufacturers. No other warranties are made.

I understand that all jobs done in this laboratory are for student learning experience, with no expressed warranty or specific completion time.

I hereby authorize the repair work listed hereon, including sublet work to be done along with the purchase of necessary materials. The automotive staff and/or students may operate the described vehicle for testing and inspection at owner's risk. An express lien is acknowledged on said vehicle to secure the amount of repairs thereto. I hereby agree to hold the School District, the Board of Trustees, and all District offices, agents, and employees free from any loss, damages, liability, cost or expense caused in any way that may arise during or as a result of this vehicle being repaired by or used in the school Auto Shop.

THOMSON
DELMAR LEARNING™

DATE	TIME REQUESTED	WRITTEN BY
NAME		
ADDRESS		
CITY	ZIP	
HOME PHONE	BUSINESS PHONE	

YEAR	MAKE	MODEL	LICENSE NO.	MILEAGE
VEHICLE ID #				

Part or Lubricant Description:

TOTAL PARTS		

Labor Description **Labor Time:** **HRS.** **MINS.**

I authorize the listed labor and materials required for this repair in an amount not to exceed

$ ________
Original estimate

You are authorized to operate this vehicle on streets for testing purposes

Customer Signature ______________________

ALL PARTS WILL BE DISCARDED UNLESS INSTRUCTED OTHERWISE SAVE ☐ DISCARD ☐

Revised Estimate	Date	Time	Person Contacted	Contacting Person	
					BY PERSON ☐ BY PHONE ☐

I acknowledge being notified and approving an increase in the original estimate.

Customer Signature ______________________

RECOMMENDED SERVICE & COMMENTS	EST. COST

TOTAL LABOR TIME		
x SHOP RATE		
LABOR TOTAL		

TOTAL LABOR AND PARTS		
LABOR		
PARTS		
SALES TAX		
TOTAL		

STUDENT TECHNICIANS

DISCLAIMER STATEMENT

The School District does not assume responsibility or accept liability for vehicles and vehicle parts or work performed by students. The proper performance of tasks by students is a slow, methodical process which cannot be accelerated. Student work is done as an educational experience and is observed by the instructor to see that it is done properly. However, instances do occur where work has to be re done because the student failed to follow instructions.

Parts installed are not warranted beyond that given by respective manufacturers. No other warranties are made.

I understand that all jobs done in this laboratory are for student learning experience, with no expressed warranty or specific completion time.

I hereby authorize the repair work listed hereon, including sublet work to be done along with the purchase of necessary materials. The automotive staff and/or students may operate the described vehicle for testing and inspection at owner's risk. An express lien is acknowledged on said vehicle to secure the amount of repairs thereto. I hereby agree to hold the School District, the Board of Trustees, and all District offices, agents, and employees free from any loss, damages, liability, cost or expense caused in any way that may arise during or as a result of this vehicle being repaired by or used in the school Auto Shop.

CERTIFIED CAR CARE SERVICE

Customer Name ________________________________

Address ________________ City ________ Zip Code ________ Phone ________

Date ________________ Time ________

Vehicle ________ Year ________ Model ________ License Number ________ Odometer Reading ________

ELECTRICAL SYSTEM CHECKS

________ Wiring Visual Inspection
Battery
________ Top Off Water Level
Posts and Cables
________ Clean ________ Corroded
________ Damaged

Battery Condition
________ Good ________ Replace
________ Recharge

LIGHTS

________ Park ________ Brake
________ Signal ________ Emergency
________ Dash Lights Back-up

Headlight Operation
________ High Left
________ Low Left
________ High Right
________ Low Left
________ Horn Operation

FUEL SYSTEM CHECKS

________ Condition of Hoses
________ Gas Cap Condition
________ Air Cleaner
________ Crankcase Vent Filter
________ Fuel Filter (miles until change suggested) ______

COOLING SYSTEM CHECKS

________ Level
________ Strength of Coolant
(Protection to ________ °)
________ No Leaks

Condition of Hoses
________ Radiator
________ Heater
________ Thermostat Bypass
Hose (if so equipped)

Pressure Test
________ Radiator
________ Cap
________ Condition of Coolant
Pump Belt

BRAKE INSPECTION

________ Pedal Travel
________ Emergency Brake
________ Brake Hoses and Lines
________ Master Brake Cylinder-
Fluid Level and Condition

ON-GROUND STEERING, SUSPENSION, DRIVE LINE CHECKS

________ Steering Wheel Freeplay
________ Power Steering Fluid Level
________ Shock Absorber Bounce Test

	Good	Unsafe
Front	______	______
Rear	______	______

________ No Squeaks
________ Ride Height Check
________ Check ATF Level

VISIBILITY

________ Mirrors
________ Wiper Blades
Wiper Operation
______ fast ______ slow
________ Washer Fluid and Pump
________ Clean and Inspect all Glass

UNDERCAR SERVICE

________ Drain Crankcase (if ordered)
________ Remove and replace oil filter
________ Inspect Under Car for
Oil, Gasoline, and Coolant Leaks
________ Check Crankcase Oil Level
________ Check Oil Filter for Leaks

INFLATE AND CHECK TIRES

Inflate to ________ lb.

Tire Condition:

	Good	Fair	Unsafe
RF	______	______	______
LF	______	______	______
RR	______	______	______
LR	______	______	______

Inflate and Check Spare
Good ______ Fair ______ Unsafe ______

SUSPENSION AND STEERING

________ Inspect Steering Linkage
________ Inspect Shock Absorbers
________ Inspect Suspension
Bushings
________ Clean Lubrication Fittings
________ Lubricate Fittings
________ Ball Joints
________ Inspect Ball Joint Seals
________ Ball Joint Wear
Inspection
________ Inspect Ride Height

UNDERCAR FUEL SYSTEM CHECKS

________ Condition of Fuel Hoses
________ Condition of Fuel Tank

DRIVE LINE CHECKS

________ Check Universal
or CV Joints

Inspect Gear Cases
________ Transmission
________ Differential

Replace Drain Plugs
________ Inspect Motor Mounts
________ Lubricate Door and Hood Hinge
and Latches

EXHAUST SYSTEM CHECKS

________ Mufflers and Pipes
________ Pipe Hangers
________ Exhaust Leaks
________ Heat Riser

FINAL VEHICLE PREPARATION

________ Replace Crankcase Oil
________ Clean Windows, Vacuum Interior
________ Fill Out and Affix Door Jamb
Record to Door Post
________ Complete a Repair Order

CERTIFIED CAR CARE SERVICE

Customer Name ____________________

Address ____________________ City ____________ Zip Code ____________ Phone ____________

Date ____________________ Time ____________

Vehicle ____________ Year ____________ Model ____________ License Number ____________ Odometer Reading ____________

ELECTRICAL SYSTEM CHECKS

________Wiring Visual Inspection
Battery
________Top Off Water Level
Posts and Cables
________Clean ________Corroded
________Damaged

Battery Condition
________Good ________Replace
________Recharge

LIGHTS

________Park ________Brake
________Signal ________Emergency
________Dash Lights Back-up

Headlight Operation
________High Left
________Low Left
________High Right
________Low Left
________Horn Operation

FUEL SYSTEM CHECKS

________Condition of Hoses
________Gas Cap Condition
________Air Cleaner
________Crankcase Vent Filter
________Fuel Filter (miles until change suggested) ______

COOLING SYSTEM CHECKS

________Level
________Strength of Coolant
(Protection to ____________°)
________No Leaks

Condition of Hoses
________Radiator
________Heater
________Thermostat Bypass
Hose (if so equipped)

Pressure Test
________Radiator
________Cap
________Condition of Coolant
Pump Belt

BRAKE INSPECTION

________Pedal Travel
________Emergency Brake
________Brake Hoses and Lines
________Master Brake Cylinder-
Fluid Level and Condition

ON-GROUND STEERING, SUSPENSION, DRIVE LINE CHECKS

________Steering Wheel Freeplay
________Power Steering Fluid Level
________Shock Absorber Bounce Test

	Good	Unsafe
Front	______	______
Rear	______	______

________No Squeaks
________Ride Height Check
________Check ATF Level

VISIBILITY

________Mirrors
________Wiper Blades
Wiper Operation
______fast ______slow
________Washer Fluid and Pump
________Clean and Inspect all Glass

UNDERCAR SERVICE

________Drain Crankcase (if ordered)
________Remove and replace oil filter
________Inspect Under Car for
Oil, Gasoline, and Coolant Leaks
________Check Crankcase Oil Level
________Check Oil Filter for Leaks

INFLATE AND CHECK TIRES

Inflate to ________ lb.

Tire Condition:

	Good	Fair	Unsafe
RF	______	______	______
LF	______	______	______
RR	______	______	______
LR	______	______	______

Inflate and Check Spare
Good ______ Fair ______ Unsafe ______

SUSPENSION AND STEERING

________Inspect Steering Linkage
________Inspect Shock Absorbers
________Inspect Suspension
Bushings
________Clean Lubrication Fittings
________Lubricate Fittings
________Ball Joints
________Inspect Ball Joint Seals
________Ball Joint Wear
Inspection
________Inspect Ride Height

UNDERCAR FUEL SYSTEM CHECKS

________Condition of Fuel Hoses
________Condition of Fuel Tank

DRIVE LINE CHECKS

________Check Universal
or CV Joints

Inspect Gear Cases
________Transmission
________Differential

Replace Drain Plugs
________Inspect Motor Mounts
________Lubricate Door and Hood Hinge
and Latches

EXHAUST SYSTEM CHECKS

________Mufflers and Pipes
________Pipe Hangers
________Exhaust Leaks
________Heat Riser

FINAL VEHICLE PREPARATION

________Replace Crankcase Oil
________Clean Windows, Vacuum Interior
________Fill Out and Affix Door Jamb
Record to Door Post
________Complete a Repair Order